Electricity and Electronics Fundamentals

2nd Edition

Electricity and Electronics Fundamentals

2nd Edition

Dale R. Patrick
Stephen W. Fardo

THE FAIRMONT PRESS, INC.

CRC Press
Taylor & Francis Group

Library of Congress Cataloging-in-Publication Data

Patrick, Dale R.
   Electricity and electronics fundamentals / Dale R. Patrick, Stephen W.
Fardo. -- 2nd ed.
      p. cm.
   Includes index.
   ISBN 0-88173-601-5 (alk. paper) -- ISBN 0-88173-602-3 (electronic) -- ISBN
1-4200-8387-2 (taylor & francis distribution : alk. paper)
      1. Electric engineering. 2. Electronics. I. Fardo, Stephen W. II. Title.

   TK146.P3425 2008
   621.3--dc22

                                                                2008008988

Published by The Fairmont Press, Inc.
700 Indian Trail
Lilburn, GA 30047
tel: 770-925-9388; fax: 770-381-9865
http://www.fairmontpress.com

Distributed by Taylor & Francis Ltd.
6000 Broken Sound Parkway NW, Suite 300
Boca Raton, FL 33487, USA
E-mail: orders@crcpress.com

Distributed by Taylor & Francis Ltd.
23-25 Blades Court
Deodar Road
London SW15 2NU, UK
E-mail: uk.tandf@thomsonpublishingservices.co.uk

Printed in the United States of America
10 9 8 7 6 5 4 3 2 1

0-88173-601-5 (The Fairmont Press, Inc.)
1-4200-8387-2 (Taylor & Francis Ltd.)

While every effort is made to provide dependable information, the publisher,
authors, and editors cannot be held responsible for any errors or omissions.

Contents

Preface

Electricity and Electronics Fundamentals (formerly *Understanding Electricity and Electronics*) is an introductory text that provides basic coverage of electricity and electronics fundamentals. The key concepts presented in this book are discussed using a simplified approach that enhances learning. The use of mathematics is kept to the very minimum and is discussed through applications and illustrations.

Chapters are organized in a step-by-step progression of concepts and theory. Each chapter begins with an introduction and objectives. A discussion of important concepts and theories follows.

Definitions of important terms are listed in a comprehensive glossary in Appendix A. Appendices dealing with electronic symbols, safety, and soldering are also provided for easy reference.

This book is organized in an easy-to-understand format and is intended to be used to acquire an understanding of electricity and electronics in the home, school, or work place. The sequence of the book allows progression at a desired pace in the study of electricity and electronics, from DC, AC, electronic devices and circuits, digital/computer basics, communications and power control.

This second edition includes revised/updated illustrations and content. In addition, chapters have been added to introduce the topics of electronic communications and electronic power control. The authors hope you will find the book easy to understand and that you are successful in your pursuit of knowledge in an exciting technical area.

Dale R. Patrick
Stephen W. Fardo

Course Objectives

Upon completion of this book—*Electricity and Electronics Fundamentals*—you should be able to:

1. Understand basic electrical concepts of voltage, current, resistance, and power in dc circuits.

2. Understand dc circuits using schematic diagrams and be able to perform tests and measurements with a multimeter.

3. Understand basic ac electronic concepts such as capacitance, inductance, reactance, impedance, phase relationship, and power.

4. Understand ac circuits and be able to perform measurements with a multimeter, oscilloscope, and signal generator.

5. Explain the operation of semiconductor devices and circuits.

6. Understand the following electronic circuits: power supplies, amplifiers, and oscillators.

7. Understand digital number systems and logic circuits.

8. Understand terms usually associated with computer operation.

9. Describe communications electronic systems.

10. Describe industrial electronic control devices.

Chapter 1

Direct Current (dc) Electronics

INTRODUCTION

The advancement of science and technology has brought about important changes in the field of electronics. At one time, the field of electronics was very limited. Recent developments in solid-state electronics and microminiaturization have brought a number of significant changes. A person working in the field of electronics must be knowledgeable about many types of systems and numerous control devices in order to be successful today. Electronics is a very fascinating science that we use in many different ways. It would be difficult to list all the ways that we use electronics each day. Everyone today should have an understanding of electronics.

This chapter deals with a basic topic in the study of electronics-direct current, or DC. Topics such as basic electrical systems; energy and power, the structure of matter, electrical charges, static electricity, electrical current, voltage, resistance, and measuring instruments are discussed. When studying this unit as well as others, you should refer to *definitions* of important terms in Appendix A. Preview these terms to gain a better understanding of what is discussed as the need arises.

OBJECTIVES

Upon completion of this chapter, you will be able to:
1. Explain the parts of an electronic system.
2. Explain the composition of matter.
3. Explain the laws of electrical charges.
4. Define the terms *insulator, conductor,* and *semiconductor.*
5. Explain electric current flow.
6. Diagram a simple electronic circuit.
7. Identify schematic electronic symbols.
8. Convert electrical quantities from metric units to English units and English units to metric units.
9. Use scientific notation to express numbers.
10. Define voltage, current, and resistance.
11. Describe basic types of batteries.
12. Explain how to connect batteries in series, parallel, and configurations.
13. Explain purposes of different configurations of battery connections,
14. Explain factors determining resistance.
15. Identify different types of resistors.
16. Identify resistor value by color code and size.
17. Explain the operation of potentiometers (variable resistors).
18. Construct basic electronic circuits.
19. Explain how to connect an ammeter in a circuit and measure current.
20. Explain how the voltmeter, ammeter, and ohmmeter are connected to a circuit.
21. Explain how to measure current, voltage, and resistance of basic dc circuits.
22. Solve basic math problems using a calculator.
23. Define Ohm's law and the power equation.
24. Solve problems finding current, voltage, and resistance.
25. Calculate power using the proper power formulas.
26. Define voltage drop in a circuit.
27. Solve circuit problems with resistors in different configurations.
28. Define the various terms relative to magnetism.
29. Explain the operation of various magnetic devices.
30. State Faraday's law for electromagnetic induction.
31. List three factors that affect the strength of electromagnets.
32. Explain how a capacitor operates.
33. Define the terms inductance and inductors.
34. Define the terms capacitance and capacitors.
35. List the factors determining capacitance.
36. Describe the construction of various types of capacitors.

37. Calculate total capacitance of capacitors in various series, parallel, and combination configurations.
38. Calculate total inductance of inductors in various series, parallel, and combination configurations.
39. List factors affecting inductance.
40. Identify different types of inductors.
41. Explain the concept of mutual inductance.

THE SYSTEM CONCEPT

For a number of years, people have worked with jigsaw puzzles as a source of recreation. A jigsaw puzzle contains a number of discrete parts that must be placed together properly to produce a picture. Each part then plays a specific role in the finished product. When a puzzle is first started, it is difficult to imagine the finished product without seeing a representative picture.

Studying a complex field such as electronics by discrete parts poses a problem that is somewhat similar to the jigsaw puzzle. In this case, it is difficult to determine the role that a discrete part plays in the operation of a complex system. A picture of the system divided into its essential parts therefore becomes an extremely important aid in understanding its operation.

The system concept will serve as the "big picture"

in the study of electronics. In this approach, a system will first be divided into a number of essential blocks. The role played by each block then becomes more meaningful in the operation of the overall system. After the location of each block has been established, discrete component operation related to each block then becomes more relevant. Through this approach, the way in which some of the "pieces" of electronic systems fit together should be more apparent.

BASIC SYSTEM FUNCTIONS

The word *system* is commonly defined as an organization of parts that are connected together to form a complete unit. There are a wide variety of electronic systems used today. Each electronic system has a number of unique features, or characteristics, that distinguish it from other systems. More importantly, however, there is a common set of parts found in each system. These parts play the same basic role in all systems. The terms energy *source, transmission path, control, load,* and *indicator* are used to describe the various system parts. A block diagram of these basic parts of the system is shown in Figure 1-1.

Each block of a basic system has a specific role to play in the overall operation of the system. This role becomes extremely important when a detailed analysis of

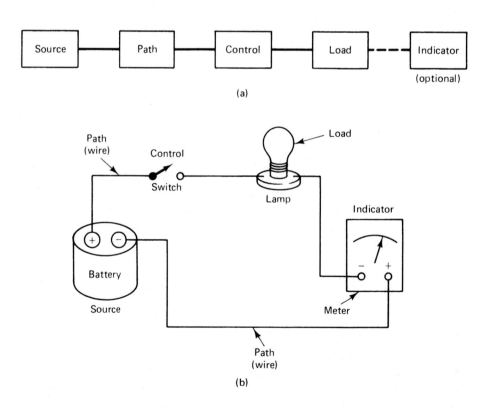

Figure 1-1. Electrical system: (a) block diagram; (b) pictorial diagram.

the system is to take place. Hundreds and even thousands of discrete components are sometimes needed to achieve a specific block function. Regardless of the complexity of the system, each block must still achieve its function in order for the system to be operational. Being familiar with these functions and being able to locate them within a complete system is a big step toward understanding the operation of the system.

The *energy source* of a system converts energy of one form into something more useful. Heat, light, sound, chemical, nuclear, and mechanical energy are considered as primary sources of energy. A primary energy source usually goes through an energy change before it can be used in an operating system.

The *transmission path* of a system is somewhat simplified when compared with other system functions. This part of the system simply provides a path for the transfer of energy. It starts with the energy source and continues through the system to the load device. In some cases, this path may be a single electrical conductor, light beam, or other medium between the source and the load. In other systems, there may be a supply line between the source and the load and a return line from the load to the source. There may also be a number of alternate or auxiliary paths within a complete system. These paths may be *series* connected to a number of small load devices or *parallel* connected to many independent devices.

The *control* section of a system is by far the most complex part of the entire system. In its simplest form, control is achieved when a system is turned on or off (see Figure 1-2). Control of this type can take place anywhere between the source and the load device. The term *full control* is commonly used to describe this operation. In addition to this type of control, a system may also employ some type of *partial control*. Partial control usually causes some type of an operational change in the system other than an on or off condition. Changes in electric current or light intensity are examples of alterations achieved by partial control.

The *load* of a system refers to a specific part or number of parts designed to produce some form of work (see Figure 1-3). Work, in this case, occurs when energy goes through a transformation or change. Heat, light, chemical action, sound, and mechanical motion are some of the common forms of work produced by a load device. As a general rule, a very large portion of all energy produced by the source is consumed by the load device during its operation. The load is typically the most prevalent part of the entire system because of its obvious work function.

The *indicator* of a system is primarily designed to display certain operating conditions at various points

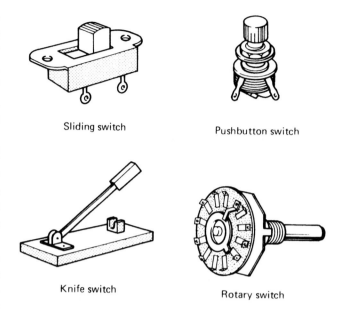

Sliding switch

Pushbutton switch

Knife switch

Rotary switch

Figure 1-2. Common control devices—switches—(full control).

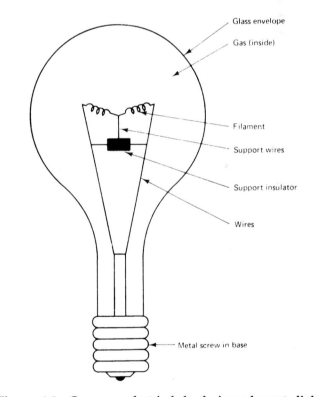

Glass envelope

Gas (inside)

Filament

Support wires

Support insulator

Wires

Metal screw in base

Figure 1-3. Common electrical load—incandescent light bulb—(converts electrical energy to light energy).

throughout the system. In some systems the indicator is an optional part, whereas in others it is an essential part in the operation of the system. In the latter case, system operations and adjustments are usually critical and are dependent upon specific indicator readings. The term *operational indicator* is used to describe this application. *Test indicators* are also needed to determine different operating values. In this role, the indicator is only temporarily attached to the system to make measurements. Test lights, meters, oscilloscopes, chart recorders, and digital display instruments are some of the common indicators used in this capacity.

A SIMPLE ELECTRONIC SYSTEM EXAMPLE

A flashlight is a device designed to serve as a light source in an emergency or as a portable light source. In a strict sense, flashlights can be classified as portable electronic systems. They contain the four essential parts needed to make this classification. Figure 1-4 is a cutaway drawing of a flashlight, with each component part shown associated with its appropriate system block.

The battery of a flashlight serves as the primary *energy source* of the system. Chemical energy of the battery must be changed into electrical energy before the system becomes operational. The flashlight is a synthesized system because it utilizes two distinct forms of energy in its operation. The energy source of a flashlight is an expendable item. It must be replaced periodically when it loses its ability to produce electrical energy.

The *transmission path* of a flashlight is commonly achieved via a metal case or through a conductor strip. Copper, brass, and plated steel are frequently used to achieve this function.

The *control* of electrical energy in a flashlight is achieved by a slide switch or a push-button switch. This type of control simply interrupts the transmission path between the source and the load device. Flashlights are primarily designed to have full control capabilities. This type of control is achieved manually by the person operating the system.

The *load* of a flashlight is a small incandescent lamp. When electrical energy from the source is forced to pass through the filament of the lamp, the lamp produces a bright glow. Electrical energy is first changed into heat and then into light energy. A certain amount of work is achieved by the lamp when this energy change takes place.

The energy transformation process of a flashlight is irreversible. It starts at the battery when chemical energy is changed into electrical energy. Electrical energy is then changed into heat and eventually into light energy by the load device. This flow of energy is in a single direction. When light is eventually produced, it consumes a large portion of the electrical energy coming from the source. When this energy is exhausted, the system becomes inoperative. The battery cells of a flashlight require periodic replacement in order to maintain a satisfactory operating condition.

Flashlights do not ordinarily employ a specific *indicator* as part of the system. Operation is indicated when the lamp produces light. In a strict sense, we could say that the load of this system also serves as an indicator. In some electrical and electronic systems the indicator is an optional system part.

A DIGITAL ELECTRONIC SYSTEM EXAMPLE

Another example of an electronic system is a digital system. Some of the most significant developments that have taken place in electronics are used for automation. Automatic fabrication methods, packaging, printing

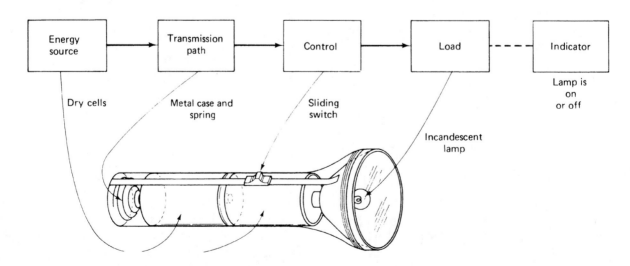

Figure 1-4. Cutaway drawing of a flashlight.

equipment, machining operations, and drafting equipment are all outgrowths of electronics in industry. In industrial systems that utilize digital control, instructions are supplied by magnetic tape, magnetic disk, or various physical changes such as pressure, temperature, or electricity. This information is then changed into digital signals and applied to the digital processor of the system. This information is decoded and directed to specific machines or machine parts, which then perform the necessary operations automatically.

A large portion of the automatic machinery being used by industry today receives instructions through digital signals. A *digital*, or numerical, system is therefore a very important part of the electronics field. In a strict sense, numerical signals are used primarily to perform the *control* function of a digital system. The *computer numerical control* (CNC) is also used to describe a specific type of digital system. A majority of the numerical systems in operation today are powered by electricity. This *source* of energy is primarily used to energize the *load* device, which in turn performs the work function of the system. The *control* function of the system must then be designed to respond to digital information. Inputs develop this information in the first operational step. Logic circuits, which provide full or partial control of the load device, are then activated.

A numerically controlled milling machine is a type of digital system. CNC machine system is shown in Figure 1-5. Operating instructions are translated into electrical signals and processed by digital logic gates housed in the machine. As an end result, these signals are used to control various physical machine operations automatically.

Figure 1-5. CNC machining Center. (Courtesy of Cellular Concepts.)

The *load* of a CNC system is typically electrical actuating motors or fluid-power cylinders designed to move the physical parts of a machine. When appropriate signals from the control unit are applied to the load, they move the two table axes or the cutting tool to a specific location. Machining operations can be performed according to the information programmed. Cutting speed, position location, clamping operations, and material flow can be controlled through this system. The operation is easily controlled by programming information into a system of this type.

ENERGY, WORK, AND POWER

An understanding of the terms energy, work, and power is necessary in the study of electronics. The first term, *energy*, means the capacity to do work. For example, the capacity to light a light bulb, to heat a home, or to move something requires energy. Energy exists in many forms, such as electrical, mechanical, chemical, and heat. If energy exists because of the movement of some item, such as a ball rolling down a hill, it is called *kinetic energy*. If energy exists because of the position of something, such as a ball that is at the top of the hill but not *yet* rolling, it is called *potential energy*. Energy has become one of the most important factors in our society.

A second important term is work. *Work* is the transferring or transforming of energy. Work is done when a force is exerted to move something over a distance against opposition, such as when a chair is moved from one side of a room to the other. An electrical motor used to drive a machine performs work. Work is performed when motion is accomplished against the action of a force that tends to oppose the motion. Work is also done each time energy changes from one form into another.

A third important term is power. *Power* is the rate at which work is done. It considers not only the work that is performed but the amount of time in which the work is done. For instance, electrical power is the rate at which work is done as electrical current flows through a wire. Mechanical power is the rate at which work is done as an object is moved against opposition over a certain distance. Power is either the rate of production or the rate of use of energy. The *watt* is the unit of measurement of power.

STRUCTURE OF MATTER

In the study of electronics, it is necessary to understand why electrical energy exists. By looking first at how certain natural materials are made, it will then be easier to see why electrical energy exists.

Here are a few basic scientific terms that are often used in the study of chemistry. They are also very important in the study of electronics. First, *matter* is anything that occupies space and has weight. Matter can be either a *solid*, a *liquid*, or a *gaseous* material. Solid matter includes such things as metal and wood; liquid matter is exemplified by water and gasoline; and gaseous matter includes such things as oxygen and hydrogen. Solids can be converted into liquids, and liquids can be made into gases. For example, water can be solid in the form of ice. Water can also be a gas in the form of steam. Matter changes state when the particles of which they are made are heated. As they are heated, the particles move and strike one another, causing them to move farther apart. Ice is converted into a liquid by adding heat. If heated to a high temperature, water becomes a gas (steam). All forms of matter exist in their most familiar states because of the amount of heat they contain. Some materials require more heat than others to become a liquid or a gas. However, all materials can be made to change from a solid to a liquid or from a liquid to a gas if enough heat is added. Also, these materials can change into liquids or solids if heat is taken from them.

The next important term in the study of the structure of matter is the *element*. An element is considered to be the basic material that makes up all matter. Materials such as hydrogen, aluminum, copper, iron, and iodine are a few of the over 100 elements known to exist. A table of elements is shown in Figure 1-6 (opposite). Some elements exist in nature and some are manufactured. Everything around us is made up of elements.

There are many more materials in our world than there are elements. Other materials are made by combining elements. A combination of two or more elements is called a *compound*. For example, water is a compound made from the elements hydrogen and oxygen. Salt is made from sodium and chlorine.

Another important term is *molecule*. A molecule is believed to be the smallest particle that a compound can be reduced to before being broken down into its basic elements. For example, one molecule of water has two hydrogen atoms and one oxygen atom.

In an even deeper look into the structure of matter, there are particles called *atoms*. Within these atoms are the forces that cause electrical energy to exist. An atom is considered to be the smallest particle to which an element can be reduced and still have the properties of that element. If an atom were broken down any further, the element would no longer exist. The smallest particles that are found in all atoms are called *electrons, protons,* and

neutrons. Elements differ from one another on the basis of the numbers of these particles found in their atoms. The relationship of matter, elements, compounds, molecules, atoms, electrons, protons, and neutrons is shown in Figure 1-7.

The simplest atom, hydrogen, is shown in Figure 1-8. The hydrogen atom has a center part called a *nucleus,* which has one proton. A proton is a particle that has a positive (+) charge. The hydrogen atom has one electron, which orbits around the nucleus of the atom. The electron has a negative (–) charge. Most atoms also have neutrons in the nucleus. A neutron has neither a positive nor a negative charge and is considered neutral. A carbon atom is shown in Figure 1-9. A carbon atom has 6 protons (+), 6 neutrons (N), and 6 electrons (–). The protons and the neutrons are in the nucleus, and the electrons *orbit* around the nucleus. The carbon atom has two orbits or circular paths. In the first orbit, there are 2 electrons. The other 4 electrons are in the second orbit.

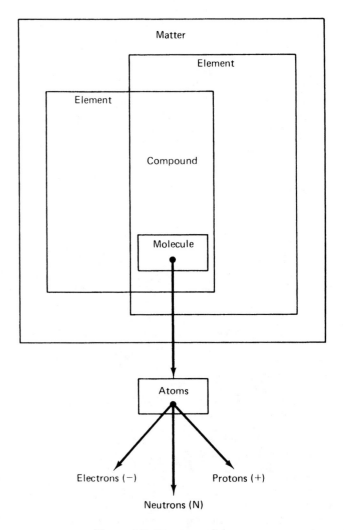

Figure 1-7. Structure of matter.

Element	Symbol	Atomic no.	Atomic weight
Actinium	Ac	89	227*
Aluminum	Al	13	26.97
Americium	Am	95	243*
Antimony	Sb	51	121.76
Argon	Ar	18	39.944
Arsenic	As	33	74.91
Astatine	At	85	210*
Barium	Ba	56	137.36
Berkelium	Bk	97	247*
Beryllium	Be	4	9.013
Bismuth	Bi	83	209.00
Boron	B	5	10.82
Bromine	Br	35	79.916
Cadmium	Cd	48	112.41
Calcium	Ca	20	40.08
Californium	Cf	98	251*
Carbon	C	6	12.01
Cerium	Ce	58	140.13
Cesium	Cs	55	132.91
Chlorine	Cl	17	35.457
Chromium	Cr	24	52.01
Cobalt	Co	27	58.94
Copper	Cu	29	63.54
Curium	Cm	96	247
Dysprosium	Dy	66	162.46
Einsteinium	E	99	254*
Erbium	Er	68	167.2
Europium	Eu	63	152.0
Fermium	Fm	100	255*
Fluorine	F	9	19.00
Francium	Fr	87	233*
Gadolinium	Gd	64	156.9
Gallium	Ga	31	69.72
Germanium	Ge	32	72.60
Gold	Au	79	197.0
Hafnium	Hf	72	178.6
Hahnium	Ha	105	262*
Helium	He	2	4.003
Holmium	Ho	67	164.94
Hydrogen	H	1	1.0080
Indium	In	49	114.76
Iodine	I	53	126.91
Iridium	Ir	77	192.2
Iron	Fe	26	55.85
Krypton	Kr	36	83.8
Lanthanum	La	57	138.92
Lawrencium	Lw	103	257*
Lead	Pb	82	207.21
Lithium	Li	3	6.940
Lutetium	Lu	71	174.99
Magnesium	Mg	12	24.32
Manganese	Mn	25	54.94
Mendelevium	Mv	101	256*
Mercury	Hg	80	200.61
Molybdenum	Mo	42	95.95
Neodymium	Nd	60	144.27
Neon	Ne	10	20.183
Neptunium	Np	93	237*
Nickel	Ni	28	58.69
Niobium	Nb	41	92.91
Nitrogen	N	7	14.008
Nobelium	No	102	253
Osmium	Os	76	190.2
Oxygen	O	8	16.000
Palladium	Pd	46	106.7
Phosphorus	P	15	30.975
Platinum	Pt	78	195.23
Plutonium	Pu	94	244
Polonium	Po	84	210
Potassium	K	19	39.100
Praseodymium	Pr	59	140.92
Promethium	Pm	61	145*
Protactinium	Pa	91	231*
Radium	Ra	88	226.05
Radon	Rn	86	222
Rhenium	Re	75	186.31
Rhodium	Rh	45	102.91
Rubidium	Rb	37	85.48
Ruthenium	Ru	44	101.1
Rutherfordium or Kurchatonium	Rf or Ku	104	260*
Samarium	Sm	62	150.43
Scandium	Sc	21	44.96
Selenium	Se	34	78.96
Silicon	Si	14	28.09
Silver	Ag	47	107.880
Sodium	Na	11	22.997
Strontium	Sr	38	87.63
Sulfur	S	16	32.066
Tantalum	Ta	73	180.95
Technetium	Tc	43	97*
Tellurium	Te	52	127.61
Terbium	Tb	65	158.93
Thallium	Tl	81	204.39
Thorium	Th	90	232.12
Thulium	Tm	69	168.94
Tin	Sn	50	118.70
Titanium	Ti	22	47.90
Tungsten	W	74	183.92
Uranium	U	92	238.07
Vanadium	V	23	50.95
Xenon	Xe	54	131.3
Ytterbium	Yb	70	173.04
Yttrium	Y	39	88.92
Zinc	Zn	30	65.38
Zirconium	Zr	40	91.22

*Mass number of the longest-lived of the known available forms of the element.

Figure 1-6. Table of elements.

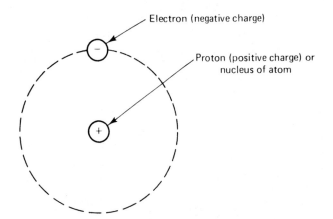

Figure 1-8. Hydrogen atom.

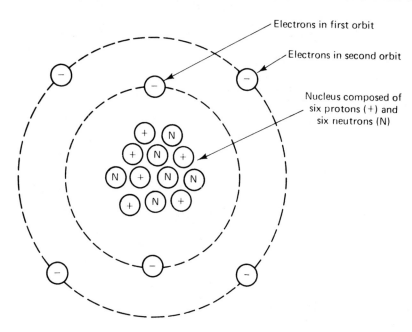

Figure 1-9. Carbon atom.

Early models of atoms showed electrons orbiting around the nucleus, in analogy with planets around the sun. This model is inconsistent with much modern experimental evidence. Atomic orbitals are very different from the orbits of satellites.

Atoms consist of a dense, positively charged nucleus surrounded by a cloud or series of clouds of electrons that occupy energy levels, which are commonly called *shells.* The occupied shell of highest energy is known as the *valence shell,* and the electrons in it are known as *valence electrons.*

Electrons behave as both particles and waves, so descriptions of them always refer to their *probability* of being in a certain region around the nucleus. Representations of orbitals are boundary surfaces enclosing the probable areas in which the electrons are found. All *s* orbitals are spherical, *p* orbitals are egg-shaped, *d* orbitals are dumbbell-shaped, and *f* orbitals are double-dumbbell-shaped.

Covalent bonding thus involves the *overlapping* of valence shell orbitals of different atoms. The electron charge then becomes concentrated in this region, thus attracting the two positively charged nuclei toward the negative charge between them. In ionic bonding, the ions are discrete units, and they group themselves in crystal structures, surrounding themselves with the ions of opposite charge.

The electron arrangement in the atoms of elements can be described from atomic mathematical theory. The first energy level, or *shell,* contains up to 2 electrons. The next shell contains up to 8 electrons. The third contains up to 18 electrons. Eighteen is the largest quantity any shell can contain. New shells are started as soon as shells nearer the nucleus have been filled with the maximum number of electrons. An atom with an incomplete outer shell is very *active.* When two unlike atoms with incomplete outer shells come together, they try to *share* their outer electrons. When their combined outer electrons are enough to make up one complete shell, *stable* atoms are formed. For example, oxygen has 8 electrons, 2 in the first shell and 6 in its outer shell. There is room for 8 electrons in the outer shell. Hydrogen has one electron in its outer shell. When two hydrogen atoms come near, oxygen combines with the hydrogen atoms by sharing the electrons of the two hydrogen atoms. Water is formed in a manner similar to the sketch of Figure 1-10. All the electrons

Look at the table of elements in Figure 1-6. Notice there are a different number of protons in the nucleus of each atom. This causes each element to be different. For example, hydrogen has 1 proton, carbon has 6, oxygen has 8, and lead has 82. The number of protons that each atom has is called its *atomic number.*

The nucleus of an atom contains protons (+) and neutrons (N). Since neutrons have no charge and protons have positive charges, the nucleus of an atom has a positive charge. Protons are believed to be about one third the diameter of electrons. The *mass,* or weight, of a proton is over 1800 times that of an electron, whose mass is about 9 $\times$ 10^{-28} g. Electrons move easily in their orbits around the nucleus of an atom. It is the movement of electrons that causes electrical energy to exist.

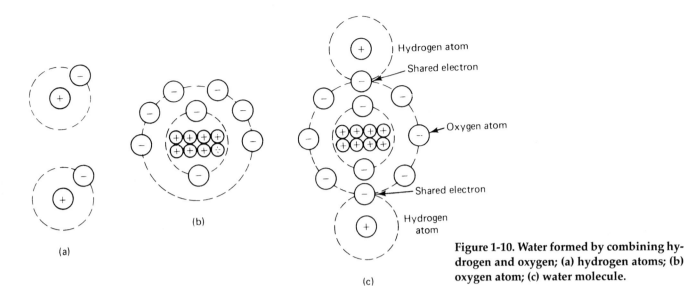

Figure 1-10. Water formed by combining hydrogen and oxygen; (a) hydrogen atoms; (b) oxygen atom; (c) water molecule.

are then bound tightly together, and a very stable water molecule is formed. The electrons in the incomplete outer shell of an atom are known as *valence electrons.* They are the only electrons that will combine with the electrons of other atoms to form compounds. They are also the only electrons that cause electric current to flow. For this reason it is necessary to understand the structure of matter.

ELECTROSTATIC CHARGES

In the preceding section, the positive and negative charges of protons and electrons were studied. Protons and electrons are parts of atoms that make up all things in our world. The positive charge of a proton is similar to the negative charge of an electron. However, a positive charge is the opposite of a negative charge. These charges are called *electrostatic charges.* Figure 1-11 shows how electrostatic charges affect each other. Each charged particle is surrounded by an *electrostatic field.*

The effect that electrostatic charges have on each other is very important. They either repel (move away) or attract (come together) each other. This action is as follows:

1. Positive charges repel each other (see Figure 1-11(a)).
2. Negative charges repel each other (see Figure 1-11(b)).
3. Positive and negative charges attract each other (see Figure 1-11c)).

Therefore, it is said that *like charges repel* and *unlike charges attract.*

The atoms of some materials can be made to gain or lose electrons. The material then becomes charged. One way to do this is to rub a glass rod with a piece of silk cloth. The glass rod loses electrons (–), so it now has a positive (+) charge. The silk cloth pulls electrons (–) away from the glass. Since the silk cloth gains new electrons, it now has a negative (–) charge. Another way to charge a material is to rub a rubber rod with fur.

It is also possible to charge other materials because some materials are charged when they are brought close to another charged object. If a charged rubber rod is touched against another material, the second material may become charged. Remember that materials are charged due to the movement of electrons and protons. Also, remember that when an atom loses electrons (–), it becomes positive (+). These facts are very important in the study of electronics.

Charged materials affect each other due to lines of force. Try to visualize these, as shown in Figure 1-11 These imaginary lines cannot be seen. However, they exert a force in all directions around a charged material. Their force is similar to the force of gravity around the earth. This force is called a *gravitational field.*

STATIC ELECTRICITY

Most people have observed the effect of *static electricity.* Whenever objects become charged, it is due to static electricity. A common example of static electricity is lightning. Lightning is caused by a difference in charge (+ and –) between the earth's surface and the clouds during a storm. The arc produced by lightning is the movement of charges between the earth and the clouds. Another

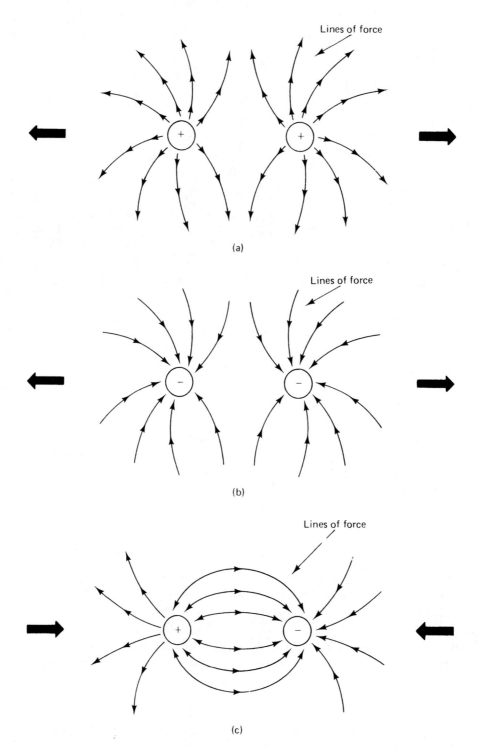

Figure 1-11. Electrostatic charges: (a) positive charges repel; (b) negative charges repel; (c) positive and negative charges attract.

common effect of static electricity is being "shocked" by a doorknob after walking across a carpeted floor. Static electricity also causes clothes taken from a dryer to cling together and hair to stick to a comb.

Electrical charges are used to filter dust and soot in devices called electrostatic filters. Electrostatic precipitators are used in power plants to filter the exhaust gas that goes into the air. Static electricity is also used in the manufacture of sandpaper and in the spray painting of automobiles. A device called an electroscope is used to detect a negative or positive charge.

ELECTRIC CURRENT

Static electricity is caused by *stationary charges.* However, electrical current is the *motion* of electrical charges from one point to another. Electric current is produced when electrons (–) are removed from their atoms. Some electrons in the outer orbits of the atoms or certain elements are easy to remove. A force or pressure applied to a material causes electrons to be removed. The movement of electrons from one atom to another is called *electric current flow.*

CONDUCTORS

A material through which current flows is called a *conductor.* A conductor passes electric current very easily. Copper and aluminum wire are commonly used as conductors. Conductors are said to have low *resistance* to electrical current flow. Conductors usually have three or fewer electrons in the outer orbit of their atoms. Remember that the electrons of an atom orbit around the nucleus. Many metals are electrical conductors. Each metal has a different ability to conduct electric current. Materials with only one outer orbit or valence electron (gold, silver, copper) are the best conductors. For example, silver is a better conductor than copper, but it is too expensive to use in large amounts. Aluminum does not conduct electrical current as well as copper, but it is commonly used, since it is cheaper and lighter than other conductors. Copper is used more than any other conductor.

INSULATORS

There are some materials that do not allow electric current to flow easily. The electrons of materials that are insulators are difficult to release. In some insulators, their valence shells are filled with eight electrons. The valence shells of others are over half-filled with electrons. The atoms of materials that are insulators are said to be stable. Insulators have high resistance to the movement of electric current. Some examples of insulators are plastic and rubber.

SEMICONDUCTORS

Materials called semiconductors have become very important in electronics. Semiconductor materials are neither conductors nor insulators. Their classification also depends on the number of electrons their atoms have in their valence shells. Semiconductors have 4 electrons in their valence shells. Remember that conductors have outer orbits *less than* half-filled and insulators ordinarily have outer orbits *more than* half-filled. Figure 1-12 compares conductors, insulators, and semiconductors. Some common types of semiconductor materials are silicon, germanium, and selenium.

CURRENT FLOW

The usefulness of electricity is due to its electric current flow. Current flow is the movement of electrical charges along a conductor. Static electricity, or electricity at rest, has some practical uses due to electrical charges. Electric current flow allows us to use electrical energy to do many types of work.

The movement of valence shell electrons of conductors produces electrical current. The outer electrons of the atoms of a conductor are called *free electrons.* Energy released by these electrons as they move allows work to be done. As more electrons move along a conductor, more energy is released. This is called an increased electric current flow. The movement of electrons along a conductor is shown in Figure 1-13.

To understand how current flow takes place, it is necessary to know about the atoms of conductors. Conductors, such as copper, have atoms that are loosely held together. Copper is said to have atoms connected together by *metallic bonding.* A copper atom has one valence shell electron, which is loosely held to the atom. These atoms are so close together that their valence shells overlap each other. Electrons can easily move from one atom to another. In any conductor the outer electrons continually move in a random manner from atom to atom.

The random movement of electrons does not result in current flow, since electrons must move in the same di-

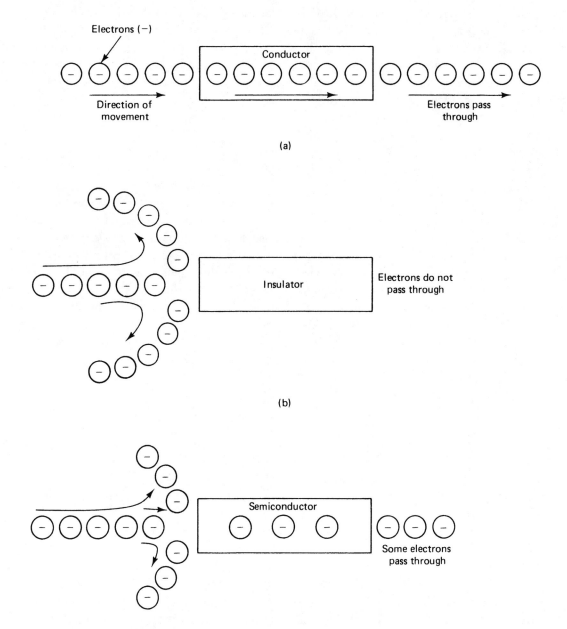

Figure 1-12. Comparisons of (a) conductors, (b) insulators, and (c) semiconductors.

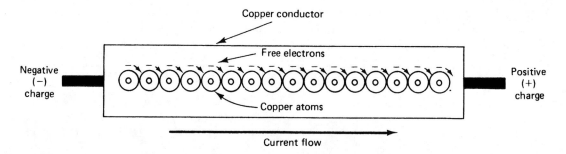

Figure 1-13. Current flow through a conductor.

rection to cause current flow. If electric charges are placed on each end of a conductor, the free electrons move in one direction. Figure 1-14 shows current flow through a conductor caused by negative and positive electrical charges. Current flow takes place because there is a difference in the charges at each end of the conductor. Remember that like charges repel and unlike charges attract.

When an electrical charge is placed on each end of the conductor, the free electrons move. Free electrons have a negative charge, so they are repelled by the negative charge on the left of Figure 1-13. The free electrons are attracted to the positive charge on the right and move to the right from one atom to another. If the charges on each end of the conductor are increased, more free electrons will move. This increased movement causes more electric current flow.

Current flow is the result of electrical energy caused as electrons change orbits. This impulse moves from one electron to another. When one electron moves out of its valence shell, it enters another atom's valence shell. An electron is then repelled from that atom. This action goes on in all parts of a conductor. Remember that electric current flow produces a transfer of energy.

Electronic Circuits

Current flow takes place in electronic circuits. A *circuit* is a path or conductor for electric current flow. Electric current flows only when it has a complete, or *closed-circuit*, path. There must be a source of electrical energy to cause current to flow along a closed path. Figure 1-14 shows a battery used as an energy source to cause current to flow through a light bulb. Notice that the path, or circuit, is *complete*. Light is given off by the light bulb due to the work done as electric current flows through a closed circuit. Electrical energy produced by the battery is changed to light energy in this circuit.

Electric current cannot flow if a circuit is open. An *open circuit* does not provide a complete path for current flow. If the circuit of Figure 1-14 were to become open, no current would flow. The light bulb would not glow. Free electrons of the conductor would no longer move from one atom to another. An example of an open circuit is a "burned-out" light bulb. Actually, the filament (the part that produces light) has become open. The open filament of a light bulb stops current flow from the source of electrical energy. This causes the bulb to stop burning, or producing light.

Another common circuit term is a *short circuit*. A short circuit, which can be very harmful, occurs when a conductor connects directly across the terminals of an electrical energy source. For safety purposes, a short circuit should never happen because short circuits cause too much current to flow from the source. If a wire is placed across a battery, a short circuit occurs. The battery would probably be destroyed and the wire could get hot or possibly melt due to the short circuit.

Direction of Current Flow

Electric current flow is the movement of electrons along a conductor. Electrons are negative charges. Negative charges are attracted to positive charges and are repelled by other negative charges. Electrons move from the negative terminal of a battery to the positive terminal. This is called *electron current flow*. Electron current flow is in a direction of electron movement from negative to positive through a circuit.

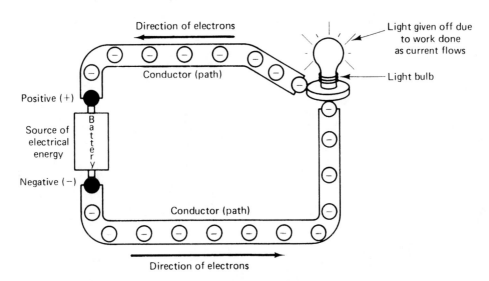

Figure 1-14. Current flow in a closed circuit.

Another way to look at electric current flow is in terms of charges. A high charge can be considered positive and a low charge, negative. Using this method, an electrical charge is considered to move from a high charge to a low charge. This is called *conventional current flow.* Conventional current flow is the movement of charges from positive to negative.

Electron and conventional current flow should not be confusing. They are two different ways of looking at current flow. One deals with electron movement and the other deals with charge movement. *In this book, electron current flow is used.*

Amount of Current Flow (The Ampere)

The amount of electric current that flows through a circuit depends on the number of electrons that pass a point in a certain time. The *coulomb* (C) is a unit of measurement of electric current. It is estimated that 1 C is 6,280,000,000,000,000,000 electrons (6.28×10^{18} in scientific notation). Since electrons are very small, it takes many to make one unit of measurement. When 1 C passes a point on a conductor in one second, 1 ampere (A) of current flows in the circuit. The unit is named for A.M. Ampere, an 18-century scientist who studied electricity. Current is commonly measured in units called *milliamperes* (mA) and *microamperes* (μA). These are smaller units of current. A milliampere is one thousandth (1/1000) of an ampere and a microampere is one millionth (1/1,000,000) of an ampere.

Current Flow Compared to Water Flow

An electrical circuit is a path in which an electric current flows. Current flow is similar to the flow of water through a pipe. Water flow can be used to show how current flows in an electrical circuit. When water flows in a pipe, something causes it to move. The pipe offers opposition, or resistance, to the flow of water. If the pipe is small, it is more difficult for the water to flow.

In an electric circuit, current flows through wires (conductors), The wires of an electric circuit are similar to the pipes through which water flows. If the wires are made of a material that has high resistance, it is difficult for current to flow. The result is the same as water flow through

a pipe with a rough surface. If the wires are large, it is easier for current to flow in an electrical circuit. In the same way, it is easier for water to flow through a large pipe. Electric current and water flow are compared in Figure 1-15. Current flows from one place to another in an electrical circuit. Similarly, water that leaves a pump moves from one place to another. The rate of water flow through a pipe is measured in gallons per minute. In an electronic circuit, the current is measured in amperes. The flow of electric current is measured by the number of coulombs that pass a point on a conductor each second.

ELECTRICAL FORCE (VOLTAGE)

Water pressure is needed to force water along a pipe. Similarly, electrical pressure is needed to force current along a conductor. Water pressure is usually measured in pounds per square inch (psi). Electrical pressure is measured in *volts* (V). If a motor is rated at 120 V, it requires 120 V of electrical pressure applied to the motor to force the proper amount of current through it. More pressure would increase the current flow and less pressure would not force enough current to flow. The motor would not operate properly with *too* much or *too* little voltage. Water pressure produced by a pump causes water to flow through pipes. Pumps produce pressure that causes water to flow. The same is true of an electrical energy source. A source such as a battery or generator produces cur-

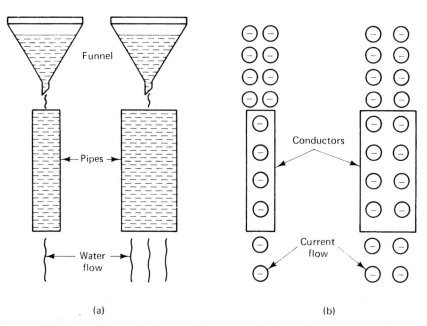

Figure 1-15. Comparison of electrical current and water flow; (a) water pipes; (b) electrical conductors.

rent flow through a circuit. As voltage is increased, the amount of current in the circuit is also increased. Voltage is also called *electromotive force* (*EMF*).

RESISTANCE

The opposition to current flow in electrical circuits is called *resistance.* Resistance is not the same for all materials. The number of free electrons in a material determines the amount of opposition to current flow. Atoms of some materials give up their free electrons easily. These materials offer low opposition to current flow. Other materials hold their outer electrons and offer high opposition to current flow.

Electric current is the movement of free electrons in a material. Electric current needs a source of electrical pressure to cause the movement of free electrons through a material. An electric current will not flow if the source of electrical pressure is removed. A material will not release electrons until enough force is applied. With a constant amount of electrical force (voltage) and more opposition (resistance) to current flow, the number of electrons flowing (current) through the material is smaller. With constant voltage, current flow is increased by decreasing resistance. Decreased current results from more resistance. By increasing or decreasing the amount of resistance in a circuit, the amount of current flow can be changed.

Even very good conductors have some resistance, which limits the flow of electric current through them. The resistance of any material depends on four factors:

1. The material of which it is made
2. The length of the material
3. The cross-sectional area of the material
4. The temperature of the material

The material of which an object is made affects its resistance. The ease with which different materials give up their outer electrons is very important in determining resistance. Silver is an excellent conductor of electricity. Copper, aluminum, and iron have more resistance but are more commonly used, since they are less expensive. All materials conduct an electric current to some extent, even though some (insulators) have very high resistance. Length also affects the resistance of a conductor. The longer a conductor, the greater the resistance. A material resists the flow of electrons because of the way in which each atom holds onto its outer electrons. The more material that is in the path of an electric current, the less current flow the circuit will have. If the length of a conductor is doubled, there is twice as much resistance in the circuit.

Another factor that affects resistance is the cross-sectional area of a material. The greater the cross-sectional area of a material, the lower the resistance. If two conductors have the same length but twice the cross-sectional area, there is twice as much current flow through the wire with the larger cross-sectional area.

Temperature also affects resistance. For most materials, the higher the temperature, the more resistance it offers to the flow of electric current. This effect is produced because a change in the temperature of a material changes the ease with which a material releases its outer electrons. A few materials, such as carbon, have lower resistance as the temperature increases. The effect of temperature on resistance varies with the type of material. The effect of temperature on resistance is the least important of the factors that affect resistance.

VOLTAGE, CURRENT, AND RESISTANCE

We depend on electricity to do many things that are sometimes taken for granted. There are three basic electrical terms which must be understood, *voltage, current*, and *resistance*. The term voltage is best understood by looking at a flashlight battery. The battery is a source of voltage. It is capable of supplying electrical energy to a light connected to it. The voltage the battery supplies should be thought of as electrical pressure. The battery has positive (+) and negative (–) terminals.

For a battery to supply electrical pressure, a circuit must be formed. A simple electrical circuit has a source, a conductor, and a load. An electrical circuit is shown in Figure 1-14. The battery is the source. The conductor is a path that allows the electric current to pass the load. The lamp is called a load because it changes electrical energy to light energy. When the source, conductors, and load are connected together, a complete circuit is made. When the battery or voltage source is connected to the light bulb by conductors, a current flows. *Current* flows when the conductor is connected to the lamp because of the electrical pressure produced by the battery. The current flow causes the lamp to light.

The movement of electrons through a conductor takes place at a rate based on the *resistance* of a circuit. A lamp offers resistance to the flow of electrical current. Resistance is opposition to the flow of electrical current. More resistance in a circuit causes less current to flow. Resistance can be explained by using the example of two water pipes shown in Figure 1-16. If a water pump is connected to a large pipe such as pipe 1, water flows easily.

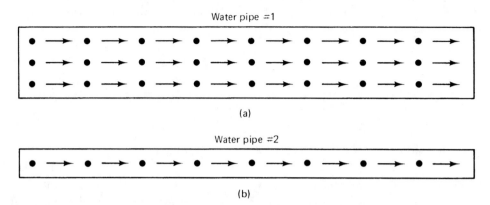

Figure 1-16. Water pipes showing the effect of resistance: (a) water pipe 1—many drops of water flow through; (b) water pipe 2—only a few drops of water flow through.

The pipe offers a small amount of resistance to the flow of water. However, if the same water pump is connected to a small pipe, such as pipe 2, there is more opposition to the flow of water. The water flow through pipe 2 is less.

Inside a lamp bulb, the part that glows is called a filament. The filament is a wire that offers resistance to the flow of electrical current. If the filament wire of a lamp is made of large wire, a great deal of current flows, as shown in circuit la} of Figure 1-17. The filament offers a small amount of resistance to the flow of current. However, circuit (b) shows a lamp with a filament of small wire. The small wire has more resistance. Therefore, less current flows in circuit (b).

In summary, *voltage* is electrical pressure that causes current to flow in a circuit. *Current* is the movement of electrons through a conductor in a circuit. *Resistance* is opposition to the flow of current in a circuit.

When an electrical pressure (voltage) is applied to an electrical circuit, the number of electrons (current) that flows is limited by the resistance of the path. Resistance is measured with a meter called an ohmmeter, since the basic unit of resistance is the ohm (Ω).

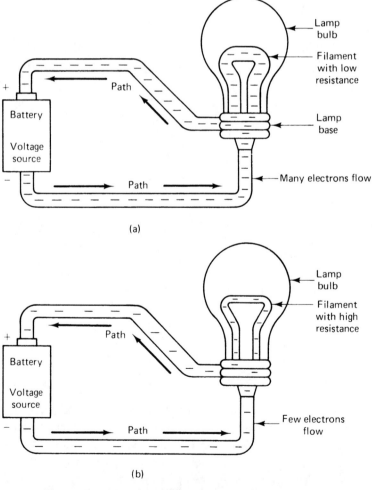

Figure 1-17. How lamp filament size affects current.

VOLTS, OHMS, AND AMPERES

Since the volt is a unit of electrical pressure, it is always measured across two points. An electrical pressure of 120 V exists across the terminals of electrical outlets in the home. This value is measured with an electrical instrument called a voltmeter.

The ampere, or amp, is a measure of the rate of flow of electric current. An ampere is a number of electrons per unit of time flowing in an electrical circuit. An ammeter is used to measure the number of electrons that flow in a circuit.

COMPONENTS, SYMBOLS, AND DIAGRAMS

Most electronic equipment is made of several parts, or components, that work together. It would be almost impossible to explain how equipment operates without using symbols and diagrams. Electronic diagrams show how the component parts of equipment fit together to form a system. Common electronic components are easy to identify. *Components* are represented by symbols. *Symbols* are used to make diagrams. A diagram shows how the components are connected together in a circuit. For example, it is easier to show symbols for a battery connected to a lamp than to draw a picture of the battery and the lamp connected together. There are several common symbols used in electronic diagrams that you should learn to recognize. Diagrams are used for installing, troubleshooting, and repairing electronic equipment. The use of symbols makes it easy to draw diagrams and to understand the purpose of each circuit. Common electronic symbols are listed in Appendix B.

Most electronic equipment uses wires (conductors) to connect its components. The symbol for a conductor is a narrow line. Figure 1-18) shows two conductors crossing one another. If two conductors are connected together, they may also be identified by symbols, as shown in Figure 1-18(b).

The symbols for two lamps connected across a battery are shown in Figure 1-19 using symbols for the battery and lamps. Notice the part of the diagram where the conductors are connected together.

A common electronic component is a switch, such as those shown in Figure 1-20. The simplest switch is a single-pole, single-throw (SPST) switch. This switch turns a circuit on or off. In Figure 1-21(a), the symbol for a switch in the "off" or open position is shown. There is no path for current to flow from the battery to the lamp. The lamp is off when the switch is open. Figure 1-21(b) shows a switch in the "on," or closed, position. This switch position completes the circuit and allows current to flow.

In many electronic circuits, a resistor is used. Resistors are usually small cylindrical components. They are used to control the flow of electrical current. The symbol for a resistor is shown in Figure 1-22. The most common type of resistor uses color coding to mark its value. Resistor values are always in ohms. For instance, a resistor might have a value of 100 Ω. Each color on the resistor represents a specific number. Resistor color-code values are easy to learn.

Another type of resistor is called a potentiometer, or pot. A pot is a variable resistor whose value can be

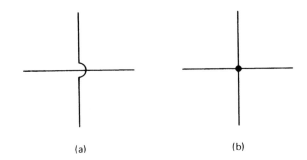

Figure 1-18. Symbols for electrical conductors: (a) conductors crossing; (b) conductors connected.

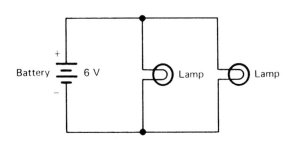

Figure 1-19. Symbols for a battery connected across two lamps.

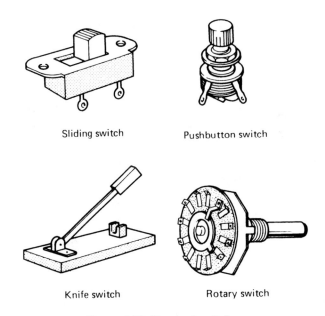

Sliding switch   Pushbutton switch

Knife switch   Rotary switch

Figure 1-20. Types of switches.

changed by adjusting a rotary shaft. For example, a 1000-Ω pot can be adjusted to any resistance value from 0 to 1000 Ω by rotating the shaft. The symbol of this component is shown in Figure 1-23(a) and 1-23(b). In the example shown in Figure l-23(c), potentiometer 1 is adjusted so that the resistance between points *A* and *B* is zero. The resistance between points B and C is 1000 Ω. By turning the

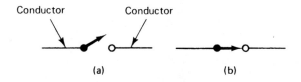

Figure 1-21. Symbol for a single-pole, single-throw (SPST) switch: (a) open or off condition; (b) closed or on condition.

Figure 1-22. The symbol for a resistor.

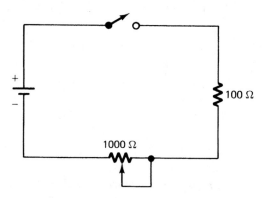

Figure 1-24. Simple circuit diagram.

shaft as far in the opposite direction as it will go, the resistance between points B and C becomes zero (see potentiometer 2). Between points *A* and *B*, the resistance is now 1000 Ω. By rotating the shaft to the center of its movement, as shown by potentiometer 3, the resistance is split in half, so the resistance from point *A* to point *B* is now about 500 Ω and the resistance from point *B* to point C is also about 500 Ω.

The symbol for a battery is shown in Figure 1-19. The symbol for any battery over 1.5 V is indicated by two sets of lines. A 1.5-V battery or cell is shown with one set of lines. The voltage of a battery is marked near its symbol. The long line in the symbol is always the positive (+) side, and the short line is the negative (–) side of the

A simple circuit diagram using symbols is shown in Figure 1-24. This diagram shows a 1.5-V battery connected to an SPST switch, a 100-Ω resistor, and a 1000-Ω potentiometer. Symbols are used to replace words. Anyone using this diagram should recognize the components represented and how they fit together to form a circuit.

RESISTORS

There is some resistance in all electronic circuits. Resistance is added to a circuit to control current flow. Devices that are used to cause proper resistance in a circuit are called *resistors*. A wide variety of resistors ire used. Some have a fixed value, and others are variable. Resistors are made of either special carbon material or of metal film. Wire-wound resistors are ordinarily used to control large currents, and carbon resistors control smaller currents.

Wire-wound resistors are constructed by winding resistance wire on an insulating material. The wire ends are attached to metal terminals. An enamel coating is used to protect the wire and to conduct heat away from it. Wire-wound resistors may have fixed taps, which can be used to change the resistance value in steps. They may also have sliders, which can be adjusted to change the resistance to any fraction of their total resistance. Precision-wound resistors are used where the resistance value must

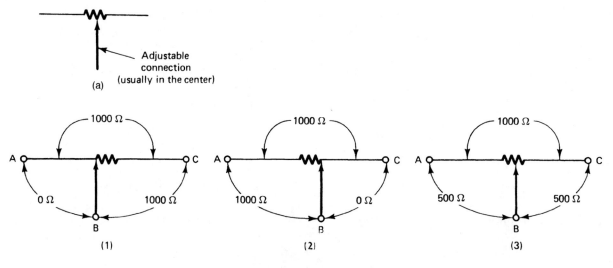

Figure 1-23. Potentiometers: (a) symbol; (b) examples.

be very accurate, such as in measuring instruments.

Carbon resistors are constructed of a small cylinder of compressed material. Wires are attached to each end of the cylinder. The cylinder is then covered with an insulating coating.

Variable resistors are used to change resistance while equipment is in operation. They are ordinarily called potentiometers or rheostats. Both carbon and wire-wound variable resistors are made on a circular form (see Figure 1-25(a)). A contact arm is attached to the form to make contact with the wire. The contact arm can be adjusted to any position on the circular form by a rotating shaft. A wire connected to the movable contact is used to vary the resistance from the contact arm to either of the two outer wires of the variable resistor. For controlling smaller currents, carbon variable resistors are made by using a carbon compound mounted on a fiber disk (see Figure 1-25(b)). A contact on a movable arm varies the resistance as the arm is turned by rotating a metal shaft.

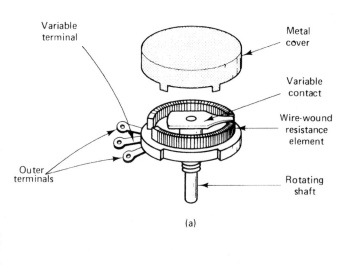

(a)

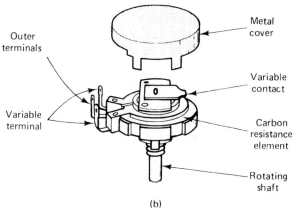

(b)

Figure 1-25. Variable resistor construction: (a) wire-wound variable resistor; (b) carbon variable resistor.

Resistor Color Codes

It is usually easy to find the value of a resistor by its color code or marked value. Most wire-wound resistors have resistance values (in ohms) printed on the resistor. If they are not marked in this way, an ohmmeter must be used to measure the value. Most carbon resistors use colored bands to identify their value, as shown in Figure 1-26. Resistors are made with the body of the resistor and the carbon cylinder coated with an insulating material.

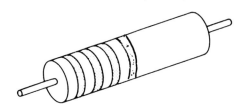

Figure 1-26. Carbon resistors.

Most resistors are color coded by an end-to-center color-band system of marking. Three colors are used to indicate the resistance value in ohms. A fourth color is used to indicate the tolerance of the resistor. The colors are read in the correct order from the end of a resistor. Numbers from the resistor color code, shown in Figure 1-27 are substituted for the colors. Through practice using the resistor color code, the value of a resistor may be determined at a glance.

It is difficult to manufacture a resistor to the exact value required. For many uses, the actual resistance value can be as much as 20% higher or lower than the value marked on the resistor without causing any problem. In most cases, the actual resistance does not need to be any closer than 10% higher or lower than the marked value. Resistors with tolerances of lower than 5% are called *precision resistors*.

Resistors are marked with color bands at one end of the resistor. For example, when a resistor that has three color bands (yellow, violet, and brown) at one end, the color bands are read from the end toward the center, as shown in Figure 1-28. The resistance value is 470 Ω. Remember that black as the center dot or third color means that no zeros are to be added to the digits. A resistor with a green band, a red band, and a black band has a value of 52 Ω.

An example of a resistor with the band color-code system is shown in Figure 1-29. Read the colors from left to right from the end of the resistor where the bands begin. Use the resistor color chart to determine the value of the resistor in ohms, its tolerance, and its failure rate.

Color	1st digit	2nd digit	Number of zeros or multiplier	Tolerance (%)
Black	0	0		
Brown	1	1	10	
Red	2	2	100	
Orange	3	3	1,000	
Yellow	4	4	10,000	
Green	5	5	100,000	
Blue	6	6	1,000,000	
Violet	7	7		
Gray	8	8		
White	9	9		
Gold*	—	—	0.1	5
Silver*	—	—	0.01	10
No color	—	—		20

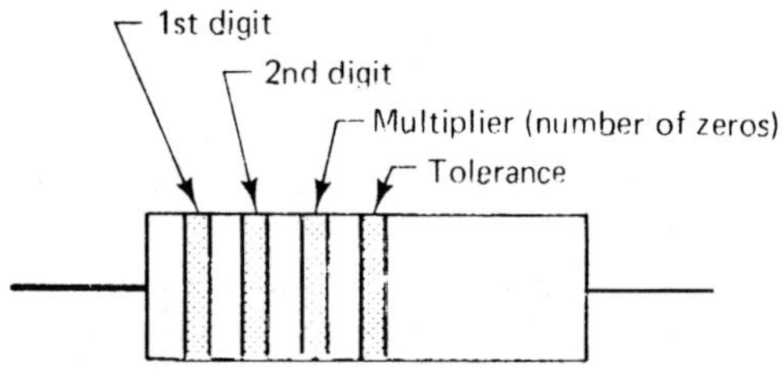

1st digit
2nd digit
Multiplier (number of zeros)
Tolerance

*When resistors have a value of less than 10 Ω, the third color band is a decimal multiplier. The two colors used are: gold = × 0.1 and silver = × 0.01.

Figure 1-27. Resistor color code.

The first color is orange, so the first digit in the value of the resistor is 3. The second color is black, so the second digit in the value of the resistor is 0. The first two digits of the resistor value is 30. The third color is yellow, which indicates the number by which the first two digits are to be multiplied. Sometimes it is easier to think of the third color as the number of zeros to add to the first two digits. The color yellow indicates a multiplier of 10,000.

Multiply 30 × 10,000 to obtain the value of the resistor (30 × 10,000 = 300,000 Ω). A yellow color means that four zeros are added to the first two digits to find the value of the resistor (30 with 4 zeros added equals 300,000 Ω).

The fourth color is silver. This band gives the tolerance of the resistor. The tolerance shows how close the actual value of the resistor should be to the color-code value. Tolerance is a percentage of the actual value. In this example, silver shows a tolerance of ±10%. This means

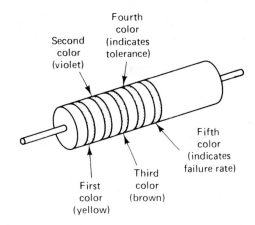

Figure 1-28. End-to-center system for axial resistors, resistor value = 470 Ω.

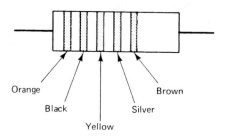

Figure 1-29. Resistor color-code example.

that the resistor should be within 10% of 300,000 Ω in either direction. The value of the resistor could be as low as 270,000 Ω (300,000 × 0.10 = 30,000 and 300,000 − 30,000 = 270,000), or the value could be as high as 330,000 (300,000 × 0.10 = 30,000 and 300,000 + 30,000 = 330,000).

The fifth color is brown, which shows the percentage of failure per 1000 h of use. The color brown shows a failure rate of 1% per 1000 h of resistor use.

Sometimes there is no fourth color on the body of the resistor. Then the tolerance is 20%. Frequently, no fifth color appears on the resistor. Using the color code, it is easy to list the value in ohms, tolerance, and failure rate of resistors. Standard values of 5% and 10% tolerance for color-coded resistors are listed in Figure 1-30.

Another way to determine the value of color-coded resistors is to remember the following: Big Brown Rabbits Often Yield Big Vocal Groans When Gingerly Slapped. The first letter of each word in the mnemonic is the same as the first letter in each of the colors used in the color code. The words of the sentence are counted (beginning with zero) to find the word corresponding to each digit or the number of zeros to be added (refer to Figure 1-31). Some method such as this mnemonic should be used to remember the color code.

0.27	1.2	5.6	27	120	560	2700	12,000	56,000	0.27M	1.2M	5.6M
0.33	1.5	6.8	33	150	680	3300	15,000	68,000	0.33M	1.5M	6.8M
0.39	1.8	8.2	39	180	820	3900	18,000	82,000	0.39M	1.8M	8.2M
0.47	2.2	10	47	220	1000	4700	22,000	0.1M	0.47M	2.2M	10M
0.56	2.7	12	56	270	1200	5600	27,000	0.12M	0.56M	2.7M	12M
0.68	3.3	15	68	330	1500	6800	33,000	0.15M	0.68M	3.3M	15M
0.82	3.9	18	82	390	1800	8200	39,000	0.18M	0.82M	3.9M	18M
1.0	4.7	22	100	470	2200	10,000	47,000	0.22M	1.0M	4.7M	22M

10% tolerance

0.24	1.1	5.1	24	110	510	2400	11,000	51,000	0.24M	1.1M	5.1M
0.27	1.2	5.6	27	120	560	2700	12,000	56,000	0.27M	1.2M	5.6M
0.30	1.3	6.2	30	130	620	3000	13,000	62,000	0.30M	1.3M	6.2M
0.33	1.5	6.8	33	150	680	3300	15,000	68,000	0.33M	1.5M	6.8M
0.36	1.6	7.5	36	160	750	3600	16,000	75,000	0.36M	1.6M	7.5M
0.39	1.8	8.2	39	180	820	3900	18,000	82,000	0.39M	1.8M	8.2M
0.43	2.0	9.1	43	200	910	4300	20,000	91,000	0.43M	2.0M	9.1M
0.47	2.2	10	47	220	1000	4700	22,000	0.1M	0.47M	2.2M	10M
0.51	2.4	11	51	240	1100	5100	24,000	0.11M	0.51M	2.4M	11M
0.56	2.7	12	56	270	1200	5600	27,000	0.12M	0.56M	2.7M	12M
0.62	3.0	13	62	300	1300	6200	30,000	0.13M	0.62M	3.0M	13M
0.68	3.3	15	68	330	1500	6800	33,000	0.15M	0.68M	3.3M	15M
0.75	3.6	16	75	360	1600	7500	36,000	0.16M	0.75M	3.6M	16M
0.82	3.9	18	82	390	1800	8200	39,000	0.18M	0.82M	3.9M	18M
0.91	4.3	20	91	430	2000	9100	43,000	0.20M	0.91M	4.3M	20M
1.0	4.7	22	100	470	2200	10,000	47,000	0.22M	1.0M	4.7M	22M

5% tolerance

Figure 1-30. Standard values of color-coded resistors (Ω).

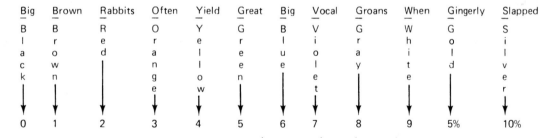

Big	Brown	Rabbits	Often	Yield	Great	Big	Vocal	Groans	When	Gingerly	Slapped
B	B	R	O	Y	G	B	V	G	W	G	S
l	r	e	r	e	r	l	i	r	h	o	i
a	o	d	a	l	e	u	o	a	i	l	l
c	w		n	l	e	e	l	y	t	d	v
k	n		g	o	n		e		e		e
			e	w			t				r
↓	↓	↓	↓	↓	↓	↓	↓	↓	↓	↓	↓
0	1	2	3	4	5	6	7	8	9	5%	10%

Figure 1-31. Quote used to memorize resistor color code.

Power Rating of Resistors

The size of a resistor helps to determine its power rating. Larger resistors are used in circuits that have high power ratings. Small resistors will become damaged if they are put in high-power circuits. The power rating of a resistor indicates its ability to give off or dissipate heat. Common power (wattage) ratings of color-coded resistors are 1/8, 1/4, 1/2, 1, and 2. Resistors that are larger in physical size will give off more heat and have higher power ratings.

SCHEMATIC DIAGRAMS

Schematic diagrams are used to show how the components or parts of each circuit or piece of electronic equipment fit together. They show the details of the electrical connections of any type of circuit or system. Schematics are used by manufacturers of electronic equipment to show operation and as an aid in servicing the equipment. A typical schematic diagram is shown in Figure 1-32. Standard electrical symbols used by all equipment manufacturers represent electrical components in schematic diagrams. Some common basic electrical symbols are shown in Appendix A These symbols should be memorized.

BLOCK DIAGRAMS

Another way to show how electrical equipment operates is to use *block diagrams.* Block diagrams show the functions of the subparts of any electrical system. A block diagram of an electrical system was shown in Figure 1-1(a). The same type of diagram is used to show the parts of a radio in Figure 1-33. Inside the blocks, symbols or words are used to describe the function of the block. Block diagrams usually show the operation of the whole system. They provide an idea of how a system operates. However, they do not show detail as does a schematic diagram. It is easy to see the major subparts of a system by looking at a block diagram.

WIRING DIAGRAMS

Another type of electrical diagram is called a *wiring diagram* (sometimes called a cabling diagram). Wiring diagrams show the actual location of parts and wires on equipment. The details of each connection are shown on a wiring diagram. Schematic and block diagrams show only how parts fit together electrically. Wiring diagrams show the details of actual connections. A simple wiring diagram is shown in Figure 1-34.

ELECTRICAL UNITS

There are four common units of electrical measurement. Voltage is used to indicate the force that causes electron movement. Resistance is the opposition to electron movement. The amount of work done or energy used in the movement of electrons in a given period of time is called power. Figure 1-35 shows the four basic units of electrical measurement.

Small Units

The measurement of a value is often less than a whole unit, such as 0.6 V. 0.025 A, and 0.0550 W. Some of the prefixes used in such measurements are shown in Figure 1-36.

For example, a millivolt (mV) is 0.001 V and a microampere (μA) is 0.000001 A. The prefixes of Figure 1-36

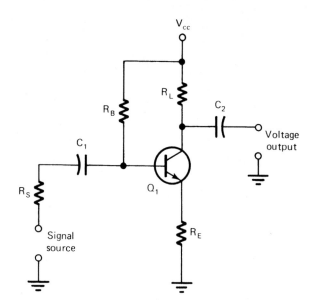

Figure 1-32. Schematic diagram of a transistor amplifier circuit.

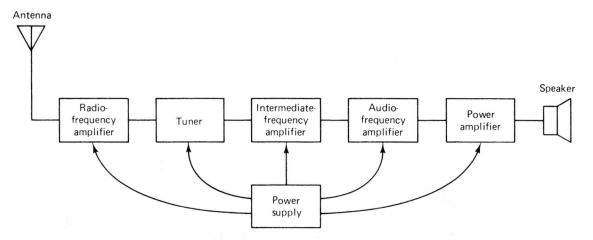

Figure 1-33. Block diagram of the parts of a radio.

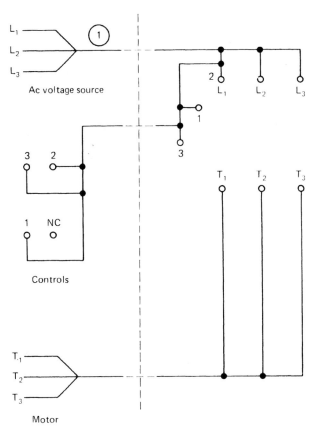

Figure 1-34. Simple wiring diagram.

may be used with any electrical unit of measurement. The unit is divided by the fractional part of the unit. For example, to change 0.6 V to millivolts, divide by the fractional part indicated by the prefix. So 0.6 V equals 600 mV, or 0.6 V ÷ 0.001 = 600 mV. To change 0.0005 A to microamperes, divide by 0.000001: 0.0005 A = 500 μA. When changing a basic electrical unit to a unit with a prefix, move the decimal point of the unit to the right by the same number of places in the fractional prefix. To change 0.8 V to millivolts, the decimal point of 0.8 V is moved three places to the right (8 0 0) since the prefix milli has three decimal places. So 0.8 V equals 800 mV. A similar method is used for converting any electrical unit to a unit with a smaller prefix.

When a unit with a prefix is converted back to a basic unit, the prefix must be multiplied by the fractional value of the prefix. For example, 68 mV is equal to 0.068 V. When 68 mV is multiplied by the fractional value of the prefix (0.001 for the prefix milli), this gives 68 mV × 0.001 = 0.0068 V. That is, to change a unit with a prefix into a basic electrical unit, move the decimal in the prefix unit to the left by the same number of places as the value of the prefix. To change 225 mV to volts, move the decimal point in 225 three places to the left (2 2 5), since the value of the prefix milli has three decimal places. So 225 m V equals 0.225 V.

Electrical quantity	Unit of measurement	Symbol	Description
Voltage	Volt (V)	*V*	Electrical "pressure" that causes current flow
Current	Ampere (A)	*I*	Amount of electron movement through a circuit
Resistance	Ohm (Ω)	*R*	Opposition to current flow
Power	Watt (W)	*P*	Amount of work done as current flows through a circuit

Figure 1-35. Basic units of electrical measurement.

Prefix	Abbreviation	Fractional part of a whole unit
milli	m	1 1000 or 0.001 (3 decimal places)
micro	μ	1 1,000,000 or 0.000001 (6 decimal places)
nano	n	1 1,000,000,000 or 0.000000001 (9 decimal places)
pico	p	1 1,000,000,000,000 or 0.000000000001 (12 decimal places)

Figure 1-36. Prefixes of units smaller than 1.

Large Units

Sometimes electrical measurements are very large, such as 20,000,000 W, 50,000 Ω, or 38,000 V. When this occurs, prefixes are used to make these numbers more manageable. Some prefixes used for large electrical values are shown in Figure 1-37. To change a large value to a smaller unit divide the large value by the value of the prefix. For example, 48,000,000 Ω is changed to 48 megohms (MΩ) by dividing: 48,000,000 Ω ÷ 1,000,000 = 48 MΩ. To convert 7000 V to 7 kilovolts (kV), divide: 7000 V ÷ 1000 = 7 kV. To change a large value to a unit with a prefix, move the decimal point in the large value to the left by the number of zeros of the prefix. Thus 3600 V equals 3.6 kV (3 6 0 0). To convert a unit with a prefix back to a standard unit, the decimal point is moved to the right by the same number of places in the unit, or, the number may be multiplied by the value of the prefix. To convert 90 MΩ to ohms, the decimal point is moved six places to the right (90,000,000). The 90-MΩ value may also be multiplied by the value of the prefix, which is 1,000,000. This 90 MΩ × 1,000,000 = 90,000,000 Ω.

Prefix	Abbreviation	Number of times larger than 1
Kilo	k	1000
Mega	M	1,000,000
Giga	G	1,000,000,000

Figure 1-37. Prefixes of large units.

The simple conversion scale shown in Figure 1-38 is useful when converting standard units to units of measurement with prefixes. This scale uses either powers of 10 or decimals to express the units.

SCIENTIFIC NOTATION

Using scientific notation greatly simplifies arithmetic operations. Any number written as a multiple of a power of 10 and a number between 1 and 10 is said to be expressed in scientific notation. For example:

$$81,000,000 = 8.1 \times 10,000,000,$$
$$\text{or } 8.1 \times 10^7$$

$$500,000,000 = 5 \times 100,000,000,$$
$$\text{or } 5 \times 10^8$$

$$0.0000000004 = 4 \times 0.0000000001,$$
$$\text{or } 4 \times 10^{-10}$$

Figure 1-39 lists some of the powers of 10. In a whole-number power of 10, the power to which 10 is raised is positive and equals the number of zeros following the 1. In decimals, the power of 10 is negative and equals the number of places the decimal point is moved to the left of the 1.

Scientific notation simplifies multiplying and dividing large numbers of small decimals. For example:

$$4800 \times 0.000045 \times 800 \times 0.0058$$
$$= (4.8 \times 10^3) \times (4.5 \times 10^{-5}) \times (8 \times 10^2)$$
$$\times 5.8 \times 10^{-3})$$
$$= (4.8 \times 4.5 \times 8 \times 5.8) \times (10^{3-5+2-3})$$
$$= 1002.24 \times 10^{-3}$$
$$= 1.00224$$

$$95,000 \div 0.0008 \quad = \quad \frac{9.5 \times 10^4}{8 \times 10^4}$$

$$= \quad \frac{9.5 \times 10^{4(-4)}}{8}$$

$$= \quad 9.5 \times 10^8$$
$$= \quad 1.1875 \times 10^8$$
$$= \quad 118,750,000$$

BATTERIES

Conversion of chemical energy into electrical energy is done by chemical cells. When two or more cells are connected in series or parallel (or a combination of both), they form a *battery*. A cell is made of two different metals immersed in a liquid or paste called an *electrolyte*. Chemical cells are either primary or secondary cells. *Primary cells* are usable only for a certain time period. *Secondary cells* are renewed after being used to produce electrical energy. This is known as *charging*. Both primary and secondary cells have many uses.

The operation of a primary cell involves the placing of two unlike materials called *electrodes* into a solution called an *electrolyte*. When the materials of the cell are brought together, their molecular structures are changed. During this chemical change, atoms may either gain additional electrons or leave behind some electrons. These atoms then have either a positive or a negative electrical charge and are called ions. Ionization of atoms allows a chemical solution of a cell to produce an electrical current.

A load device such as a lamp may be connected to a cell. Electrons flow from one of the cell's electrodes to

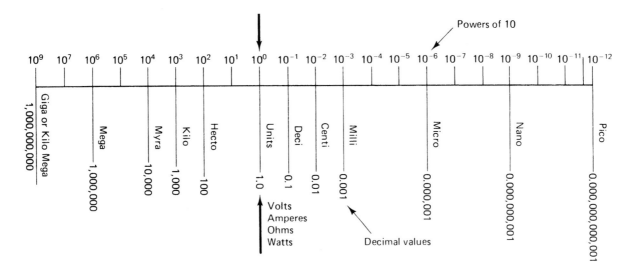

Figure 1-38. Simple conversion scale for large or small numbers.

	Number	Power of 10
Whole numbers	1,000,000	10^6
	100,000	10^5
	10,000	10^4
	1000	10^3
	100	10^2
	10	10^1
	1.0	10^0
Decimals	0.1	10^{-1}
	0.01	10^{-2}
	0.001	10^{-3}
	0.0001	10^{-4}
	0.00001	10^{-5}
	0.000001	10^{-6}

Figure 1-39. Powers of 10.

the other through the electrolyte material. This creates an electrical current flow through the load. Current leaves the cell through its negative electrode. It passes through the load device and then goes back to the cell through its positive electrode. A complete circuit exists between the cell (source) and the lamp (load).

The voltage output of a *primary* cell depends on the electrode materials used and the type of electrolyte. The familiar carbon-zinc cell produces approximately 1.5 V. The negative electrode of this cell is the zinc container. The positive electrode is a carbon rod. A paste material acts as the electrolyte. It is placed between the two electrodes. This type of cell is called a *dry cell.*

There are many types of primary cells used today. The carbon-zinc cell is the most common type. This cell is low in cost and is available in many sizes. Applications include portable equipment and instruments. For uses that require higher voltage or current than one cell can deliver, several cells are combined in series, parallel, or series-parallel connections, as shown in Figure 1-40. Carbon-zinc batteries are available in many voltage ratings.

Another type of primary cell is the mercury (zinc-mercuric oxide) cell. This cell is an improvement over the carbon-zinc cell because it has a more constant voltage output, a longer life, and a smaller size. Mercury cells are more expensive. They produce a voltage of about 1.35 V, which is slightly less than carbon-zinc cells.

An alkaline (zinc-manganese dioxide) cell has a voltage per cell of 1.5 V. They supply higher-current electrical loads. They have much longer lives than carbon-zinc cells of the same types.

Chemical cells that may be reactivated by charging are called *secondary cells*, or *storage cells*. Common types are the lead-acid, nickel-iron (Edison), and nickel-cadmi-

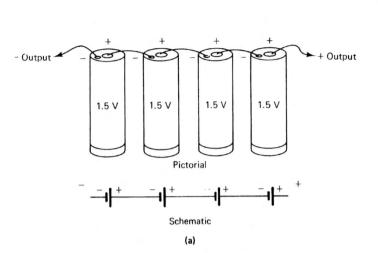

Figure 1-40(a). Series voltage connection.

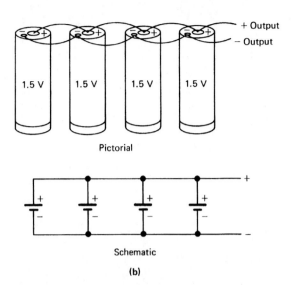

Figure 1-40(b). Parallel voltage connection.

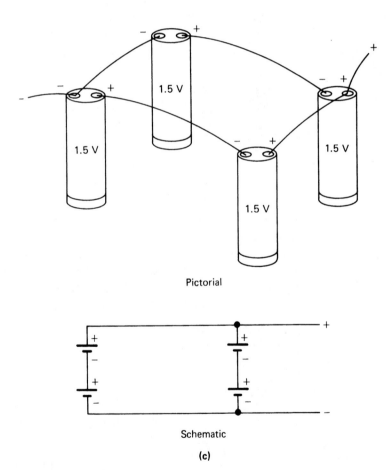

Figure 1-40(c). Series-parallel voltage connection.

um cells. The latter two cells serve as common electrical sources. The lead-acid cell is a *secondary* cell. The electrodes of lead-acid cells are made of lead and lead peroxide. The positive plate is lead peroxide, and the negative plate is lead. The electrolyte is sulfuric acid (H_2SO_4).

There are other types of secondary cells used today, many of which have specialized applications. Among these are silver oxide-zinc cells and nickel-cadmium cells. These cells have a high output and long life and are more expensive than other types of secondary cells of the same size.

Secondary cells have many uses. Storage batteries are used in some buildings to provide emergency power when a power failure occurs. Standby systems are needed, especially for lighting when power is off. Automobiles use storage batteries for their everyday operation. Many types of instruments and portable equipment use batteries for power.

MEASURING VOLTAGE, CURRENT, AND RESISTANCE

Another important activity in the study of electronics is measurement. Measurements are made in many types of electronic circuits. The proper ways of measuring resistance, voltage, and current should be learned.

Measuring Resistance

Many important electrical tests may be made by measuring resistance. Resistance is opposition to the flow of current in an electrical circuit. The current that flows in a circuit depends upon the amount of resistance in that circuit. You should learn to measure resistance in an electronic circuit by using a meter.

A volt-ohm-milliammeter (VOM), or multimeter, is used to measure resistance. VOMs are among the most-used meters for doing electronic work. A VOM is used to measure resistance, voltage, or current. The type of measurement is changed by adjusting the "function-select switch" to the desired measurement.

The test leads used with the VOM are ordinarily black and red. These colors are used to help identify which lead is the positive and which is the negative side of the meter. This is important when measuring direct-current (dc) values. Red indicates positive polarity (+) and black indicates the negative (–) polarity.

The red test lead is put in the hole, or jack, marked V-Ω-A, or volts-ohms-amperes. The black test lead is put in the hole, or jack, labeled –COM, or negatance ranges. When the test leads are touched together, or "shorted," the meter is operational, indicating zero (0) ohms.

Never measure the resistance of a component until it has been disconnected, or the reading may be wrong. Voltage should never be applied to a component when measuring resistance.

A VOM may also be used to measure the resistance of a potentiometer, as shown in Figure 1-41. If the shaft of the pot is adjusted while the ohmmeter is connected to points *A* and *C*, no resistance change will take place. The resistance of the potentiometer is measured in this way. Connecting to points *B* and *C* or to points *B* and *A* allows changes in resistance as the shaft is turned. The potentiometer shaft may be adjusted both clockwise and counterclockwise. This adjustment affects the measured resistance across points B and C or B and A. The resistance varies from zero to maximum and from maximum back to zero as the shaft is adjusted.

How to Measure Resistance with a VOM

Remember that resistance is the opposition to flow of electric current. For example, a lamp connected to a battery has resistance, which is determined by the size of the filament wire. The filament wire opposes the flow of electrical current from the battery, and the battery causes current to flow through the lamp's filament. The amount of current through the lamp depends on the filament resistance. If the resistance offers little opposition to current flow from the battery, a large current *flows* in the circuit.

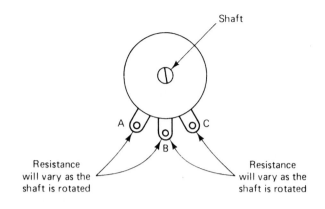

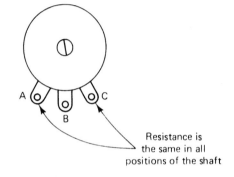

Figure 1-41. Measuring the resistance of a potentiometer.

If the lamp filament has high resistance, it offers a great deal of opposition to current *flow* from the battery. Then a small current *flows* in the circuit.

Resistance tests are sometimes called *continuity checks*. A continuity check is made to see if a circuit is open or closed (a "continuous" path). An ohmmeter is also used to measure exact values of resistance. Resistance must always be measured with no voltage applied to the component being measured. The ohmmeter ranges of a VOM are used to measure resistance. This *type* of meter is often used by electronics technicians for measuring resistance, voltage, or current. By adjusting the rotary function select switch, the meter can be set to measure either resistance, voltage, or current.

To learn to measure resistance, it is easy to use color-coded resistors. These resistors are small and easy to handle.

Measuring Voltage

Voltage is applied to electrical equipment to cause it to operate. It is important to be able to measure voltage to check the operation of equipment. Many electrical problems develop when either too much or too little voltage is applied to the equipment. A voltmeter is used to measure voltage in an electrical circuit.

When making voltage measurements, adjust the function-select switch to the highest range of dc voltage. Connect the red and black test leads to the meter by putting them into the proper jacks. The red test lead should be put into the jack labeled V-Ω-A. The black test lead should be put into the jack labeled –COM.

Before making any measurements, the proper dc voltage range is chosen. The value of the range being used is the maximum value of voltage that can be measured on that range. For example, when the range selected is 12 V, the maximum voltage the meter can measure is 12 V. Any voltage above 12 V could damage the meter. To measure a voltage that is unknown (no indication of its value), start by using the highest range on the meter. Then slowly adjust the range downward until a voltage reading is indicated on the right side of the meter scale.

Matching the meter polarity to the voltage polarity is very important when measuring de voltage. Meter polarity is simple to determine. The positive (+) red test lead is connected to the positive side of the dc voltage being measured. The negative (–) black test lead is connected to the negative side of the dc voltage being measured. The meter is always connected *across* (in parallel with) the dc voltage being measured.

How to Measure dc Voltage with a VOM

Voltage is the electrical pressure that causes current to flow in a circuit. A common voltage source is a battery. Batteries come in many sizes and voltage values. The voltage applied to a component determines how much current will flow through it.

A dc voltmeter or the dc voltage ranges of a VOM are used to measure dc voltage.

When measuring voltage with a VOM, first put the test leads into the meter. The red test lead is plugged into the hole marked V-Ω-A. The black test lead is plugged into the hole marked –COM. The scale of the meter is used to indicate voltage (in volts).

To measure a dc voltage, first select the proper range. If the meter's range setting is on the 3-V range, this means that the voltage being measured cannot be larger than 3 V. If the voltage is greater than 3 V, the meter would probably be damaged. You must be careful to use a meter range that is larger than the voltage being measured.

When measuring direct current volt age, polarity is very important. The proper matching of meter polarity and voltage source polarity must be assured. The negative (black) test lead of the meter is connected to the negative polarity of the voltage being measured. The positive (red) test lead is connected to the positive polarity of the voltage.

A certain amount of voltage, called *voltage drop*, is needed to cause electrical current to flow through a resistance in a circuit. Voltage drop is measured across any component through which current flows. The polarity of a voltage drop depends on the direction of current flow. Current flows from the negative polarity of a battery to the positive polarity. In Figure 1-42 the bottom of each resistor is negative. The top of each resistor is positive. The negative test lead of the meter is connected to the bottom of the resistor, and the positive test lead is connected to the top. The meters are connected as shown to measure each of the voltage drops in the circuit.

Measuring Current

Current flows through a complete electrical circuit when voltage is applied. Many important tests are made by measuring current flow in electrical circuits. The current values in an electrical circuit depend on the amount of resistance in the circuit. Learning to use an ammeter to measure current in an electrical circuit is very important.

Most VOMs will also measure dc current. The function-select switch may be adjusted to any of the ranges of direct current. For example, when the function-select switch is placed in the 120-mA range, the meter is capable of measuring up to 120 mA of current. The value of the current set on the range is the maximum value that can be measured on that range. The function-select switch should first be adjusted to the highest range of direct current. Current is measured by connecting the meter into a circuit as shown in Figure 1-43. This is referred to as connecting the meter in series with the circuit.

Current flows from a voltage source when some device that has resistance is connected to the source. When a lamp is connected to a battery, a current flows from the battery through the lamp. In the circuit of Figure 1-43, electrons flow from the negative battery terminal through the lamp and back to the positive battery terminal.

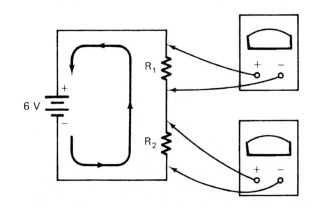

Figure 1-42. Measuring voltage drop in a dc circuit.

As the voltage applied to a circuit increases, the current also increases. So, if 12 V is applied to the lamp in Figure 1-43, a larger current will flow through the lamp.

How to Measure dc Current with a VOM

When measuring current, always start with the meter set on its highest range. Then move the range setting to a lower value if the current is a small amount.

When the meter is set on the 120-mA range, the meter will measure up to 120 mA of direct current.

The test-lead polarity of the VOM is also important when measuring direct current. The VOM is connected to allow current to flow through the meter in the right direction. The negative test lead is connected nearest to the negative side of the voltage source. The meter is then connected into the circuit. To measure current, a wire is removed from the circuit to place the meter into the circuit. No voltage should be applied to the circuit when connecting the current meter. The meter is placed in series with the circuit. Series circuit have one path for current flow.

The proper procedure for measuring current through point A in the circuit of Figure 1-43 is as follows:

1. Turn off the circuit's voltage source by opening the switch.

2. Set the meter to the highest current range.

3. Remove the wire at point A.

4. Connect the negative test lead of the meter to the negative side of the voltage source.

5. Connect the positive test lead to the end of the wire that was removed from point A.

6. Turn on the switch to apply voltage to the circuit.

Always remember the following safety tips when measuring current:

1. Turn off the voltage before connecting the meter in order not to get an electrical shock. This is an important habit to develop. Always remember to turn off the voltage before connecting the meter.

2. Set the meter to its highest current range.

3. Disconnect a wire from the circuit and put the meter in series with the circuit. Always remember to disconnect a wire and reconnect the wire to one of the meter test leads. If a wire is not removed to put the meter into the circuit, the meter will not be connected properly.

4. Use the proper meter polarity. The negative test lead is connected so that it is nearest the negative side of the voltage source. Similarly, the positive test lead is connected so that it is nearest the positive side of the voltage source.

DIGITAL METERS

Many digital meters are now in use. They employ numerical readouts to simplify the measurement process and to make more accurate measurements. Digital meters rely on the operation of digital circuitry to produce a numerical readout of the measured quantity.

The readout of a digital meter is designed to transform electric signals into numerical data. Both letter and number readouts are available, as are seven-segment, discrete-number, and bar matrix displays. Each method has a device designed to change electric energy into light energy on the display.

OHM'S LAW

Ohm's law is the most basic and most used of all electrical theories. Ohm's law explains the relationship of voltage (the force that causes current to flow), current (the movement of electrons), and resistance (the opposition to current flow). Ohm's law is stated as follows: An increase

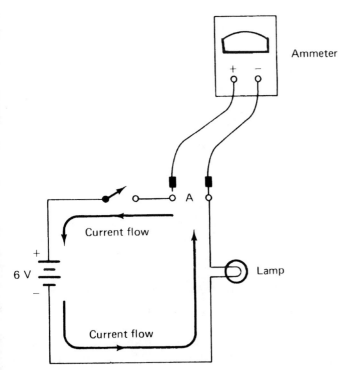

Figure 1-43. Meter connection for measuring dc current.

in voltage increases current if resistance remains the same. Ohm's law stated in another way is: An increase in resistance causes a decrease in current if voltage remains the same. The electrical values used with Ohm's law are usually represented with capital letters. For example, voltage is represented with the letter *V*, current with the letter *I*, and resistance with the letter *R*. The mathematical relationship of the three electrical units is shown in the following formulas, which should be memorized. The Ohm's law circle of Figure 1-44 is helpful in remembering the formulas.

$$V = I \times R$$

$$R = \frac{V}{R}$$

$$R = \frac{V}{I}$$

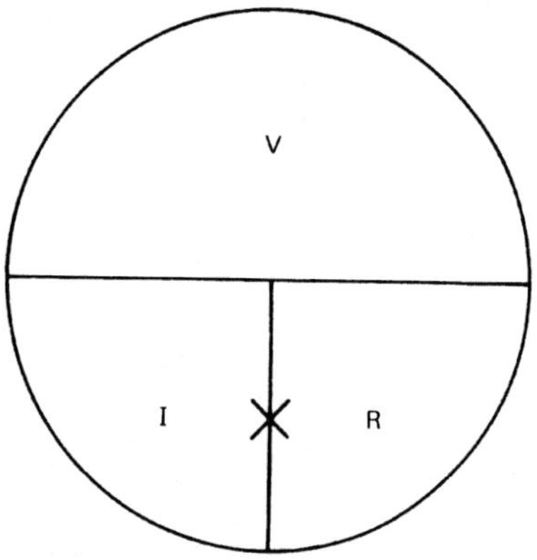

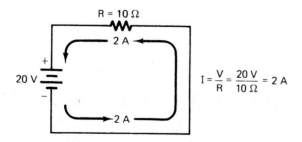

Figure 1-44. Ohm's law circle; *V*, voltage; *I*, current; *R*, resistance. To use the circle, cover the value you want to find and read the other values as they appear in the formula: *V = I × R*, *I = V/R*, *R = V/I*.

Voltage (*V*) is measured in volts. Current (*I*) is measured in amperes. Resistance (*R*) is measured in ohms. If two values are known, the third value can be calculated by using one of the formulas. Look at Figure 1-45. Using the Ohm's law current formula, *I = V/R*, the calculated value of *I* in the circuit is

Figure 1-45. Ohm's law example.

$$I = \frac{V}{R} = \frac{10\ V}{10\ \Omega} = 1\ A$$

If the voltage in the circuit is doubled, as shown in Figure 1-46 the calculated current using the Ohm's law current formula (*I = V/R*), is

$$I = \frac{V}{R} = \frac{20\ V}{10\ \Omega} = 2\ A$$

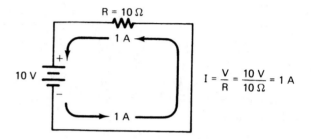

$$I = \frac{V}{R} = \frac{10\ V}{10\ \Omega} = 1\ A$$

Figure 1-46. Effect of doubling the voltage.

From this example, notice that as voltage is increased, current also increases if resistance remains the same. Also, if voltage is doubled, current is doubled. If voltage is 10 times as large, the current becomes 10 times as large.

Now look at Figure 1-47(a). The calculated current flow in the circuit is

$$I = \frac{V}{R} = \frac{10\ V}{100\ \Omega} = 0.1\ A$$

In Figure 1-48, the 100-Ω resistor in the circuit is replaced with a 1-kΩ resistor. The calculated value of the current with 10 V applied to the 1-kΩ resistance is

$$I = \frac{V}{R} = \frac{10\ V}{1000\ \Omega} = 0.01\ A$$

As resistance increases, current decreases.

Ohm's law explains the relationship of voltage, current, and resistance in electrical circuits. The circle shown in Figure 1-44 can be used to help remember this relationship. To calculate the voltage in a circuit, cover the *V* on the circle. Note that *V* is equal to *I* times *R*. To find current, cover the *I* and note that *I* is equal to *V* over *R*. To find resistance, cover the *R* and note that *R* is equal to *V* over *I*.

Another example of using Ohm's law is shown in Figure 1-48(a). To find the value of current that flows in this circuit, use the Ohm's law circle. Cover the *I* and find

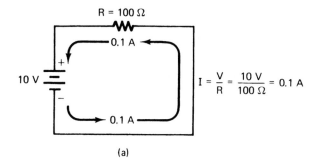

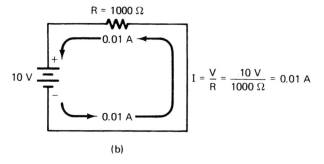

Figure 1-47. Effect of increasing resistance.

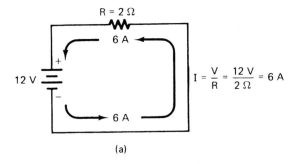

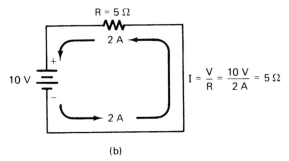

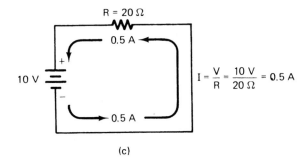

Figure 1-48. Ohm's law examples.

that *I* equals *V* over *R*. In this circuit, *I* = 12 V divided by 2 Ω, so the current is equal to 6 A. The voltage of the circuit in Figure 1-48(b) is equal to 10 V, and the current is equal to 2 A. The resistance of the circuit is equal to 10 V divided by 2 A, or 5 Ω. Ohm's law also states that if the resistance of a circuit increases, the current will decrease if the voltage stays the same. An example of this is shown in Figure 1-48(c). The resistance of the circuit is increased to 20 Ω. The current of the circuit is now equal to 10 V divided by 20 Ω, or 0.5 A. The current in the previous circuit was 2 A. If the resistance of a circuit is increased by a factor of four, the current flow decreases to one-fourth of its original value.

In a circuit in which current and resistance are known, Ohm's law is used to find voltage. In the circuit of Figure 1-49, assume that resistance is equal to 20 Ω and the current is 5 A. Use the circle and cover the *V*. This shows that *V* equals *I* times *R*. So the voltage required to cause 5 A of current through a 20 Ω resistance is equal to 5 A times 20 Ω (100 V).

Ohm's law is also used to find the value of resistance in a circuit. Assume that a circuit has a known value of voltage and current. The value of resistance required to cause this value of current flow may be found. In the example of Figure 1-50, the voltage is equal to 70 V and the current equals 10 A. To find the resistance of the circuit, cover the *R* and find that *R* is equal to *V* over *I*. So the resistance of this circuit is equal to 70 V divided by 10 A, or 7 Ω.

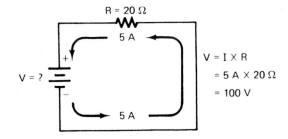

Figure 1-49. Using Ohm's law to find voltage.

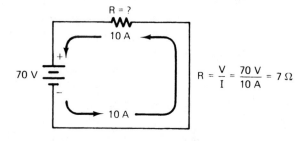

Figure 1-50. Using Ohm's law to find resistance.

PROBLEM-SOLVING
WITH CALCULATORS

Simple electronic circuit problems involving small whole numbers are often solved in a few seconds by pencil-and-paper arithmetic. We can determine that the application of 12 V to a 6-Ω resistor results in a current of 2 A through the resistor because Ohm's law instructs us to divide voltage by resistance to get current. With greater numbers, calculations become more difficult and time-consuming. An electronic calculator is a great labor-saving device that also reduces errors and increases accuracy of answers. Multidigit numbers may be added, subtracted, multiplied, divided, squared, and more by a few simple strokes on the buttons of a calculator keypad.

SERIES ELECTRICAL CIRCUITS

One type of electrical circuit is referred to as a series circuit. In a series circuit, there is only one path for current to flow. Since there is only one current path, the current flow is the same value in any part of the circuit. The voltages in the circuit depend on the resistance of the components in the circuit. When a series circuit is opened, there is no path for current flow. Thus, the circuit does not operate.

There are three types of electrical circuits. They are called *series* circuits, *parallel* circuits, and *combination* circuits. The easiest type of circuit to understand is the series circuit. Series circuits are different from other types of electrical circuits. It is important to remember the characteristics of a series circuit.

In the circuit examples that follow, subscripts (such as in R_T, V_T, I_1) are used to identify electrical components in circuit diagrams. The circuit shown in Figure 1-51 has two resistors and a battery. The resistors are labeled R_1 and R_2. The subscripts identify each of these resistors. Subscripts also aid in making measurements. The voltage drop across resistor R_1 is called voltage drop V_1. The term *total* is represented by the subscript T, such as in V_T, which is total voltage applied to a circuit. The current measurement I_2 is the current through resistor R_2 measured at point B. Total current (I_T) is measured at point A. The voltage drop across R_2 is called V_2.

Subscripts are also valuable in troubleshooting and repair of electronic equipment. It would be impossible to isolate problems in equipment without components that are easily identified.

The main characteristic of a series circuit is that it has only one path for current flow. In the circuit shown

in Figure 1-51, current flows from the negative side of the voltage source through resistor R_1, through resistor R_2, and then to the positive side of the voltage source. Another characteristic of a series circuit is that the current is the same everywhere in the circuit. A wire at point A could be removed and a VOM placed into the circuit to measure current. The value of current at point A is the same as the current at points B and C of the circuit.

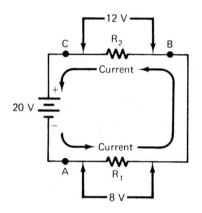

Figure 1-51. Series electrical circuit.

In any series circuit, the sum of the voltage drops is equal to the voltage applied to the circuit. The circuit shown in Figure 1-51 has voltage drops of 12 V plus 8 V, which is equal to 20 V. Another characteristic of a series circuit is that its total resistance to current flow is equal to the sum of all resistance in the circuit. In the circuit shown in Figure 1-52, the total resistance of the circuit is the sum of the two resistances, so the total resistance is equal to 20 Ω plus 10 Ω, or 30 Ω.

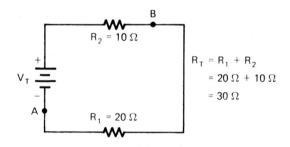

Figure 1-52. Finding total resistance in a series circuit.

When a series circuit is opened, there is no longer a path for current flow. The circuit will not operate. In the circuit of Figure 1-53, if lamp 1 is burned out, its filament is open. Since a series circuit has only one current path, that path is broken. No current flows in the circuit. Lamp 2 will not work either. If one light burns out, the others go out also.

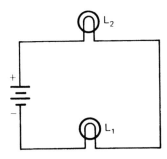

Figure 1-53. Two lamps connected in series.

Ohm's law is used to explain how a series circuit operates. In the circuit of Figure 1-54, the total resistance is equal to 2 Ω plus 3 Ω, or 5 Ω. The applied voltage is 10 V. Current is equal to voltage divided by resistance, or $I = V/R$. In the circuit shown, current is equal to 10 V divided by 5 Ω, which is 2 A. If a current meter were connected into this circuit, the current measurement would be 2 A. Voltage drops across each of the resistors may also be found. Voltage is equal to current times resistance ($V = I \times R$). The voltage drop across R_1 (V_1) is equal to the current through R_1 (2 A) times the value of R_1 (2 Ω), which is 2 A × 2 Ω, or 4 V. The voltage drop across R_2 (V_2) equals 2 A × 3 Ω or 6 V. The sum of these voltage drops is equal to the applied voltage. To check these values, add 4 V plus 6 V, which is equal to 10 V.

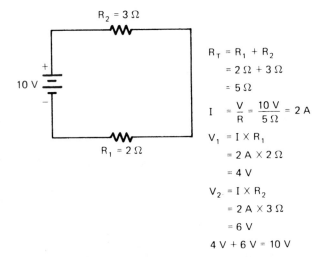

Figure 1-54. Using Ohm's law for a series circuit.

If another resistance is added to a series circuit, as shown in Figure 1-55 resistance increases. Since there is more resistance, the current flow becomes smaller. The circuit now has R_3 (a 5-Ω resistor) added in series to R_1 and R_2. The total resistance is now 2 Ω + 3 Ω + 5 Ω, or 10 Ω, compared to 5 Ω in the previous example. The current is now 1 A, compared to 2 A in the other circuit:

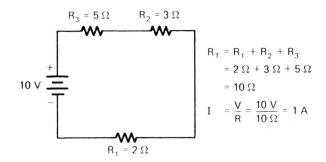

Figure 1-55. Effect of adding resistance to a series circuit.

Summary of Series Circuits

There are several important characteristics of series circuits:

1. The same current flows through each part of a series circuit.

2. The total resistance of a series circuit is equal to the sum of the individual resistances.

3. The voltage applied to a series circuit is equal to sum of the individual voltage drops.

4. The voltage drop across a resistor in a series circuit is directly proportional to the size of the resistor.

5. If the circuit is broken at any point, no current will flow.

Examples of Series Circuits

There is only one path for current flow in a series circuit. The same current flows through each part of the circuit. To find the current through a series circuit, only one value must be found. Kirchhoff's current law states that the sum of the voltage drops across the parts of a series circuit is equal to the applied voltage. In a series circuit, the sum of the voltage drops always equals the source voltage. This can be shown by looking at the circuit shown in Figure 1-56. In this circuit, a voltage (V_T) of 30 V is applied to a series circuit that has two 10-Ω resistors. The total resistance (R_T) of the circuit is equal to the sum of the resistor values (20 Ω). Using Ohm's law, the total current of the circuit is

$$I_T = \frac{V_T}{R_T} = \frac{30\ V}{20\ \Omega} = 1.5\ A$$

The resistors each equal 10 Ω and the current through them is 1.5 A. The voltage drop across the resistors is then found. The voltage (V_1) across resistor R_1 is

$$V_1 = I \times R_1 = 1.5\,A \times 10\,\Omega = 15\,V$$

Resistor R_2 is the same value as resistor R_1 and the same current flows through it. The voltage drop across resistor R_2 also equals 15 V. Adding these two 15-V drops gives a total voltage of 30 V. The sum of the voltage drops is equal to the applied voltage (15 V + 15 V = 30 V).

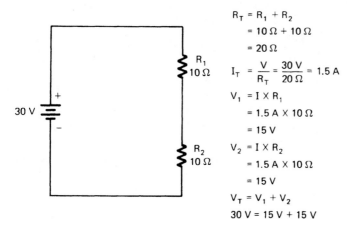

$R_T = R_1 + R_2$
$\quad = 10\,\Omega + 10\,\Omega$
$\quad = 20\,\Omega$

$I_T = \dfrac{V}{R_T} = \dfrac{30\,V}{20\,\Omega} = 1.5\,A$

$V_1 = I \times R_1$
$\quad = 1.5\,A \times 10\,\Omega$
$\quad = 15\,V$

$V_2 = I \times R_2$
$\quad = 1.5\,A \times 10\,\Omega$
$\quad = 15\,V$

$V_T = V_1 + V_2$
$30\,V = 15\,V + 15\,V$

Figure 1-56. Series-circuit example.

The series circuit of Figure 1-57 has three resistors with values of 10, 20, and 30 Ω. The applied voltage (V_T) may be found, since the current through the circuit is known to be 1 A. The circuit is a series circuit, so the same current flows through each resistor. Using Ohm's law ($V = I \times R$), the voltage drop across each resistor can be found:

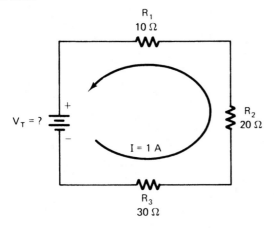

Figure 1-57. Series-circuit example.

$V_1 = I \times R_1 = 1\,A \times 10\,\Omega = 10\,V$
$V_2 = I \times R_2 = 1\,A \times 20\,\Omega = 20\,V$
$V_3 = I \times R_3 = 1\,A \times 30\,\Omega = 30\,V$

When the individual voltage drops are known, they may be added to find the applied voltage (V_T):

$$\begin{aligned} V_T &= V_1 + V_2 + V_3 \\ &= 10\,V + 20\,V + 30\,V \\ &= 60\,V \end{aligned}$$

Notice that the voltage drops in a series circuit are in direct proportion (when one increases, the other does too) to the resistance across which they appear. This is true since the same current flows through each resistor. The larger the value of the resistor, the larger the voltage drop across it will be.

PARALLEL ELECTRICAL CIRCUITS

It is also important to understand the characteristics of *parallel circuits*. Parallel circuits are different from series circuits in several ways. A parallel circuit has two or more paths for current to flow from the voltage source. In Figure 1-58, path 1 is from the negative side of the voltage source through R_1 and back to the positive side of the voltage source. Path 2 is from the negative side of the voltage source through R_2 and back to the positive side of the voltage source. Also, path 3 is from the negative side of the power supply through R_3 and back to the voltage source.

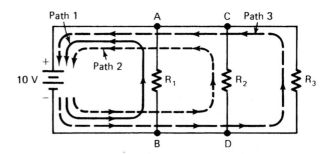

Figure 1-58. Parallel electrical circuit.

In a parallel circuit, the voltage is the same across every component of the circuit. In Figure 1-58, the voltage across points A and B is equal to 10 V. This is the value of the voltage applied to the circuit. By following point A to point C, it can be seen that these two points are connected. Points B and D are also connected together. So the voltage from point A to point B will be the same as the voltage from point C to point D.

Another characteristic of a parallel circuit is that the sum of the currents through each path equals the total current that flows from the voltage source. In Figure 1-59, the currents through the paths are 1 A, 2 A, and 3 A. One ampere of current could be measured through R_1 at point *A* or point *B*. Two amperes of current could be measured through R_2 and 3 A through R_3. The total current is equal to 6 A. This value of total current could be measured at point *C* or point *D* in the circuit. Remember that the current is the same in every part of a series circuit, and current divides through each branch in a parallel circuit. More resistance causes less current to flow through a *branch*, or parallel path of a circuit.

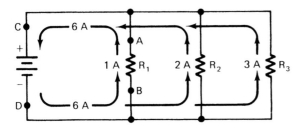

Figure 1-59. Current flow in a parallel circuit.

The *total resistance* (R_T) of a parallel circuit is found by the formula

$$\frac{1}{R_T} = \frac{1}{R_1} + \frac{1}{R_2} + \ldots + \frac{1}{R_n}$$

1 divided by total resistance $(1/R_T)$ is equal to $1/R_1 + 1/R_2 + 1/R_3 + 1$ divided by as many other resistances as there are in the circuit. This is called an inverse or reciprocal formula (see Figure 1-60). When trying to find the total resistance of a parallel circuit, first write the formula. Notice that each value is divided into 1. Next, divide each resistance value into 1 and write the values. Then add these values to get a value of 1 divided by total resistance. Do not forget to divide the value obtained by 1 to find the total resistance of the circuit. Also, always remember that the total resistance in a parallel circuit is less than any individual resistance in the circuit. In the example shown, 1.33 is less than 2 Ω or 4 Ω.

If there are *only two paths* in a parallel circuit, the following formula can be used:

$$R_T = \frac{R_1 \times R_2}{R_1 \times R_2}$$

The two resistances are multiplied and then added. When using this formula, it is not necessary to find the reciprocals of the resistance values. This formula may be

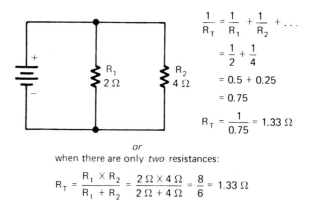

$$\frac{1}{R_T} = \frac{1}{R_1} + \frac{1}{R_2} + \ldots$$
$$= \frac{1}{2} + \frac{1}{4}$$
$$= 0.5 + 0.25$$
$$= 0.75$$
$$R_T = \frac{1}{0.75} = 1.33\ \Omega$$

or

when there are only *two* resistances:

$$R_T = \frac{R_1 \times R_2}{R_1 + R_2} = \frac{2\,\Omega \times 4\,\Omega}{2\,\Omega + 4\,\Omega} = \frac{8}{6} = 1.33\ \Omega$$

Figure 1-60. Finding total resistance of a parallel circuit.

used *only* when there are two resistances in a parallel circuit. If there are more than two resistances, the reciprocal formula must be used.

Another simple method of finding total resistance in a parallel circuit is when all the resistance values are the same. An example of a parallel circuit with all resistances the same is a string of lights connected in parallel. Each lamp has the same resistance. When all resistances are equal, to find total resistance, divide the resistance value of each resistor by the number of paths (refer to Figure 1-61). If five 20-Ω resistors are connected in parallel, the total resistance is equal to 20 divided by 5, or 4 Ω.

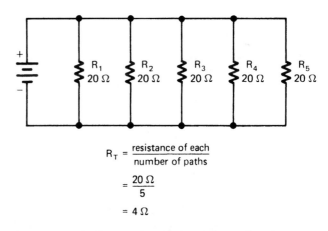

$$R_T = \frac{\text{resistance of each}}{\text{number of paths}}$$
$$= \frac{20\ \Omega}{5}$$
$$= 4\ \Omega$$

Figure 1-61. Finding total resistance when all resistances are the same.

When one of the components of a parallel path is opened, the rest of the circuit will continue to operate. Remember that in a series circuit, when one component is opened, no current will flow in the circuit. Since the same voltage is applied to all parts of a parallel circuit, the circuit will operate unless the path from the voltage source is broken (refer to Figure 1-62). If lamp 3 has a filament burned out, this causes an open circuit. No current flows

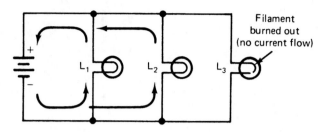

Figure 1-62. Three lamps connected in parallel.

through lamp 3. However, lamps 1 and 2 continue to operate. Some types of lights are connected in parallel. If one of these lights burns out, the rest of them will still burn.

A sample problem with a parallel circuit is shown in Figure 1-63. This circuit has a source voltage of 10 V and resistors of 5, 10, and 20 Ω. The total resistance of this circuit is 2.85 Ω. The total opposition to current flow of this circuit is 2.85 Ω. Next, find the current that flows from the voltage source. In this circuit the total current is found by dividing the voltage by the total resistance, so 10 V ÷ 2.85 Ω = 3.5 A. This means that 3.5 A of current flows from the voltage source and divides into the three paths of the circuit.

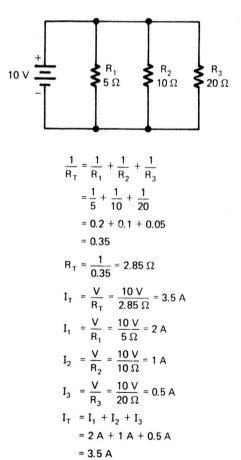

$$\frac{1}{R_T} = \frac{1}{R_1} + \frac{1}{R_2} + \frac{1}{R_3}$$

$$= \frac{1}{5} + \frac{1}{10} + \frac{1}{20}$$

$$= 0.2 + 0.1 + 0.05$$

$$= 0.35$$

$$R_T = \frac{1}{0.35} = 2.85 \ \Omega$$

$$I_T = \frac{V}{R_T} = \frac{10 \ V}{2.85 \ \Omega} = 3.5 \ A$$

$$I_1 = \frac{V}{R_1} = \frac{10 \ V}{5 \ \Omega} = 2 \ A$$

$$I_2 = \frac{V}{R_2} = \frac{10 \ V}{10 \ \Omega} = 1 \ A$$

$$I_3 = \frac{V}{R_3} = \frac{10 \ V}{20 \ \Omega} = 0.5 \ A$$

$$I_T = I_1 + I_2 + I_3$$

$$= 2 \ A + 1 \ A + 0.5 \ A$$

$$= 3.5 \ A$$

Figure 1-63. Simple parallel-circuit problem.

Now, find the current through each path of the circuit. The current from the voltage source divides into each path of the circuit. A lower resistance in a path causes more current to flow through the path. Current is equal to voltage divided by resistance. The current through each path is found by dividing the voltage (10 V) by the resistance of the path. Notice that the same voltage is across each path. The current values are 2, 1, and 0.5 A. These are added together to equal the total current. The total current is 3.5 A (2 A + 1A+ 0.5 A).

Summary of Parallel Circuits

There are several important basic rules for parallel circuits.

1.   There are two or more paths for current flow.

2.   Voltage is the same across each component of the circuit.

3.   The sum of the currents through each path is equal to the total current that flows from the source.

4.   Total resistance is found by using the formula

$$\frac{1}{R_T} = \frac{1}{R_1} + \frac{1}{R_2} \ \frac{1}{R_3} + ... + \frac{1}{R_n}$$

5.   If one of the parallel paths is broken, current will continue to flow in all the other paths.

Examples of Parallel Circuits

Parallel circuits have more than one current path connected to the same voltage source. A parallel circuit example is shown in Figure 1-64 Start at the voltage source (V_T) and look at the circuit. Two complete current paths are formed. One path is from the voltage source through resistor R_1 and back to the voltage source. The other path is from the voltage source through resistor R_2 and back to

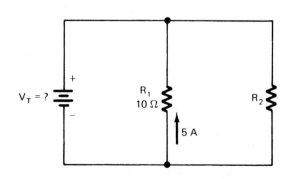

Figure 1-64. Parallel-circuit example.

the voltage source. In a parallel circuit, the same voltage occurs across all the resistors.

Assume that the current through resistor R_1 in Figure 1-64 is 5 A and the value of the resistor is 10 Ω. The voltage across resistors R_1 and R_2 and the source voltage (V_T) may be found. Using Ohm's law, the voltage across resistor R_1 is:

$$V_T = I_1 \times R_1 = 5\,A \times 10\,Ω = 50\,V$$

Since this is a parallel circuit, the voltages are the same. The source voltage in this circuit is equal to 50 V.

Ohm's law shows that the current in a circuit is *inversely* proportional to the resistance of the circuit. This means that when resistance increases, current decreases. The division of current in parallel circuit paths is based on this fact. Remember that in a series circuit, the current is the same through each part. The total current in a parallel circuit divides through the paths based on the value of resistance in each path.

The flow of current in a parallel circuit is shown by using the example in Figure 1-65. Figure 1-65(a) shows a series circuit. The total current passes through one resis-

tor (R_1). The amount of current is

$$I_1 = \frac{V_T}{R_1} = \frac{10\,V}{10\,Ω} = 1.0\,A$$

Figure 1-65(b) shows the same circuit with another resistor (R_2) connected in parallel across the voltage source. The current through R_2 is the same as through R_1, since their resistances are equal:

$$I_2 = \frac{V_T}{R_2} = \frac{10\,V}{10\,Ω} = 1.0\,A$$

Since 1 A of current flows through each of the two resistors, the total current of 2 A flows from the source. The division of currents through the resistors is shown here. Each branch carries part of the total current that flows from the source.

Changing the value of any resistor in a parallel circuit has no effect on the current in the other branches. However, a change in resistance does affect the total current. If R_2 of Figure 1-65(b) is changed to 5 Ω (Figure 1-65(c)) the total current increases:

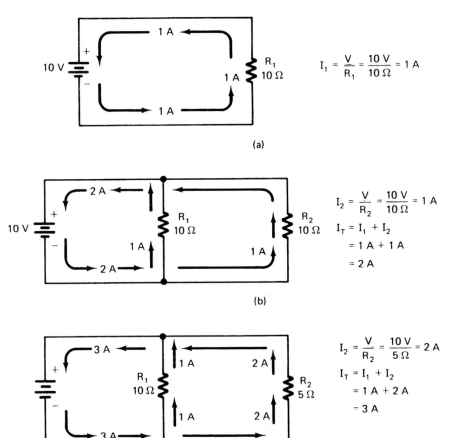

$$I_1 = \frac{V}{R_1} = \frac{10\,V}{10\,Ω} = 1\,A$$

$$I_2 = \frac{V}{R_2} = \frac{10\,V}{10\,Ω} = 1\,A$$
$$I_T = I_1 + I_2$$
$$= 1\,A + 1\,A$$
$$= 2\,A$$

$$I_2 = \frac{V}{R_2} = \frac{10\,V}{5\,Ω} = 2\,A$$
$$I_T = I_1 + I_2$$
$$= 1\,A + 2\,A$$
$$= 3\,A$$

Figure 1-65. Current flow in a parallel circuit: (a) one path; (b) two paths; (c) R_2 changed to 5 Ω.

$$I_1 = \frac{V_T}{R_1} = \frac{10 \text{ V}}{10 \text{ } \Omega} = 1.0 \text{ A}$$

$$I_2 = \frac{V_T}{R_2} = \frac{10 \text{ V}}{5 \text{ } \Omega} = 2 \text{ A}$$

$$I_T = I_1 + I_2 = 1 \text{ A} + 2 \text{ A} = 3.0 \text{ A}$$

The total resistance of a parallel circuit is not equal to the sum of the resistors, as in series circuits. When more branches are added to a parallel circuit, the total current (I_T) increases. In a parallel circuit, the total resistance is *less than any* of *the branch resistances*. As more parallel resistances are added, the total resistance of the circuit decreases.

There are several ways to find the total resistance of parallel circuits. The method used depends on the type of circuit. Some common methods for finding total parallel resistance are as follows.

1. *Equal resistors.* When two or more equal-value resistors are connected in parallel, their total resistance (R_T) is

$$R_T = \frac{\text{value of one resistance}}{\text{number of paths}}$$

If four 20-Ω resistors are connected in parallel, their total resistance is $20 \div 4 = 5 \text{ } \Omega$. For two 50-$\Omega$ resistors in parallel, $R_T = 50 \div 2 = 25 \text{ } \Omega$.

2. *Product over the sum.* Another shortcut for finding total resistance of two parallel resistors is called the product-over-sum method. This method is:

$$R_T = \frac{R_1 \times R_2}{R_1 + R_2}$$

For example, to find the total resistance (R_T) of a 10-Ω and a 20-Ω resistor connected in parallel,

$$R_T = \frac{R_1 \times R_2}{R_1 + R_2} = \frac{10 \times 20}{10 + 20} + \frac{200}{30}$$
$$= 6.67 \text{ } \Omega$$

Notice that the total resistance of a 10-Ω and a 20-Ω resistor is less than 10 Ω. The product-over-sum method can be used only for *two* resistances in parallel.

3. *Reciprocal method.* Most circuits have more than two resistors of unequal value. The reciprocal method must then be used to find total resistance. The reciprocal method is as follows:

$$\frac{1}{R_T} = \frac{1}{R_1} + \frac{1}{R_2} + \frac{1}{R_3} + \dots + \frac{1}{R_n}$$

For example, if 1-, 2-, 3-, and 4-Ω resistors are connected in parallel,

$$\frac{1}{R_T} = \frac{1}{1} + \frac{1}{2} + \frac{1}{3} + \frac{1}{4}$$
$$= 1 + 0.5 + 0.33 + 0.25 = 2.08$$
$$R_T = \frac{1}{2.08}$$
$$= 0.48 \text{ } \Omega$$

COMBINATION ELECTRICAL CIRCUITS

Combination electrical circuits are made up of both series and parallel parts. They are sometimes called *series-parallel* circuits. Almost all electrical equipment has combination circuits rather than only series circuits or only parallel circuits. However, it is important to understand series *and* parallel circuits in order to work with combination circuits.

In the circuit shown in Figure 1-66, R_1 is in series with the voltage source, and R_2 and R_3 are in parallel. There are many different types of combination circuits. Some have only one series component and many parallel components. Others have many series components and only a few parallel components. In the circuit shown in Figure 1-67, R_1 and R_2 are in series with the voltage source. R_3-R_4 and R_5-R_6 are in parallel. Notice that there are two paths in each parallel part of the circuit. The total current of the circuit flows through each series part of the circuit. In this circuit, the current is the same through resistors R_1 and R_2.

Another combination circuit is shown in Figure 1-68. R_1, R_2, and R_3 are in series with the voltage source. The total current (I_T) flows through R_1, R_2, and R_3. At point A, the current divides through the parallel paths of R_4, R_5, and R_6. The currents I_4, I_5, and I_6 flow through the parallel paths.

To find the *total resistance* of a combination circuit, the series resistance is added to the parallel resistance. In the circuit shown in Figure 1-69, R_1 is the only series part

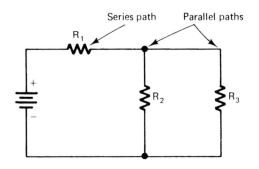

Figure 1-66. Simple combination circuit.

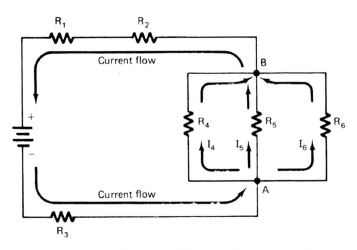

Figure 1-68. Current path in a combination circuit.

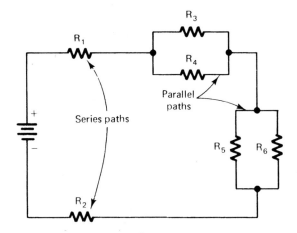

Figure 1-67. Combination Circuit.

of the circuit. R_2 and R_3 are in parallel. First, the parallel resistance of R_2 and R_3 is found. Then the series resistance and parallel resistance are added to find the total resistance. The total resistance of this circuit is 4 Ω.

To find the *total current* in the circuit, the same method that was discussed before is used. Current is equal to voltage divided by resistance. The current that flows from the voltage source in this circuit is 2.5 A. The total current also flows through each series part of the circuit, so 2.5 A also flows through R_1. In this circuit, it is easy to find the current flow through R_2 and R_3. Since R_2 and R_3 are equal resistance values, the same current flows through each of them. Here, 2.5 A of current flows to point *A* of the cir-

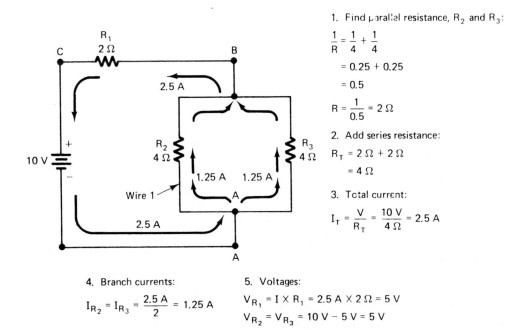

1. Find parallel resistance, R_2 and R_3:

$$\frac{1}{R} = \frac{1}{4} + \frac{1}{4}$$
$$= 0.25 + 0.25$$
$$= 0.5$$
$$R = \frac{1}{0.5} = 2\ \Omega$$

2. Add series resistance:

$$R_T = 2\ \Omega + 2\ \Omega$$
$$= 4\ \Omega$$

3. Total current:

$$I_T = \frac{V}{R_T} = \frac{10\ V}{4\ \Omega} = 2.5\ A$$

4. Branch currents:

$$I_{R_2} = I_{R_3} = \frac{2.5\ A}{2} = 1.25\ A$$

5. Voltages:

$$V_{R_1} = I \times R_1 = 2.5\ A \times 2\ \Omega = 5\ V$$
$$V_{R_2} = V_{R_3} = 10\ V - 5\ V = 5\ V$$

Figure 1-69. Combination-circuit example.

cuit. This 2.5 A divides into two paths through R_2 and R_3. To find the current through R_2 and R_3, divide the current coming into point A (2.5 A) by the number of paths (2), which gives 1.25 A.

To find voltage across R_1, multiply the current through R_1 by the value of R_1. The voltage across R_1 is equal to 2.5 A times 2 Ω, or 5 V. This is the voltage from point C to point B. The voltage across R_2 *or* R_3 is also across points B and A. The voltage across points B and A is found by subtracting the voltage across R_1 from the applied voltage, so it is equal to 10 V – 5 V, or 5 V.

Examples of Combination Circuits

Problems for combination circuits are solved by combining rules for series and parallel circuits.

Look at the circuit of Figure 1-70. The value that should first be calculated is the resistance of R_2 and R_3 in parallel. When this quantity is found, it can be added to the value of the series resistor (R_1) to find the total resistance of the circuit.

$$R_T = R_1 + R_2 \parallel R_3$$

$$= 30\ \Omega + \frac{10 \times 20}{10 + 20}$$

$$= 30\ \Omega + 6.67\ \Omega$$

$$= 36.67\ \Omega$$

(The symbol $\parallel$ means R_2 is in parallel with R_3.) When the total resistance (R_T) is found, the total current (I_T) may be found:

$$I_T = \frac{V_T}{R_T} = \frac{40\ \text{V}}{36.67\ \Omega} = 1.09\ \text{A}$$

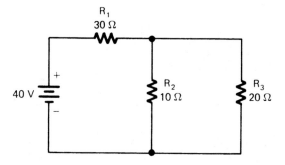

Figure 1-70. Combination-circuit example.

Notice that the total current flows through resistor R_1, since it is in series with the voltage source. The voltage drop across resistor R_1 is

$$V_3 = I_T \times R_1 = 1.09\ \text{A} \times 30\ \Omega = 32.7\ \text{V}$$

The applied voltage is 40 V, and 32.7 V is dropped across resistor R_1. The remaining voltage is dropped across the two parallel resistors R_2 and R_3; 40 V – 32.7 V = 7.3 V across R_2 and R_3. The currents through R_2 and R_3 are

$$I_2 = \frac{V_2}{R_2} = \frac{7.3\ \text{V}}{10\ \Omega} = 0.73\ \text{A}$$

$$I_3 = \frac{V_3}{R_3} = \frac{7.3\ \text{V}}{20\ \Omega} = 0.365\ \text{A}$$

Another type of combination circuit is shown in Figure 1-71. Resistors R_2 and R_3 are in series. When they are combined by adding their values, the circuit becomes a two-branch parallel circuit. Total resistance (R_T) is found by using the product-over-sum method:

$$R_T = \frac{10 \times 50}{10 + 50} = \frac{500}{60} = 8.33\ \Omega$$

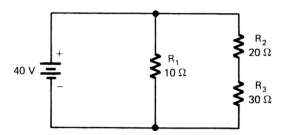

Figure 1-71. Combination-circuit example.

Total current is

$$I_T = \frac{V_T}{R_T} = \frac{40\ \text{V}}{8.33\ \Omega} = 4.8\ \text{A}$$

The voltage across R_1 is 40 V, since it is in parallel with the voltage source. Voltage drops across R_2 and R_3 are

$$I_2 = \frac{V_T}{R\ \text{of branch}} = \frac{40\text{V}}{50} = 0.8\ \text{A}$$

$$V_2 = I \times R_2$$
$$= 0.8\ \text{A} \times 20$$
$$= 16\text{V}$$
$$V_3 = I \times R_3$$
$$= 0.8\ \text{A} \times 30\ 1$$
$$= 24\text{V}$$

Notice that the sum of V_2 and V_3 (16 V + 24 V) is equal to the source voltage.

Combination circuit problems may be solved by using the following step-by-step procedure:

1. Combine series and parallel parts to find the total resistance of the circuit.

2. Find the total current that flows through the circuit.

3. Find the voltage across each part of the circuit.

4. Find the current through each resistance of this circuit.

Steps 3 and 4 must often be done in combination with each other, rather than doing one and then the other.

KIRCHHOFF'S LAWS

Ohm's law shows the relationship of voltage, current, and resistance in electrical circuits. *Kirchhoff's voltage law* is also important in solving electrical problems.

$$R_T = R_1 + R_2 + R_3 + R_4$$
$$= 5\,\Omega + 5\,\Omega + 5\,\Omega + 5\,\Omega$$
$$= 20\,\Omega$$

$$I_T = \frac{V}{R_T} = \frac{10\,V}{20\,\Omega} = 0.5\,A$$

All resistances are equal:
$$V_{R_1} = V_{R_2} = V_{R_3} = V_{R_4}$$
$$V_{R_1} = I \times R_1$$
$$= 0.5\,A \times 5\,\Omega$$
$$= 2.5\,V$$
Source voltage = sum of voltage drops
$$10\,V = 2.5\,V + 2.5\,V + 2.5\,V + 2.5\,V$$

(a)

Kirchhoff found that the sum of voltage drops around any closed-circuit loop must equal the voltage applied to that loop. Figure 1-72(a) shows a simple series circuit that illustrates this law. This law holds true for any series circuit loop.

Kirchhoff's current law is also important in solving electrical problems, especially for parallel circuits. The current law states that at any junction of electrical conductors in a circuit, the total amount of current entering the junction must equal the amount of current leaving the junction. Figure 1-72(b) shows some examples of Kirchhoff's current law.

POWER IN ELECTRICAL CIRCUITS

In terms of voltage and current, power (P) in watts (W) is equal to voltage (in volts) multiplied by current (in amperes). The formula is $P = V \times I$. For example, a 120-V electrical outlet with 4 A of current flowing from it has a power value of

$$P = V \times I = 120\,V \times 4\,A = 480\,W$$

The unit of electrical power is the watt. In the example, 480 W of power is converted by the load portion of the circuit. Another way to find power is

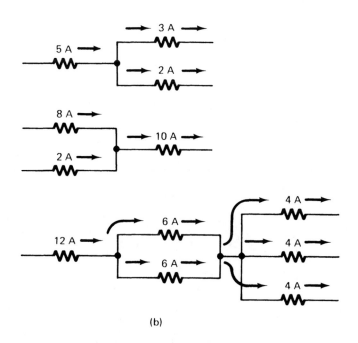

(b)

Figure 1-72. Kirchhoff's laws: (a) voltage law example; (b) current law examples.

$$P = \frac{V_2}{R}$$

This formula is used when voltage and resistance are known but current is not known. The formula $P = I_2 \times R$ is used when current and resistance are known.

Several formulas are summarized in Figure 1-73. The quantity in the center of the circle may be found by any of the three formulas along the outer part of the circle in the same part of the circle. This circle is handy to use for making electrical calculations for voltage, current, resistance, or power.

It is easy to find the amount of power converted by each of the resistors in a series circuit, such as the one shown in Figure 1-74. In the circuit shown, the amount of power converted by each of the resistors and the total power is found as follows:

1. Power converted by resistor R_1:
$P_1 = I_2 \times R_1 = 2^2 \times 20\ \Omega = 80\ \text{W}$

2. Power converted by resistor R_2:
$P_2 = I_2 \times R_2 = 2^2 \times 30\ \Omega = 120\ \text{W}$

3. Power converted by resistor P_3:
$P_3 = I_2 \times R_3 = 2^2 \times 50\ \Omega = 200\ \text{W}$

4. Power converted by the circuit:
$P_T = P_1 + P_2 + P_3$

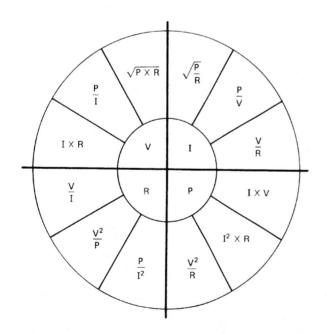

Figure 1-73. Formulas for finding voltage, current, resistance, or power.

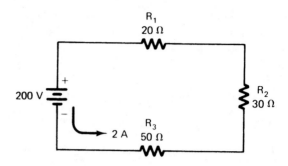

Figure 1-74. Finding power values in a series circuit.

$= 80\ \text{W} + 120\ \text{W} + 200\ \text{W} = 400\ \text{W}$

or

$P_T = V_T \times I = 200\ \text{V} \times 2\ \text{A} = 400\ \text{W}$

When working with electrical circuits, it is possible to check the results by using other formulas.

Power in parallel circuits is found in the same way as for series circuits. In the example of Figure 1-75, the power converted by each of the resistors and the total power of the parallel circuit are found as follows:

1. Power converted by resistor R_1:
$$P_1 = \frac{V_2}{R_1} = \frac{30^2}{5} = \frac{900}{9} = 180\ \text{W}$$

2. Power converted by resistor R_2:
$$P_2 = \frac{V_2}{R_2} = \frac{30^2}{10} = \frac{900}{10} = 90\ \text{W}$$

3. Power converted by resistor R_3:
$$P_3 = \frac{V_2}{R_3} = \frac{30^2}{20} = \frac{900}{20} = 45\ \text{W}$$

4. Total power converted by the circuit:
$$P_T = P_1 = P_2 = P_3$$
$$= 180\ \text{W} + 90\ \text{W} + 45\ \text{W} = 315\ \text{W}$$

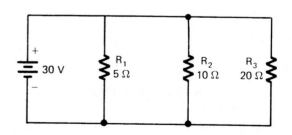

Figure 1-75. Finding power values in a parallel circuit.

MAGNETISM AND ELECTROMAGNETISM

Magnetism has been studied for many years. Some metals in their natural state attract small pieces of iron. This property is called *magnetism*. Materials that have this ability are called natural magnets. The first magnets used were called *lodestones*. Now, artificial magnets are made in many different strengths, sizes, and shapes. Magnetism is important because it is used in electric motors, generators, transformers, relays, and many other electrical devices. The earth itself has a magnetic field like a large magnet.

Electromagnetism is magnetism that is brought about due to electrical current flow. There are many electrical machines that operate because of electromagnetism. This unit deals with magnetism and electromagnetism and some important applications.

PERMANENT MAGNETS

Magnets are made of iron, cobalt, or nickel materials, usually in an alloy (or mixture). Each end of the magnet is called a pole. If a magnet is broken, each part becomes a magnet with two poles. Magnetic poles always occur in pairs. When a magnet is suspended in air so that it can turn freely, one pole will point to the North Pole of the earth, which explains why compasses can be used to determine direction. The north pole of a magnet *attracts* the south pole of another magnet. A north pole repels another north pole, and a south pole *repels* another south pole. The two laws of magnetism are (1) like poles repel, and (2) unlike poles attract.

The magnetic field patterns when two permanent magnets are placed end to end are shown in Figure 1-76. When the magnets are farther apart, a smaller force of attention or repulsion exists. This type of permanent magnet is called a *bar magnet*.

Some materials retain magnetism longer than others. Hard steel holds its magnetism much longer than soft steel. A *magnetic field* is set up around any magnetic material. The field is made up of *lines of force*, or *magnetic flux*. These magnetic flux lines are invisible. They never cross one another, but they always form individual closed loops around a magnetic material. They have a definite direction from the north to the south pole along the outside of a magnet. When magnetic flux lines are close together, the magnetic field is strong. When magnetic flux lines are farther apart, the field is weaker. The magnetic field is strongest near the poles. Lines of force pass through all materials. It is easy for lines of force to pass through iron and steel. Magnetic flux passes through a piece of iron as shown in Figure 1-77.

When magnetic flux passes through a piece of iron, the iron acts like a magnet. Magnetic poles are formed due to the influence of the flux lines. These are called induced poles. The induced poles and the magnet's poles attract and repel each other. Magnets attract pieces of soft iron in this way. It is possible to magnetize pieces of metal temporarily by using

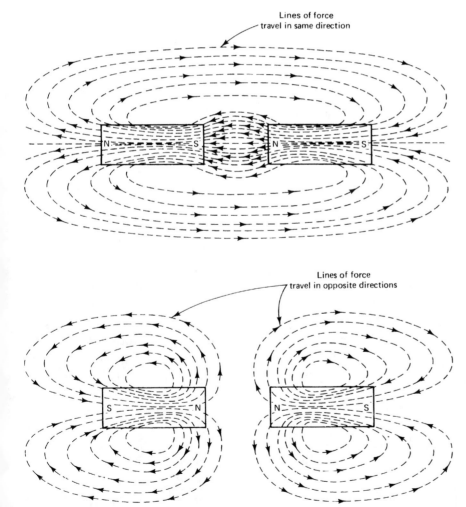

Figure 1-76. Magnetic field patterns when magnets are placed end to end.

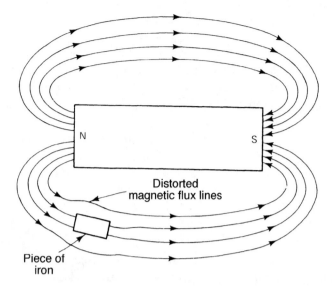

Figure 1-77. Magnetic flux lines distorted through a piece of iron.

a bar magnet. If a magnet is passed over the top of a piece of iron several times in the same direction, the soft iron becomes magnetized and stays magnetized for a short period of time.

When a compass is brought near the north pole of a magnet, the north-seeking pole of the compass is attracted to it. The polarities of the magnet may be determined by observing a compass brought near each pole. Compasses detect the presence of magnetic fields.

Horseshoe magnets are similar to bar magnets. They are bent in the shape of a horseshoe, as shown in Figure 1-78. This shape gives more magnetic field strength than a similar bar magnet, since the magnetic poles are closer. The magnetic field strength is more concentrated into one area. Many electrical devices use horseshoe magnets.

A magnetic material can lose some of its magnetism if it is jarred or heated. People must be careful when handling equipment that contains permanent magnets. A magnet also becomes weakened by loss of magnetic flux. Magnets should always be stored with a keeper, which is a soft-iron piece used to join magnetic poles. The keep-

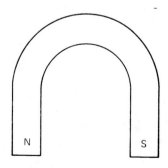

Figure 1-78. Horseshoe magnet.

er provides the magnetic flux with an easy path between poles. The magnet will retain its greatest strength for a longer period of time if keepers are used. Bar magnets should always be stored in pairs with a north pole and a south pole placed together. A complete path for magnetic flux is made in this way.

MAGNETIC FIELD AROUND CONDUCTORS

Current-carrying conductors produce a magnetic field. A compass is used to show that the magnetic flux lines are circular in shape. The conductor is in the center of the circular shape. The direction of the current flow and the magnetic flux lines can be shown by using the *left-hand rule* of magnetic flux. A conductor is held in the left hand, as shown in Figure 1-79(a). The thumb points in the direction of current flow from negative to positive. The fingers then encircle the conductor in the direction of the magnetic flux lines.

The circular magnetic field produced around a conductor is stronger near the conductor and becomes weaker farther away from the conductor. A cross-sectional end view of a conductor with current flowing toward the observer is shown in Figure 1-79(b). Current flow toward the observer is shown by a circle with a dot in the center. Notice that the direction of the magnetic flux lines is clockwise. This can be verified by using the left-hand rule.

When the direction of current flow through a conductor is reversed, the direction of the magnetic lines of force is also reversed. The cross-sectional end view of a conductor in Figure 1-79(c) shows a current flow in a direction away from the observer. Notice that the direction of the magnetic lines of force is now counterclockwise.

The presence of magnetic lines of force around a current-carrying conductor can be observed by using a compass. When a compass is moved around the outside of a conductor, the needle aligns itself tangent to the lines of force, as shown in Figure 1-79(d). The needle does not point toward the conductor. When current flows in the opposite direction, the compass polarities reverse. The compass needle aligns itself tangent to the conductor.

MAGNETIC FIELD AROUND A COIL

The magnetic field around one loop of wire is shown in Figure 1-80. Magnetic flux lines extend around the conductor as shown. Inside the loop, the magnetic flux is in one direction. When many loops are joined together to form a coil, the magnetic flux lines surround the

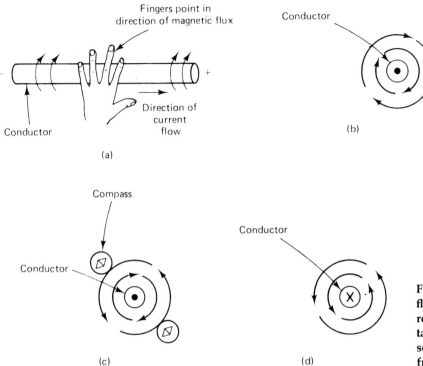

(a)

(b)

(c)

(d)

Figure 1-79.(a) Left-hand rule of magnetic flux; (b) cross section of conductor with current flow toward observer; (c) compass aligns tangent to magnetic lines of force; (d) cross section of conductor with current flow away from the observer.

coil, as shown in Figure 1-81. The field around a coil is much stronger than the field of one loop of wire. The field around the coil is the same shape as the field around a bar magnet. A coil that has an iron or steel core inside it is called an electromagnet. A core increases the magnetic flux density of a coil.

ELECTROMAGNETS

Electromagnets are produced when current flows through a coil of wire, as shown in Figure 1-82. The north pole of a coil of wire is the end where the lines of force come out. The south pole is the end where the lines of force enter the coil. This is like the field of a bar magnet. To find the north pole of a coil, use the *left-hand rule for polarity*, as shown in Figure 1-83. Grasp the coil with the left hand. Point the fingers in the direction of current flow through the coil. The thumb points to the north polarity of the coil.

When the polarity of the voltage source is reversed, the magnetic poles of the coil also reverse. The poles of an electromagnet can be checked with a compass. The compass is placed near a pole of the electromagnet. If the north-seeking pole of the compass points to the coil, that side is the north pole.

Electromagnets have several turns of wire wound around a soft-iron core. An electrical power source is then

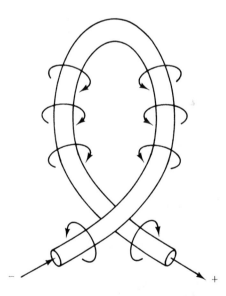

Figure 1-80. Magnetic field around a loop of wire.

connected to the ends of the turns of wire. When current flows through the wire, magnetic polarities are produced at the ends of the soft iron core. The three basic parts of an electromagnet are (1) an iron core, (2) wire windings, and (3) an electrical power source. Electromagnetism is made possible by electric current flow, which produces a magnetic field. When electrical current flows through the coil, the properties of magnetic materials are developed.

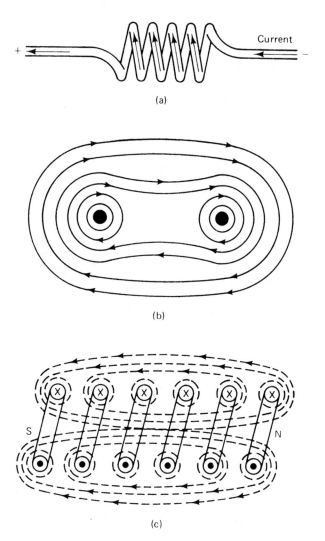

Figure 1-81. Magnetic field around a coil: (a) coil of wire showing current flow; (b) lines of force around two loops that are parallel; (c) cross section of coil showing lines of force.

Magnetic Strength of Electromagnets

The magnetic strength of an electromagnet depends on three factors: (1) the amount of current passing through the coil, (2) the number of turns of wire, and (3) the type of core material. The number of magnetic lines of force is increased by increasing the current, by increasing the number of turns of wire, or by using a more desirable type of core material. The magnetic strength of electromagnets is determined by the *ampere-turns* of each coil. The number of ampere-turns is equal to the current in amperes multiplied by the number of turns of wire ($I \times N$). For example, 200 ampere-turns are produced by 2 A of current through a 100-turn coil. One ampere of current through a 200-turn coil produces the same magnetic field strength. Figure 1-84 shows how the magnetic field strength of an electromagnet changes with the number of ampere-turns.

The magnetic field strength of an electromagnet also depends on the type of core material. Cores are usually made of soft iron or steel, since these materials transfer a magnetic field better than air or other nonmagnetic materials. Iron cores increase the flux density of an electromagnet. Figure 1-85 shows that an iron core causes the magnetic flux to be more dense.

An electromagnet loses its field strength when the current stops flowing. However, an electromagnet's core retains a small amount of magnetic strength after current stops flowing. This is called *residual magnetism*, or "leftover" magnetism. It can be reduced by using soft-iron cores or increased by using hard-steel core material. Residual magnetism is very important in the operation of some types of electrical generators.

In many ways, electromagnetism is similar to magnetism produced by natural magnets such as bar mag-

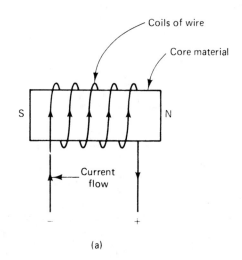

Figure 1-82. Electromagnet—pictorial

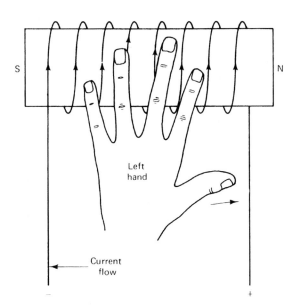

Figure 1-83. Left-hand rule for finding the polarities of an electromagnet.

terials aid in the development of magnetic lines of force to a greater extent. Other types of core materials offer greater resistance to the development of magnetic flux around an electromagnet.

OHM'S LAW FOR MAGNETIC CIRCUITS

A relationship similar to Ohm's law for electrical circuits exists in magnetic circuits. Magnetic circuits have *magnetomotive force* (MMF), *magnetic flux* (Φ), and reluctance ($\mathcal{R}$). MMF is the force that causes a magnetic flux to be developed. Magnetic flux consists of the lines of force around a magnetic material. Reluctance is the opposition to the development of a magnetic flux. These terms may be compared to voltage, current, and resistance in electrical circuits, as shown in Figure 1-86. When MMF increases, magnetic flux increases. Remember that in an electrical circuit, when voltage increases, current increases. When resistance in an electrical circuit in-

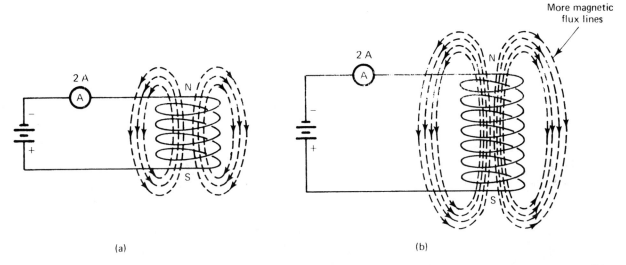

(a) (b)

Figure 1-84. Effect of ampere turns on magnetic field strength: (a) five turns, two amperes = 1 0 ampere-turns; (b) eight turns, two amperes = 1 6 ampere-turns.

nets. However, the main advantage of electromagnetism is that it is easily controlled. It is easy to increase the strength of an electromagnet by increasing the current flow through the coil. This is done by increasing the voltage applied to the coil. The second way to increase the strength of an electromagnet is to have more turns of wire around the core. A greater number of turns produces more magnetic lines of force around the electromagnet. The strength of an electromagnet is also affected by the type of core material used. Different alloys of iron are used to make the cores of electromagnets. Some ma-

creases, current decreases. When reluctance of a magnetic circuit increases, magnetic flux decreases. The relationship of magnetic and electrical terms in Figure 1-86 should be studied.

DOMAIN THEORY OF MAGNETISM

A theory of magnetism was presented in the nineteenth century by the German scientist Wilhelm Weber. Weber's theory of magnetism was called the molecular

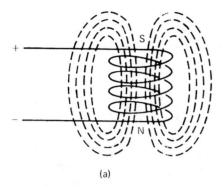

(a)

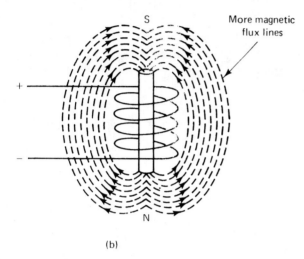

(b)

Figure 1-85. Effect of an iron core on magnetic strength: (a) coil without core; (b) coil with core.

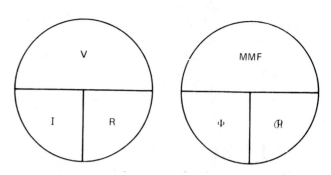

Figure 1-86. Relationship of magnetic and electrical terms.

rials, half of the electrons spin in one direction and half spin in the opposite direction. Their charges cancel each other out, so no magnetic field is produced. Electron rotation in magnetic materials is in the same direction, which causes the domains to act like tiny magnets that align to produce a magnetic field.

ELECTRICITY PRODUCED BY MAGNETISM

A scientist named Michael Faraday discovered in the early 1830s that electricity is produced from magnetism. He found that if a magnet is placed inside a coil of wire, electrical current is produced when the magnet is moved.

Faraday's law is stated as follows: When a coil of wire moves across the lines of force of a magnetic field, electrons flow through the wire in one direction. When

theory. It dealt with the alignment of molecules in magnetic materials. Weber felt that molecules were aligned in an *orderly* arrangement in magnetic materials. In non-magnetic materials, he thought that molecules were arranged in a *random* pattern.

Weber's theory has now been modified somewhat to become the *domain* theory of magnetism. This theory deals with the alignment in materials of domains rather than molecules. A domain is a group of atoms (about 10^{15} atoms). Each domain acts like a tiny magnet. The rotation of electrons around the nucleus of these atoms is important. Electrons have a negative charge. As they orbit around the nucleus of atoms, their electrical charge moves. This moving electrical field produces a magnetic field. The polarity of the magnetic field is determined by the direction of electron rotation.

The domains of magnetic materials are atoms grouped together. Their electrons are believed to spin in the same directions. This produces a magnetic field due to electrical charge movement. Figure 1-87, shows the arrangement of domains in magnetic, nonmagnetic, and partially magnetized materials. In nonmagnetic mate-

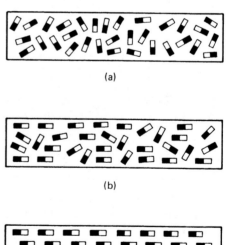

(a)

(b)

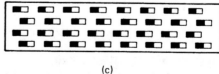

(c)

Figure 1-87. Domain theory of magnetism: (a) unmagnetized; (b) slightly magnetized; (c) fully magnetized saturation.

the coil of wire moves across the magnetic lines of force in the opposite direction, electrons flow through the wire in the opposite direction. This law is the principle of electrical power generation produced by magnetism. Figure 1-88 shows Faraday's law as it relates to electrical power generation.

Current flows in a conductor placed inside a magnetic field only when there is motion between the conductor and the magnetic field. If a conductor is stopped while moving across the magnetic lines of force, current stops flowing. The operation of electrical generators depends on conductors moving across a magnetic field. This principle is called electromagnetic induction.

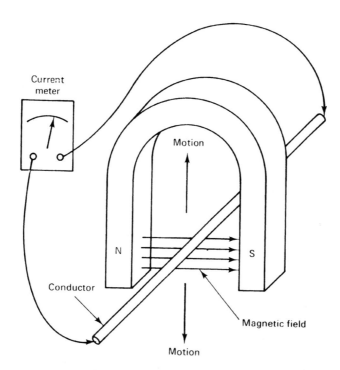

Figure 1-88. Faraday's law—electrical current is produced when there is relative motion between a conductor and a magnetic field.

Chapter 2

Alternating Current (ac) Electronics

INTRODUCTION

Alternating current (ac) electronics is somewhat more complex than direct current (dc) electronics. Alternating current circuits, like dc circuits, have a *source* of energy and *load* in which power conversion takes place. Most of the electrical energy produced in our country is alternating current; therefore, ac systems are used for many applications. In terms of electronic circuits, there are three important characteristics present in ac circuits. These characteristics are resistance, inductance, and capacitance. Also, there are two types of ac voltage—single-phase and three-phase.

OBJECTIVES

1. Explain the difference between ac and dc.
2. Define the process of electromagnetic induction.
3. Describe factors affecting induced voltage.
4. Draw a simple ac generator and explain ac voltage generation.
5. Convert peak, peak-peak, average, and RMS/effective values from one to the other.
6. Explain the relationship between period and frequency of an ac waveform.
7. Recognize the different types of ac waveforms.
8. Measure alternating current with an ac ammeter.
9. Use a multi-meter to measure ac voltage.
10. Explain the basic operation of an oscilloscope.
11. Explain how an oscilloscope can be used to measure the amplitude, period, frequency, and phase relationships of ac waveforms.
12. Explain how an ac voltmeter is used to measure ac voltages.
13. Perform basic ac circuit calculations.
14. Define and calculate impedance.
15. Draw diagrams illustrating the phase relationship between current and voltage in a capacitive circuit.
16. Define capacitive reactance.
17. Solve problems using the capacitive reactance formula.
18. Define impedance.
19. Calculate impedance of series and parallel resistive/capacitive circuits.
20. Determine current in *RC* circuits.
21. Explain the relationship between ac voltages and current in a series resistive circuit.
22. Describe the effect of capacitors in series and parallel.
23. Explain the characteristics of a series *RC* circuit.
24. Solve Ohm's law problems for ac circuits.
25. Solve problems involving true power, apparent power, power factor and reactive power.
26. List several purposes of transformers.
27. Describe the construction of a transformer.
28. Explain transformer action.
29. Calculate turns ratio, voltage ratio, current ratio, power, and efficiency of transformers.
30. Explain the purpose of isolation transformers and autotransformers.
31. Explain factors that cause losses in transformer efficiency.
32. Identify types of filter circuits.
33. Draw response curves for basic filter circuits.
34. Investigate the characteristics of series resonant and parallel circuits.
35. Investigate the frequency response of high-pass, low-pass, and band-pass filters.
36. Define resonance.
37. Calculate resonant frequency.
38. List the characteristics of series and parallel resonant circuits.
39. Define and calculate *Q* and bandwidth.

ALTERNATING CURRENT
(AC) VOLTAGE

When an ac source is connected to some type of load, current direction changes several times in a given unit of time. Remember that dc flows in one direction only. A diagram of one cycle of ac is compared to a dc waveform in Figure 2-1. This waveform is called an ac sine wave. When the ac generator shaft rotates one complete revolution, or 360°, one ac sine wave is produced. Note that the sine wave has a positive peak at 90° and then decreases to zero at 180°. It then increases to a peak negative voltage at 270° and then decreases to zero at 360°. The cycle

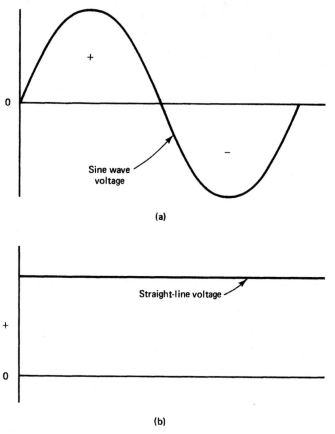

(a)

(b)

Figure 2-1. Comparison of (a) ac and (b) dc waveforms.

then repeats itself. Current flows in one direction during the positive part and in the opposite direction during the negative half-cycle.

Figure 2-2 shows five cycles of alternating current. If the time required for an ac generator to produce five cycles were 1 s, the frequency of the ac would be 5 cycles per second. Ac generators at power plants in the United States operate at a frequency of 60 cycles per second, or 60 hertz (Hz). The hertz is the international unit for frequency measurement. If 60 ac sine waves are produced every second, a speed of 60 revolutions per second is needed. This produces a frequency of 60 cycles per second.

Ac voltage is measured with a VOM and polarity of the meter leads is not important because ac changes direction. (Remember that polarity is important when measuring dc, since direction current flows only in one direction.)

Figure 2-3 shows several voltage values associated with ac. Among these are peak positive, peak negative, and peak-to-peak ac values. Peak positive is the maximum positive voltage reached during a cycle of ac. Peak negative is the maximum negative voltage reached. Peak-to-peak is the voltage value from peak positive to peak negative. These values are important to know when working with radio and television amplifier circuits. For example, the most important ac value is called the *effective*, or *measured*, value. This value is less than the peak positive value. A common ac voltage is 120 V, which is used in homes. This is an effective value voltage. Its peak value is about 170 V. Effective value of ac is defined as the ac voltage that will do the same amount of work as a dc voltage of the same value. For instance, in the circuit of Figure 2-4, if the switch is placed in position 2, a 10-V ac effective value is applied to the lamp. The lamp should produce the same amount of brightness with a 10-V ac effective value as with 10 V dc applied. When ac voltage is measured with a meter, the reading indicated is effective value.

In some cases, it is important to convert one ac value to another. For instance, the voltage rating of electronic devices must be greater than the peak ac voltage applied to them. If 120 V ac is the measured voltage applied to a

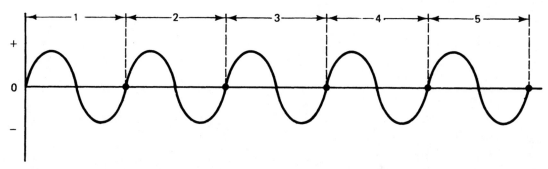

Figure 2-2. Five cycles of alternating current.

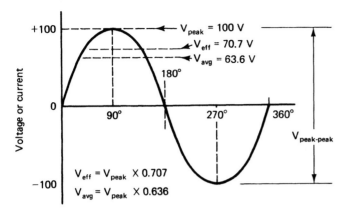

Figure 2-3. Voltage values of an ac waveform.

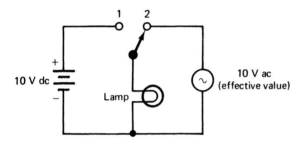

Figure 2-4. Comparison of effective ac voltage and dc voltage.

device, the peak voltage is about 170 V, so the device must be rated over 170 V rather than 120 V.

To determine peak ac, when the measured or effective value is known, the formula

$$\text{Peak} = 1.41 \times \text{effective value}$$

is used. When 120 V is multiplied by the conversion factor 1.41, the peak voltage is found to be about 170 V.

Two other important terms are RMS value and average value. RMS stands for *root mean square* and is equal to 0.707 times peak value. RMS refers to the mathematical method used to determine effective voltage. RMS voltage and effective voltage are the same. Average voltage is the mathematical average of all instantaneous voltages that occur at each period of time throughout an alternation. The average value is equal to 0.636 times the peak value.

SINGLE-PHASE AND THREE-PHASE AC

Single-phase ac voltage is produced by single-phase ac generators or is obtained across two power lines of a three-phase system. A single-phase ac source has a hot wire and a neutral wire, which is grounded to help pre-

vent electrical shocks. Single-phase power is the type of power distributed to our homes. A three-phase ac source has three power lines. Three-phase voltage is produced by three-phase generators at power plants and is a combination of three single-phase voltages, which are electrically connected together. This voltage is similar to three single-phase sine waves separated in phase by 1200. Three-phase ac is used to power large equipment in industry and commercial buildings. It is not distributed to homes. There are three power lines on a three-phase system. Some three-phase systems have a neutral connection and others do not.

The term *phase* refers to time, or the difference between one point and another. If two sine-wave voltages reach their zero and maximum values at the same time, they are in phase. Figure 2-5 shows two ac voltages that are in phase. If two voltages reach their zero and maximum values at different times, they are out of phase. Figure 2-6 shows two ac voltages that are out of phase. Phase difference is given in degrees. The voltages shown are out of phase by 90°.

Remember that single-phase ac voltage is in the form of a sine wave. Single-phase ac voltage is used for

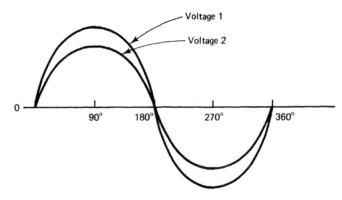

Figure 2-5. Two ac voltages that are in phase.

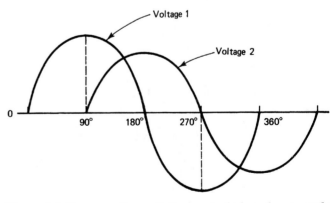

Figure 2-6. Two ac voltages that are out of phase by an angle of 90°.

low-power applications, primarily in the home. Almost all electrical power is generated and transmitted over long distances as three-phase ac. Three coils are placed 1200° apart in a generator to produce three-phase ac voltage. Most ac motors over 1 horsepower (hp) in size operate with three-phase ac power applied.

Three-phase ac systems have several advantages over single-phase systems. In a single-phase system, the power is said to be pulsating. The peak values along a single-phase ac sine wave are separated by 360°, as shown in Figure 2-7(a). This is similar to a one-cylinder gas engine. A three-phase system is some what like a multi-cylinder gas engine. The power is more steady, since one cylinder is compressing when the others are not. This is similar to the voltages in three-phase ac systems. The power of one separate phase is pulsating, but the total power is more constant. The peak values of three-phase ac are separated by 120°, as shown in Figure 2-7(b). This makes three-phase ac power more desirable to use.

The power ratings of motors and generators is greater when three-phase ac power is used. For a certain frame size, the rating of a three-phase ac motor is almost 50% larger than a similar single-phase ac motor.

ELECTROMAGNETIC INDUCTION

Alternating current electrical energy is produced by placing a conductor inside a magnetic field. An experi-ment by a scientist named Michael Faraday showed the following important principle: When a conductor moves across the lines of force of a magnetic field, electrons in the conductor tend to flow in a certain direction. When the conductor moves across the lines of force in the opposite direction, electrons in the conductor tend to flow in the opposite direction. This is the principle of electrical power generation. Most of the electrical energy used today is produced by using magnetic energy.

Figure 2-8 shows the principle of electromagnetic induction. Electrical current is produced only when there is motion. When the conductor is brought to a stop while crossing lines of force, electrical current stops.

If a conductor or a group of conductors is moved through a strong magnetic field, induced current will flow and a voltage will be produced. Figure 2-9 shows a loop of wire rotated through a magnetic field. The position of the loop inside the magnetic field determines the amount of induced current and voltage. The opposite sides of the loop move across the magnetic lines of force in opposite directions. This movement causes an equal amount of electrical current to flow in *opposite directions* through the two sides of the loop. Notice each position of the loop and the resulting output voltage in Figure 2-9. The electrical current flows in one direction and then in the opposite direction with every complete revolution of the conductor. This method produces alternating current. One complete rotation is called a cycle. The num-

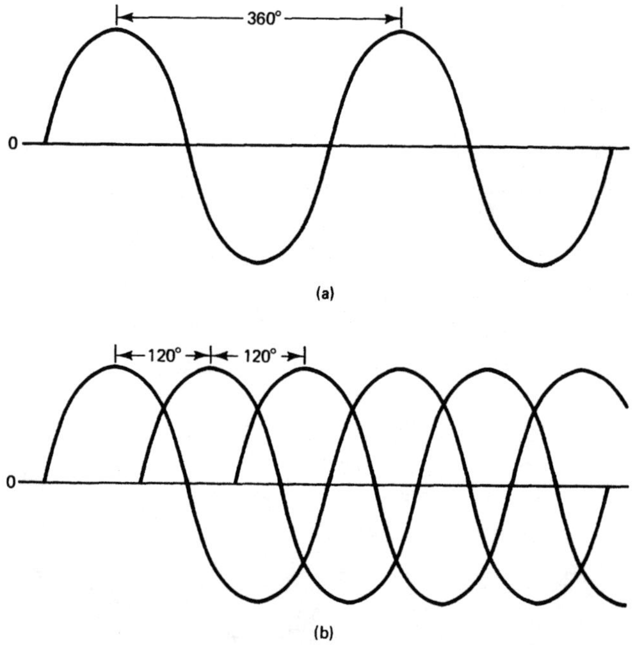

Figure 2-7. Comparison of (a) single-phase and (b) three-phase ac voltages.

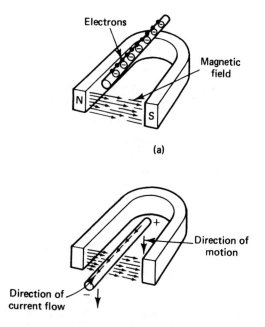

Figure 2-8. Principle of electromagnetic induction: (a) conductor placed inside a magnetic field; (b) conductor moved downward through the magnetic field.

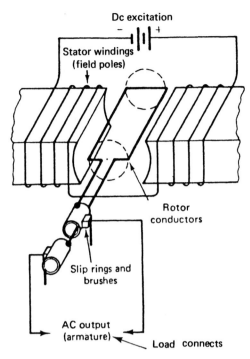

Figure 2-9. Basic ac generator.

GENERATING AC VOLTAGE

Electromagnetic induction takes place when a conductor passes through a magnetic field and cuts across lines of force. As a conductor is passed through a magnetic field, it cuts across the magnetic flux lines. As the conductor cuts across the flux lines, the magnetic field develops a force on the electrons of the conductor. The direction of the electron movement determines the polarity of the induced voltage. The left-hand rule is used to determine the direction of electron flow. This rule for generators is stated as follows: Hold the thumb, forefinger, and middle finger of the right hand perpendicular to each other. Point the forefinger in the direction of the magnetic field from north to south. Point the thumb in the direction of the motion of the conductor. The middle finger will then point in the direction of electron current flow.

The amount of voltage induced into a conductor cutting across a magnetic field depends on the number of lines of force cut in a given time. This is determined by these three factors:

1. The *speed* of the relative motion between the magnetic field and the conductors

2. The *strength* of the magnetic field

3. The *length* of the conductor passed through the magnetic field

If the speed of the conductor cutting the magnetic lines of force is increased, the generated voltage increases. If the strength of the magnetic field is increased, the induced voltage is also increased. A longer conductor allows the magnetic field to induce more voltage into the conductor. The induced voltage increases when each of the three quantities just listed is increased.

ELECTRICAL GENERATOR BASICS

Electrical generators are used to produce electrical energy. They require some form of mechanical energy input. The mechanical energy is used to move electrical conductors (turns of wire) through a magnetic field inside the generator. All generators operate due to electromagnetic induction. A generator has a stationary part and a rotating part housed inside a machine assembly. The stationary part is called the *stator* and the rotating part is called the *rotor*. The generator has *magnetic field poles* of north and south polarities. Generators must have a method of

ber of cycles per second is known as the *frequency*. Most ac generators produce 60 cycles per second.

The ends of the conductor that move across the magnetic field of the generator shown in Figure 2-9 are connected to slip rings and brushes. The slip rings are mounted on the same shaft as the conductor. Carbon brushes are used to make contact with the slip rings. The electrical current induced into the conductor flows through the slip rings to the brushes. When the conductor turns half a revolution, electrical current flows in one direction through the slip rings and the load. During the next half revolution of the coil, the positions of the two sides of the conductor are opposite. The direction of the induced current is reversed. Current now flows through the load in the opposite direction.

The conductors that make up the rotor of a generator have many turns. The generated voltage is determined by the number of turns of wire used, the strength of the magnetic field, and by the speed of the prime mover used to rotate the machine.

Direct current can also be produced by electromagnetic induction. A simple dc generator has a *split-ring* commutator instead of two slip rings. The split rings resemble one full ring, except that they are separated by small openings. Induced electrical current still flows in opposite directions to each half of the split ring. However, current flows in the same direction in the load circuit due to the action of the split rings.

producing rotary motion (mechanical energy). This system is called a *prime mover* and is connected to the generator shaft. There must also be a method of electrically connecting the rotating conductors to an external circuit. This is done by a *slip ring* or *split ring* and brush assembly. Slip rings are used with ac generators, and split rings are used with dc generators. Slip rings are made of solid sections of copper and split rings are made of several copper sections which are separated from each other. The rings are permanently mounted on the shaft of a generator and connect to the ends of the conductors of the rotor. When a load is connected to a generator, a complete circuit is made. With all the generator parts working together, electrical power is produced. A basic ac generator is shown in Figure 2-10.

Single-Phase AC Generation

Single-phase electrical power is often used, particularly in homes. However, very little electrical power is produced by single-phase generators. Alternating current generators are usually referred to as *alternators*. Single-phase electrical power used in homes is usually produced by three-phase generators at power plants. It is then converted to single-phase electrical energy before it is distributed to homes.

The current produced by single-phase generators is in the form of a sine wave. This waveform is called a sine wave due to its mathematical origin. It is based on the trigonometric sine function used in mathematics.

The voltage induced into the conductors of the armature varies as the sine of the angle of rotation between the conductors and the magnetic field (see Figure 2-11).

The voltage induced at a specific time is called *instantaneous voltage* (V_i). Voltage induced into an armature conductor at a specific time is found by using the formula

$$V_i = V_{max} \times \sin \theta$$

V_{max} is the maximum voltage induced into the conductor. The symbol θ is the angle of conductor rotation.

For example, at the 60° position, assume that the maximum voltage output is 100 V. The instantaneous voltage induced at 60° is $V_i = 100 \times \sin \theta$. A calculator with trigonometric functions gives sin 60° = 0.866. The induced voltage at 60° is 86.6 V ($V_i = 100 \times 0.866 = 86.6$ V).

The frequency of the voltage produced by alternators is usually 60 Hz. A cycle of ac is generated when the rotor moves one complete revolution (360°). Cycles per second or hertz refers to the number of revolutions per second. For example, a speed of 60 revolutions per second (3600 rpm) produces a frequency of 60 Hz. The frequency (f) of an alternator is found by using the formula

$$f(Hz) = \frac{\text{no. of poles} \times \text{speed (rpm)}}{120}$$

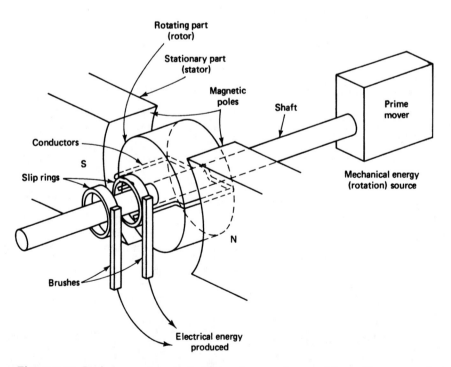

Figure 2-10. Basic generator construction. A generator must have (1) a magnetic field, (2) conductors, and (3) a source of mechanical energy (rotation).

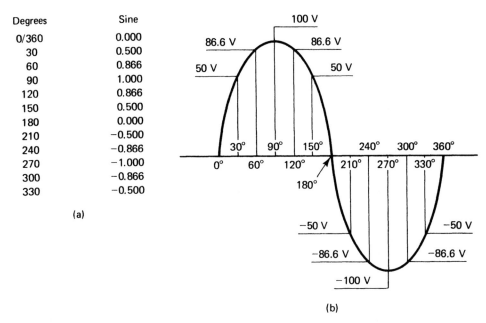

Degrees	Sine
0/360	0.000
30	0.500
60	0.866
90	1.000
120	0.866
150	0.500
180	0.000
210	−0.500
240	−0.866
270	−1.000
300	−0.866
330	−0.500

(a)

(b)

Figure 2-11. Generation of an ac sine wave: (a) sine values of angles from 0 to 3600; (b) sine wave produced.

If the number of poles (field coils) is increased, the speed of rotation can be reduced and still produce a 60-Hz frequency.

For a generator to convert mechanical energy into electrical energy, three conditions must exist: (1) There must be a magnetic field. (2) There must be conductors cutting the magnetic field. (3) There must be relative motion between the magnetic field and the conductors.

Three-phase AC Generation

Most electrical power produced in the United States is three-phase ac produced at power plants. Power distribution systems use many three-phase generators (alternators) connected in parallel. A simple diagram of a three-phase alternator and a three-phase voltage diagram are shown in Figure 2-12. The alternator output windings are connected in either of two ways. These methods are called the wye connection and the delta connection; they are shown with schematics in Figure 2-13. These methods are also used for connecting the windings of three-phase transformers, three-phase motors, and other three-phase equipment.

In the wye connection of Figure 2-13(a), the beginnings or ends of each winding are connected together. The other sides of the windings are the ac lines which extend from the alternator. The voltage across the power lines is called line voltage (V_L). Line voltage is higher than the voltage across each phase. Line voltage (V_L) is equal to the square root of 3 (1.73) multiplied by the voltage across the phase windings (V_p), or $V_L = V_p \times 1.73$.

The line current (I_L) is equal to the phase current (I_p), or $I_L = I_p$.

In the delta connection, shown in Figure 2-13(b), the end of the phase windings is connected to the beginning of the next phase winding. Line voltage (V_L) is equal to the phase current (I_p) multiplied by 1.73, or $I_L = I_p \times 1.73$. The differences between voltages and currents in wye and delta systems should be remembered.

Three-phase power is used mainly for high-power industrial and commercial equipment. The power produced by three-phase generators is a more constant output than that of single-phase power. Three-phase power is more economical to supply energy to the large motors that are often used in industries. Three separate single-phase voltages can be delivered from a three-phase power system. It is more economical to distribute three-phase power from power plants to homes, cities, and industries. Three conductors are needed to distribute three-phase voltage. Six conductors are necessary for three separate single-phase systems. Equipment that uses three-phase power is smaller in size than similar single-phase equipment. It saves energy to use three-phase power whenever possible.

One type of a three-phase alternator is used in automobiles. The three-phase ac it produces is converted to dc by a rectifier circuit. The dc voltage is then used to charge the automobile battery. The charging time and voltage is controlled by a voltage regulator circuit. The parts of an automobile alternator are shown in Figure 2-14.

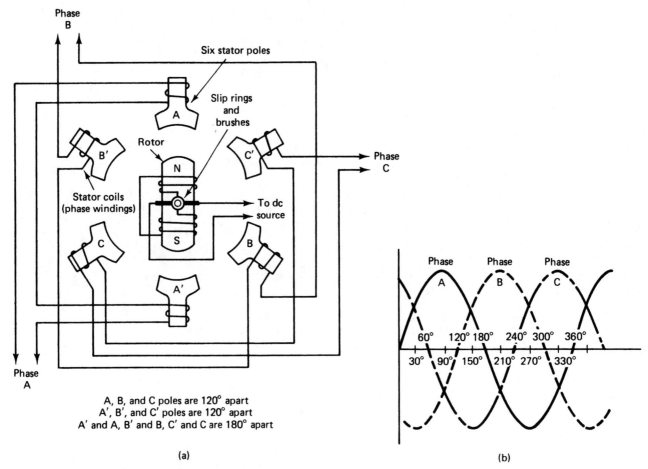

A, B, and C poles are 120° apart
A', B', and C' poles are 120° apart
A' and A, B' and B, C' and C are 180° apart

(a)   (b)

Figure 2-12. (a) Diagram of a three-phase alternator; WI diagram of three-phase voltage.

MEASURING AC VOLTAGE WITH A VOM

A VOM may be used to measure ac voltage in the same way as dc voltage with two exceptions:

1. When measuring ac voltage, proper polarity does not have to be observed.

2. When measuring ac voltage, the ac voltage ranges and scales of the meter must be used.

USING AN OSCILLOSCOPE

Another way to measure ac voltage is with an oscilloscope. Oscilloscopes or scopes, are also used to measure a wide range of frequencies with precision and to examine wave shapes. For electronic servicing, it is necessary to be able to observe the voltage waveform while troubleshooting.

An oscilloscope permits various voltage waveforms to be analyzed visually. The controls must be properly adjusted to produce an image on its screen. The image, called a trace, is usually a line on the screen or cathode-ray tube (CRT). A stream of electrons strikes the phosphorescent coating on the inside of the screen, causing the screen to produce light.

The oscilloscope displays voltage waveforms on two axes, like a graph. The horizontal axis on the screen is the time axis. The vertical axis is the voltage axis. Figure 2-15 shows an ac waveform displayed on the CRT. Oscilloscopes are slightly different, but most scopes have some of the following controls:

1. *Intensity*: Controls the brightness of the trace and sometimes the on-off control also.

2. Focus: Adjusts the thickness of the trace so that it is clear and sharp.

3. *Vertical position:* Adjusts the entire trace up or down.

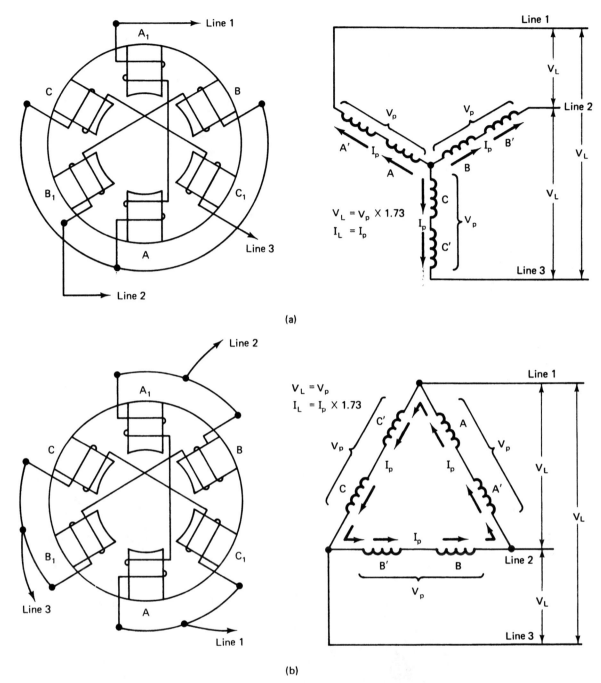

Figure 2-13. Three-phase connections: (a) wye connection, sometimes called star connection; (b) delta connection.

4. *Horizontal position:* Adjusts the entire trace to the left or right.

5. *Vertical gain:* Controls the height of the trace.

6. *Horizontal gain:* Controls the horizontal size of the trace.

7. *Vertical attenuation, or variable volts/cm:* Acts as a "coarse" adjustment to reduce the trace vertically.

8. *Horizontal sweep, or variable time/cm:* Controls the speed at which the trace moves across (sweeps) the CRT horizontally. This control determines the number of waveforms displayed on the screen.

9. *Synchronization select:* Controls how the input to the scope is "locked in" with the circuitry of the scope.

10. *Vertical input:* External connections used to apply

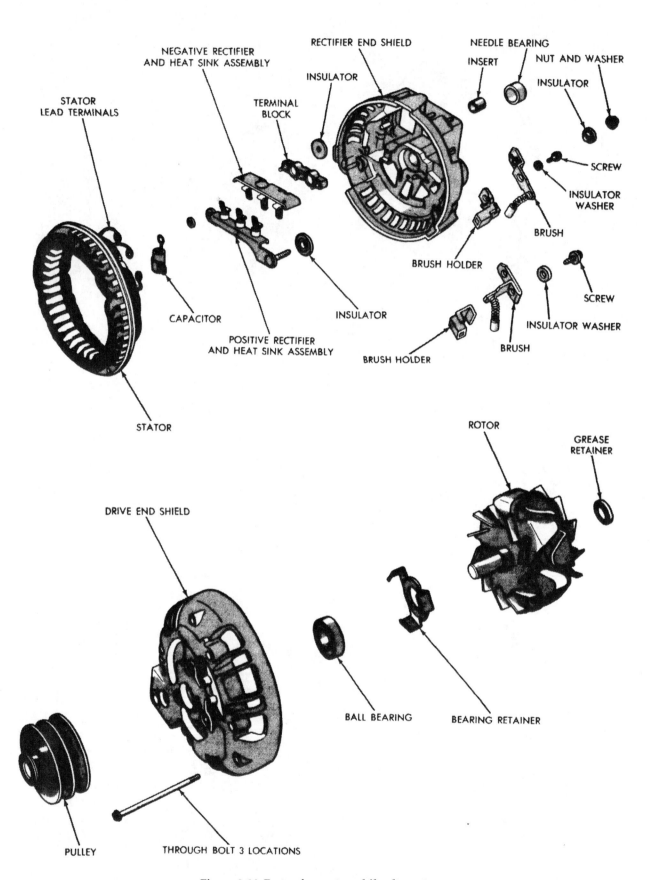

Figure 2-14. Parts of an automobile alternator.

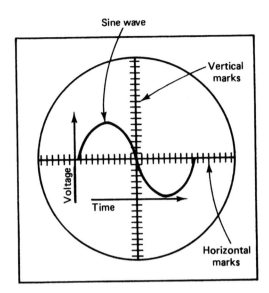

Figure 2-15. Ac waveform displayed on the screen of an oscilloscope.

the input to the vertical circuits of the scope.

11. *Horizontal input:* External connections used to apply an input to the horizontal circuits of the scope.

The following procedure is used to adjust the oscilloscope controls to measure ac voltage. The names of some of the controls vary on different types of oscilloscopes.

1. Turn on the oscilloscope and adjust the intensity and focus controls until a bright, narrow, straight-line trace appears on the screen. Use the horizontal position and vertical position controls to position the trace in the center of the screen. Adjust the horizontal gain and the variable time/cm until the trace extends from the left side of the screen to the right side of the screen. This allows the entire waveform to be displayed.

2. Connect the proper test probes into the oscilloscope's vertical input connections.

3. The scope is now ready to measure ac voltage.

4. After a waveform is displayed, adjust the vertical attenuation (volts/cm) and vertical gain controls until the height of the trace equals about 2 in. or 4 cm. Most scopes have scales that are marked in centimeters. Adjust the vernier or stability control until the trace becomes stable. One ac waveform or more should appear on the screen of the scope.

Oscilloscope Operation

Figure 2-16 shows the construction of the CRT and its internal electron gun assembly. A beam of electrons is produced by the cathode of the tube. Electrons are given off when the filament is heated by a filament voltage. The electrons have a negative charge. They are attracted to the positive potential of anode 1. The number of electrons that pass to anode 1 is changed by the amount of negative voltage applied to the control grid. Anode 2 has a higher positive voltage applied. The higher voltage accelerates the electron beam toward the screen of the CRT. The difference in voltage between anodes 1 and 2 sets the point where the beam strikes the CRT screen. The screen has a phosphorescent coating. When electrons strike the CRT, light is produced. The phosphorescent coating of the screen allows an electrical charge to produce light. The horizontal and vertical movement of the electron beam is controlled by deflection plates. With no voltage applied to either set of plates, the electron beam appears as a dot in the center of the CRT screen. The movement of the electron beam caused by the change of the plates is called *electrostatic deflection*. When a potential is placed on the horizontal and vertical deflection plates, the electron beam is moved. Horizontal deflection is produced by a circuit called a sweep oscillator circuit inside the oscilloscope. A voltage "sweeps" the electron beam back and forth across the CRT screen. The horizontal setting of an oscilloscope is adjusted to match the frequency of the voltage being measured. Vertical deflection is caused by the voltage being measured. The voltage to be measured is applied to the vertical deflection of the oscilloscope. A block diagram of an oscilloscope circuit is shown in Figure 2-17. Oscilloscopes are used to measure ac and dc voltages, frequency, phase relationships, distortion in amplifiers, and various timing and special-purpose applications. They also have some important medical uses, such as to monitor heartbeat.

INDUCTANCE AND CAPACITANCE

Two very important electronic properties are inductance and capacitance. This section discusses these properties as well as inductors and capacitors.

Inductance

When energized with dc voltages, a magnetic field is produced around a coil. Dc current flow produces a constant magnetic field around a coil. When the same coil is supplied with ac voltage, the constantly changing ac current produces a constant changing magnetic flux. This

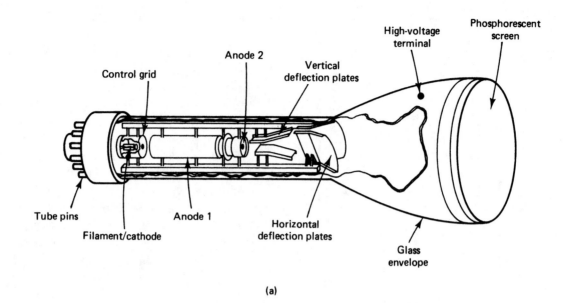

Figure 2-16. Construction of a CRT: (a) cutaway view; (b) electron beam detail.

changing flux sets up a magnetic field with a constantly reversing polarity and changing strength. It also induces a counter EMF or counter voltage. This counter electromotive force (CEMF) opposes the source voltage. CEMF limits current flow from the source.

The opposition of the flow of ac current by a magnetic field is due to a property called *inductance* (L). The opposition to current flow of an inductive device depends on the resistance of the wire and the magnetic properties of the circuit. The opposition due to the magnetic effect in ac circuits is called *inductive reactance* (X_L). X_L varies with the applied frequency and is found by using the formula $X_L = 2\pi fL$, where $2\pi = 6.28$, f is the applied frequency in hertz, and L is the inductance in henries. The basic unit of

inductance is the henry (H).

At zero frequency (or dc), there is no opposition due to inductance. Only a coil's resistance limits current flow. As ac frequency increases, the inductive effect becomes greater. Many ac machines use magnetic circuits in one form or another. The inductive reactance of an ac circuit usually has more effect on current flow than resistance. An ohmmeter measures dc resistance only. Inductive reactance must be calculated or determined experimentally.

Figure 2-18 shows some symbols used to indicate various types of inductors. Often they are called *choke* coils. Choke coils are used to pass dc current and oppose ac current flow.

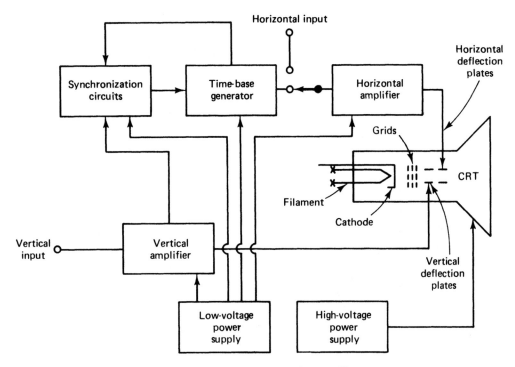

Figure 2-17. Functional parts of an oscilloscope.

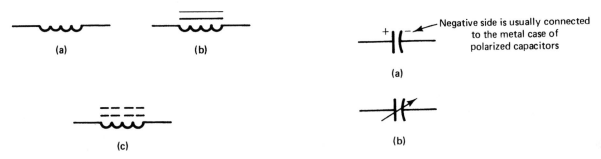

Figure 18. Symbols used for inductors: (a) air core; (b) iron core; (c) powered metal core.

Figure 2-19. Capacitor symbols; (a) fixed capacitor; (b) variable capacitor.

Capacitance

When two conductors are separated by an insulator (dielectric), an electrostatic charge may be set up on the conductors. This charge becomes a source of stored energy. The strength of the charge depends on the applied voltage, the size of the conductors, which are called plates, and the quality or dielectric strength of the insulation. The closer the two plates are placed together, the more charge may be set up on them. This type of device is called a capacitor. The size of capacitors is measured in units called farads, microfarads, and picofarads. Some capacitor symbols are shown in Figure 2-19.

When a dc voltage is applied to the plates of a capacitor, it will charge to the value of the source voltage. The dielectric between the plates then stops current flow. The capacitor remains charged to the value of the dc volt-

age source. When the voltage source is removed, the capacitor charge will leak away. Dc current flows to or from a capacitor only when the source voltage is turned on or off.

If the same capacitor is connected to an ac voltage source, the voltage constantly changes. The capacitor receives energy from the source during one-quarter cycle and very little current flows through the dielectric of a capacitor. Ac voltage applied to a capacitor causes it constantly to change its amount of charge. Capacitors have the ability to pass ac current because of their charging and discharging action. They can be used to pass ac current and block dc current flow. The opposition of a capacitor to a source voltage depends on frequency. The faster the applied voltage changes across a capacitor, the more easily the capacitor passes current.

Inductive Effects in Circuits

Inductance is the property of a circuit that opposes changes of current flow. Any coil of wire has inductance. Coils oppose current changes by producing a counter voltage (CEMF). The term reactive is used due to a coil's reaction to changes in applied voltage. The opposition of a coil to ac is called *reactance* (X) and is measured in ohms. The subscript L is added to represent inductive reactance (X_L). Inductive reactance depends on the "speed," or frequency, of the *ac* source.

Inductance does not change with ac frequency changes. Coils have inductance due to their ability to oppose current change. This property stays the same for any coil. The factors that determine inductance are the number of turns of the coil, area and length of the core material, and the type of core used. The main factor is the number of coil turns. The inductive reactance (X_L) does change with changes in frequency. This can be verified by looking at the formula $X_L = 2\pi f L$. Notice that as frequency is increased, inductive reactance increases. Inductors have a low *dc resistance*. This value can be measured with an ohmmeter.

MUTUAL INDUCTANCE

When inductors are connected together, a property called *mutual inductance* (M) must be considered. Mutual inductance is the magnetic field interaction or flux linkage between coils. The amount of flux linkage is called the *coefficient of coupling* (k). If all the lines of force of one coil cut across a nearby coil, it is called *unity coupling*. There are many possibilities, determined by coil placement, of coupling between coils. The amount of mutual inductance between coils is found by using the formula

$$M = k \times L_1 \times L_2$$

The term k is the coefficient of coupling, which gives the amount of coupling, and L_1 and L_2 are the inductance values of the coils. Mutual inductance should also be considered when two or more coils are connected together.

INDUCTORS IN SERIES AND PARALLEL

Inductors may be connected in series or parallel. When inductors are connected to prevent the magnetic field of one from affecting the others, these formulas are used to find total inductance (L_T). L_1, L_2, L_3, and so on, are inductance values in henries.

1. Series inductance:

$$L_T = L_1 + L_2 + L_3 + \dots + L_n$$

2. Parallel inductance:

$$\frac{1}{L_T} = \frac{1}{L_1} + \frac{1}{L_2} + \frac{1}{L_3} + \dots + \frac{1}{L_n}$$

When inductors are connected so that the magnetic field of one affects the other, mutual inductance increases or decreases the total inductance. The effect of mutual inductance increases or decreases the total inductance. The effect of mutual inductance depends on the physical positioning of the inductors. Their distance apart and the direction in which they are wound effects mutual inductance. Inductors are connected in series or parallel with an aiding or opposing mutual inductance. The formulas used to find total inductance (L_T) are:

1. *Series aiding:*
$$L_T = L_1 + L_2 + 2M$$

2. *Series opposing:*
$$L_T = L_1 + L_2 - 2M$$

3. *Parallel aiding:*
$$\frac{1}{L_T} = \frac{1}{L_1 + M} + \frac{1}{L_2 + M}$$

4. *Parallel opposing:*
$$\frac{1}{L_T} = \frac{1}{L_1 - M} + \frac{1}{L_2 M}$$

L_1 and L_2 are the inductance values and M is the value of mutual inductance.

INDUCTIVE CURRENT RELATIONSHIPS

A change in current through a coil causes a change in the magnetic flux around the coil. The CEMF of the coil is at maximum value when current value changes. Since CEMF opposes source voltage, it is directly opposite applied voltage.

Lenz's law states that the counter voltage (CEMF) always opposes a change in current. CEMF opposes the rise in current that causes current to decrease. When the values of the counter voltage and the applied voltage are maximum, the circuit current is maximum. The dc resistance of a coil is small. Ohm's law can be used with inductive dc circuits to find current as follows:

$$I_L = \frac{V_L}{R}$$

where I_L is the current through the coil, V_L is the voltage across the coil, and R is the dc resistance of the coil (in ohms).

CAPACITIVE EFFECTS IN CIRCUITS

Inductance is defined as the property of a circuit to oppose changes in current. Capacitance is the property of a circuit to oppose changes in voltage. Inductance stores energy in an electromagnetic field, whereas capacitance stores energy in an electrostatic field. A capacitor is measured in units called *farads*. One farad is the amount of capacitance that permits a current of 1 A to flow when the voltage change across the plates of a capacitor is 1 V per second. The farad is too large for practical use. The microfarad (one-millionth of a farad, 0.000001 F, abbreviated µF or MFD) is the most common subunit of capacitance. For high-frequency ac circuits, the microfarad is also too large. The subunit micro-microfarad (one-millionth of a microfarad, 0.000001 µF, abbreviated µF or MMFD) is then used. This subunit is called the picofarad (pF) to avoid confusion.

The three factors that determine the capacitance of a capacitor are the following:

1. *Plate area:* Increasing plate area increases capacitance.

2. *Distance between plates:* Capacitance is decreased when the distance between the plates increases.

3. *Dielectric material:* Dielectrics, including air, are used as insulators between capacitor plates. They are rated with a dielectric constant (K). Capacitors with higher dielectric constants have higher capacitance.

Capacitors in Series

Adding capacitors in series has the same effect as increasing the distance between plates. This reduces total capacitance. The total capacitance of capacitors connected in series is less than any individual capacitance. Total capacitance (C_T) is found in the same way as parallel resistance. The reciprocal formula is used:

$$\frac{1}{C_T} = \frac{1}{C_1} + \frac{1}{C_2} + \frac{1}{C_3} + \ldots + \frac{1}{C_N}$$

Capacitors in Parallel

Capacitors in parallel are similar to one capacitor with its plate area increased. Doubling the plate area would double capacitance. Total capacitance (C_T) for capacitors in parallel is found by adding individual values, just as with series resistors, or

$$C_T = C_1 + C_2 + C_3 + \ldots + C_N$$

Capacitor Charging and Discharging

Refer to the circuit of Figure 2-20. A capacitor is shown connected to a two-position switch. With the switch in position 1, the capacitor is uncharged. No voltage is applied to the capacitor. Each plate of the capacitor is neutral, since no voltage is applied. When a voltage is placed across the capacitor, an electrostatic field or charge is developed between its plates. When the switch is placed in position 2, this places the capacitor across a 6-V battery. This causes a displacement of electrons in the circuit. Electrons accumulate on the negative plate of the capacitor. At the same time, electrons leave the positive plate. This causes a difference of potential to develop across the capacitor. When electrons move onto the negative plate, its becomes more negative. Also, as electrons leave the positive plate, it becomes more positive.

The polarity of the potential that exists across the capacitor opposes the source voltage. As the capacitor continues to charge, the voltage across the capacitor increases. Current flow stops when the voltage across the capacitor is equal to the source voltage. At this time, the two voltages cancel each other out. No current actually flows through the capacitor because the material between the plates of a capacitor is an insulator.

An important safety factor to remember is that a capacitor may hold a charge for a long period of time. Capacitors can be an electrical shock hazard if not handled properly.

When a capacitor is discharged, the charges on the

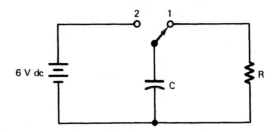

Figure 2-20. Capacitor charge and discharge circuit.

plates become neutralized. A current path between the two plates must be developed. When the switch in Figure 2-20 is moved to position 1, the electrons on the negative plate move to the positive plate and neutralize the charge. The capacitor releases the energy it has absorbed during its charging as it discharges through the resistor.

Types of Capacitors

Capacitors are classified as either fixed or variable. Fixed capacitors have one value of capacitance. Variable capacitors are constructed to allow capacitance to be varied over a range of values. Variable capacitors often use air as the dielectric. The capacitance is varied by changing the position of the movable plates. This changes the plate area of the capacitor. When the movable plates are fully meshed together with the stationary plates, the capacitance is maximum.

Fixed capacitors come in many types. Some types of fixed capacitors are as follows:

1. *Paper capacitors*: Paper capacitors use paper as their dielectric. As shown in Figure 2-21, they are made of flat strips of metal foil plates separated by a dielectric, which is usually waxed paper. Paper capacitors have values in the picofarad and low microfarad ranges. The voltage ratings are usually less than 600 V. Paper capacitors are usually sealed with wax to prevent moisture problems.

    The voltage rating of capacitors is very important. A typical set of values marked on a capacitor might be "10 µF, 50 DCWV." Such a capacitor has a capacitance of 10 µF and a "dc working voltage" of 50 V. This means that a voltage in excess of 50 V could damage the plates of the capacitors.

2. Mica capacitors: Mica capacitors have a layer of mica and then a layer of plate material. Their capacitance is usually small (in the picofarad range). They are small in physical size but have high voltage ratings.

3. *Oil-filled capacitors*: Oil-filled capacitors are used when high capacitance and high voltage ratings are needed. They are like paper capacitors immersed in oil. The paper, when soaked in oil, has a high dielectric constant.

4. *Ceramic capacitors*: Ceramic capacitors use a ceramic dielectric. The plates are thin films of metal deposited on ceramic material or made in the shape of a disk. They are covered with a moisture-proof coating and have high voltage ratings.

5. *Electrolytic capacitors*: Electrolytic capacitors are used when very high capacitance is needed. Electrolytic capacitors contain a paste electrolyte. They have two metal plates with the electrolyte between them and are usually housed in a cylindrical aluminum can. The aluminum can is the negative terminal of the capacitor. The positive terminal (or terminals) is brought out of the can at the bottom. The size and voltage rating are usually printed on the capacitor. Electrolytic capacitors often have two or more capacitors housed in one unit and are called multisection capacitors.

The positive plate of an electrolytic capacitor is aluminum foil, covered with a thin oxide film. The film is formed by an electrochemical reaction and acts as the dielectric. A strip of paper that contains a paste electrolyte is placed next to the positive plate. The electrolyte is the negative plate of the capacitor. Another strip of aluminum foil is placed next to the electrolyte. These three layers are then coiled up and placed into a cylinder.

Electrolytic capacitors are said to be polarized. If the positive plate is connected to the negative terminal of a voltage source, the capacitor will become short-circuited. Capacitor polarity is marked on the capacitor to prevent this from happening. Some special high-value nonpolarized electrolytic capacitors, such as motor-starting capacitors, are used for ac applications.

RESISTANCE, INDUCTANCE, AND CAPACITANCE IN AC CIRCUITS

Ac electronic circuits are similar in many ways to dc circuits. Ac circuits are classified by their electrical characteristics (resistive, inductive, or capacitive). All electronic

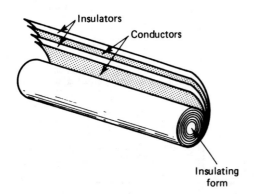

Figure 2-21. Paper capacitor construction.

ac circuits are either resistive, inductive, capacitive, or a combination of these. The operation of each type of electronic circuit is different. The nature of alternating current causes certain circuit properties to exist.

Resistive AC Circuits

The simplest type of ac circuit is a resistive circuit, such as the one shown in Figure 2-22. The purely resistive circuit offers the same type of opposition to ac as it does to pure dc sources. In dc circuits,

$$\text{Voltage } (V) = \text{Current } (I) \times \text{Resistance } (B)$$

$$\text{Current } (I) = \frac{\text{voltage } (V)}{\text{resistance } (B)}$$

$$\text{Resistance } (B) = \frac{\text{voltage } (V)}{\text{current } (I)}$$

$$\text{Power } (P) = \text{voltage } (V) \times \text{current } (I)$$

These basic electrical relationships show that when voltage is increased, the current in the circuit increases proportionally. Also, as resistance is increased, the current in the circuit decreases. By looking at the waveforms of Figure 2-23, it can be seen that the voltage and current in a purely resistive circuit, with ac applied, are in phase. An in-phase relationship exists when the minimum and maximum values of both voltage and current occur at the same time interval. Also, the power converted by the circuit is a product of voltage times current ($P = V \times I$). The power curve is shown in Figure 2-24. Thus, when an ac circuit contains only resistance, its behavior is very similar to a dc circuit. Purely resistive circuits are seldom encountered in the design of electrical power systems, although some devices are primarily resistive in nature.

Inductive Circuits

The property of inductance (L) is very commonly encountered in electronic circuits. This circuit property, shown in Figure 2-25 adds more complexity to the relationship between voltage and current in an ac circuit. All motors, generators, and transformers exhibit the property of inductance. The magnetic flux produced around coils affects circuit action. Thus, the inductive property (CEMF) produced by a magnetic field offers an opposition to change in the current flow in a circuit.

The opposition to change of current is evident in the diagram of Figure 2-25. In an inductive circuit, we can say that voltage leads current or that current lags voltage. If the circuit were purely inductive (contains no resistance),

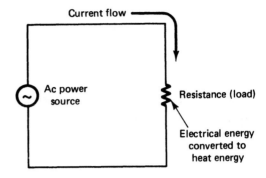

Figure 2-22. Resistive ac circuit.

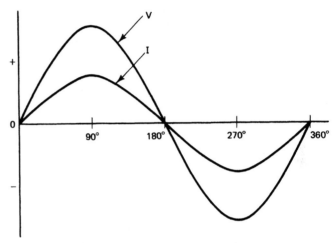

Figure 2-23. Power curve for a resistive ac circuit.

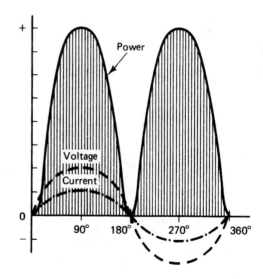

Figure 2-24. Power curve for a purely inductive circuit.

the voltage would lead the current by 90° (Figure 2-25(b) and no power would be converted in the circuit (Figure 2-26. However, since all actual circuits have resistance, the inductive characteristic of a circuit might typically cause the condition shown in Figure 2-26 to exist. Here the voltage is leading the current by 30°. The angular separation between voltage and current is called the *phase angle*. The phase angle increases as the inductance of the circuit increases. This type of circuit is called a resistive-inductive (*RL*) circuit.

In terms of power conversion, a purely inductive circuit would not convert any power in a circuit. All ac power would be delivered back to the power source. Refer back to Figure 2-26 and note points *A* and *B* on the waveforms. These points show that at the peak of each waveform, the corresponding value of the other waveform is zero. The power curves shown are equal and opposite in value and cancel each other out. Where both voltage and current are positive, the power is also positive, since the

product of two positive values is positive. When one value is positive and the other is negative, the product of the two values is negative; therefore, the power converted is negative. Negative power means that electrical energy is being returned from the load device to the power source without being converted to another form of energy. Therefore, the power converted in a purely inductive circuit (90° phase angle) is equal to zero.

Compare the purely inductive waveforms of Figure 2-26 with those of Figure 2-27. 1 In the practical resistive-inductive (*RL*) circuit, part of the power supplied from the source is converted in the circuit. Only during the intervals from 0° to 30° and from 180° to 210° does negative power result. The remainder of the cycle produces posi-

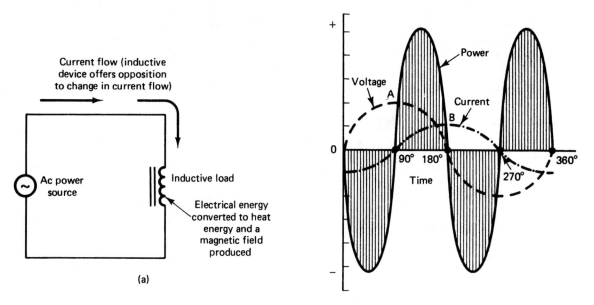

Figure 2-25. Voltage and current waveforms of a purely inductive ac circuit: (a) circuit; (b) waveforms.

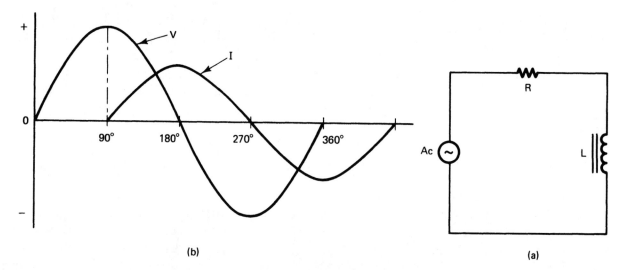

Figure 2-26. Power curve for a purely inductive circuit.

tive power; therefore, most of the electrical energy supplied by the source is converted to another form of energy.

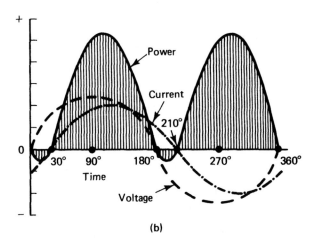

(b)

Figure 2-27. Resistive-inductive (RL) circuit and its waveform: (a) RL ac circuit; (b) voltage and current waveforms.

Any inductive circuit exhibits the property of inductance (L), which is the opposition to a change in current flow in a circuit. This property is found in coils of wire (which are called inductors) and in rotating machinery and transformer windings. Inductance is also present in electrical power transmission and distribution lines to some extent. The unit of measurement for inductance is the henry. A circuit has a 1-H inductance if a current changing at a rate of 1 A per second produces an induced CEMF of 1 V.

In an inductive circuit with ac applied, an opposition to current flow is created by the inductance. This type of opposition is known as inductive reactance (X_L). The inductive reactance of an ac circuit depends upon the inductance (L) of the circuit and the rate of change of current. The frequency of the applied ac establishes the rate of change of the current. Inductive reactance (X_L) may be expressed as:

$$X_L = 2\pi f L$$

where X_L is the inductive reactance in ohms, 2π is approximately 6.28, the mathematical expression for one sine wave of alternating current (0°–360°), f is the frequency of the ac source in hertz, and L is the inductance of the circuit in henrys.

Capacitive Circuits

Figure 2-28 shows a capacitive device connected to an ac source. We know that whenever two conductive materials (plates) are separated by an insulating (dielectrical) material, the property of capacitance is exhibited. Capacitors have the capability of storing an electrical charge. They also have many applications in electronic systems.

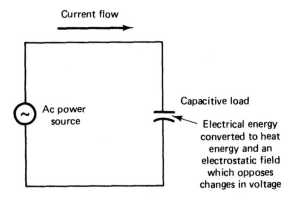

Figure 2-28. Capacitive ac circuit.

The operation of a capacitor in a circuit is dependent upon its ability to charge and discharge. When a capacitor charges, an excess of electrons (negative charge) is accumulated on one plate and a deficiency of electrons (positive charge) is created on the other plate. Capacitance (C) is determined by the size of the conductive material (plates) and by their separation (determined by the thickness of the dielectric or insulating material). The type of insulating material is also a factor in determining capacitance. Capacitance is directly proportional to the plate size and inversely proportional to the distance between the plates. A capacitance of 1 F results when a potential of 1 V causes an electrical charge of 1 C (a specific mass of electrons) to accumulate on a capacitor.

If a direct current is applied to a capacitor, the capacitor will charge to the value of that dc voltage. After the capacitor is fully charged, it will block the flow of direct current. However, if ac is applied to a capacitor, the changing value of current causes the capacitor alternately to charge and discharge. In a purely capacitive circuit, the situation shown in Figure 2-34 would exist. The greatest amount of current flows in a capacitive circuit when the voltage changes most rapidly, which occurs at the 0° and 180° positions where the polarity changes. At these positions, maximum current is developed in the circuit. When the rate of change of the voltage value is slow, such as near the 90° and 270° positions, a small amount of current flows. In examining Figure 2-29, we can observe that current leads voltage by 90° in a purely capacitive circuit or the voltage lags the current by 90°. Since a 90° phase angle exists, no power would be converted in this circuit, just as no power was developed in the purely inductive circuit.

As shown in Figure 2-30, the positive and negative power waveforms cancel one another.

Since all circuits contain some resistance, a more practical circuit is the *RC* circuit shown in Figure 2-31. In an *RC* circuit, the current leads the voltage by some phase angle between 0° and 90°. As capacitance increases, with no corresponding increase in resistance, the phase angle becomes greater. The waveforms of Figure 2-32 show an *RC* circuit in which current leads voltage by 30°. This circuit is similar to the *RL* circuit of Figure 2-27. Power is converted in the circuit except during the 0° to 30° interval and the 180° to 210° interval. In the *RC* circuit shown, most of the electrical energy supplied by the source is converted to another form of energy in the load.

Due to the electrostatic field that is developed around a capacitor, an opposition to the flow of alternating current exists. This opposition is known as capacitive

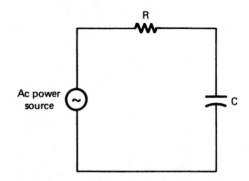

Figure 2-31. Resistive-capacitive (*RC*) circuit.

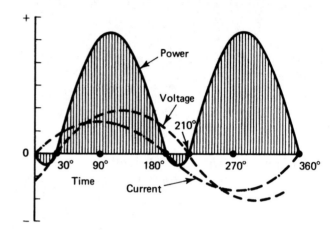

Figure 2-32. Waveforms of an *RC* circuit.

reactance (X_C). Capacitive reactance is expressed as

$$X_C = \frac{1}{2\pi f C}$$

where X_C is the capacitive reactance in ohms, 2π is the mathematical expression of one sine wave (0° to 360°), f is the frequency of the source in hertz, and C is the capacitance in farads.

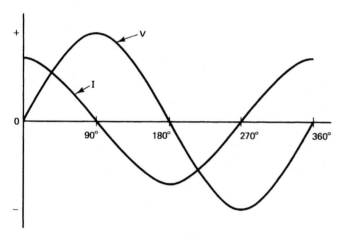

Figure 2-29. Voltage and current waveform of a purely capacitive ac circuit.

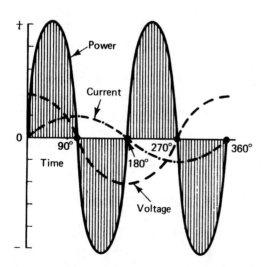

Figure 2-30. Power curves for a purely capacitive ac circuit.

Vector (Phasor) Diagrams

In Figure 2-33 a vector diagram is shown for each circuit condition that was illustrated. Vectors are straight lines that have a specific direction and length. They may be used to represent voltage or current values. An understanding of vector diagrams (sometimes called phasor diagrams) is important when dealing with alternating current. Rather than using waveforms to show phase relationships, it is possible to use a vector or phasor representation.

Ordinarily, when beginning a vector diagram, a horizontal line is drawn with its left end as the reference point. Rotation in a counterclockwise direction from the

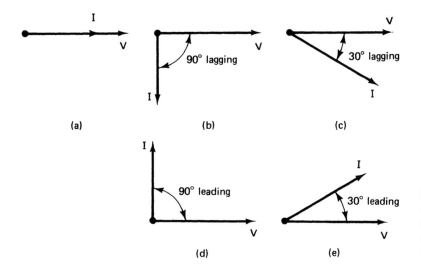

Figure 2-33. Vector diagrams showing voltage and current relationship of ac circuits: (a) resistive (R) circuit; (b) purely inductive (L) circuit; (c) RL circuit; (d) purely capacitive (C) circuit; (e) AC circuit; (f) rectangular and polar coordinates.

reference point is then considered to be a positive direction. Note that in the preceding diagrams, the voltage vector was the reference. For the inductive circuits, the current vector was drawn in a clockwise direction, indicating a lagging condition. A leading condition is shown for the capacitive circuits by the use of a current vector drawn in a counterclockwise direction from the voltage vector.

Series AC Circuits

In any series ac circuit, the current is the same in all parts of the circuit. The voltages must be added by using a voltage triangle. Impedance (Z) of a series ac circuit is found by using an impedance triangle. Power values are found by using a power triangle.

Series RL Circuits

Series RL circuits are often found in electronic equipment. When an ac voltage is applied to a series RL circuit, the current is the same through each part. The voltage drops across each component distributed according to the values of resistance (R) and inductive reactance (X_L) in the circuit.

The total opposition to current flow in any ac circuit is called impedance (Z). Both resistance and reactance in an ac circuit oppose current flow. The impedance of a series RL circuit is found by using either of these formulas:

$$Z = \frac{V}{I} \quad \text{or} \quad Z = \sqrt{R^2 + X_L^2}$$

The impedance of a series RL circuit is found by using an impedance triangle. This right triangle is formed by the three quantities that oppose ac. A triangle is also used to compare voltage drops in series RL circuits. Inductive

voltage (V_L) leads resistive voltage (V_R) by 90°. V_A is the voltage applied to the circuit. Since these values form a right triangle, the value of V_A may be found by using the formula

$$V_A = \sqrt{V_R^2 + V_L^2}$$

An example of a series RL circuit is shown in Figure 2-34.

Series RC Circuits

Series RC circuits also have many uses. This type of circuit is similar to the series RL circuit. In a capacitive circuit, current leads voltage. The reactive values of RC circuits act in the opposite direction as RL circuits. An example of a series RC circuit is shown in Figure 2-35.

Series RLC Circuits

Series RLC circuits have resistance (R), inductance (L), and capacitance (C). The total reactance (X_T) is found by subtracting the smaller reactance (X_L or X_C) from the larger one. Reactive voltage (Vs) is found by obtaining the difference between V_L and V_C. The effects of the capacitance and inductance are 180° out of phase with each other. Right triangles are used to show the simplified relationships of the circuit values. An example of a series RLC circuit is shown in Figure 2-36.

Parallel AC Circuits

Parallel ac circuits have many applications. The basic formulas used with parallel ac circuits are different from those of series circuits. Impedance (Z) of a parallel circuit is less than individual branch values of resistance, inductive reactance, or capacitive reactance. There is no impedance triangle for parallel circuits, since Z is smaller than R, X_L, or X_C. A right triangle is drawn to show the

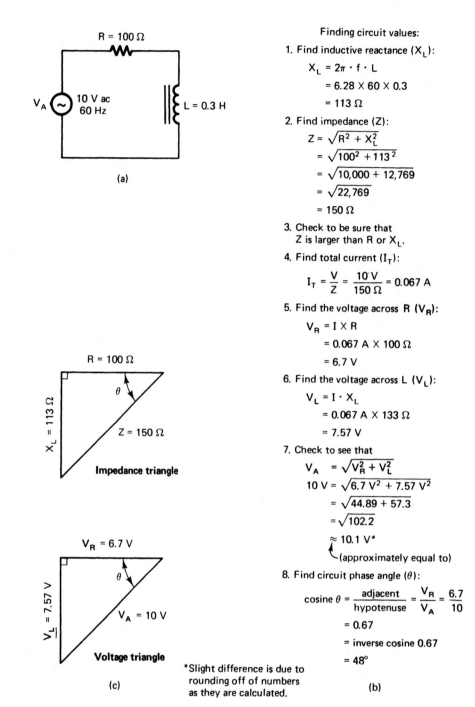

Finding circuit values:

1. Find inductive reactance (X_L):

$$X_L = 2\pi \cdot f \cdot L$$
$$= 6.28 \times 60 \times 0.3$$
$$= 113 \ \Omega$$

2. Find impedance (Z):

$$Z = \sqrt{R^2 + X_L^2}$$
$$= \sqrt{100^2 + 113^2}$$
$$= \sqrt{10{,}000 + 12{,}769}$$
$$= \sqrt{22{,}769}$$
$$= 150 \ \Omega$$

3. Check to be sure that Z is larger than R or X_L.

4. Find total current (I_T):

$$I_T = \frac{V}{Z} = \frac{10 \ V}{150 \ \Omega} = 0.067 \ A$$

5. Find the voltage across R (V_R):

$$V_R = I \times R$$
$$= 0.067 \ A \times 100 \ \Omega$$
$$= 6.7 \ V$$

6. Find the voltage across L (V_L):

$$V_L = I \cdot X_L$$
$$= 0.067 \ A \times 133 \ \Omega$$
$$= 7.57 \ V$$

7. Check to see that

$$V_A = \sqrt{V_R^2 + V_L^2}$$
$$10 \ V = \sqrt{6.7 \ V^2 + 7.57 \ V^2}$$
$$= \sqrt{44.89 + 57.3}$$
$$= \sqrt{102.2}$$
$$\approx 10.1 \ V^*$$
(approximately equal to)

8. Find circuit phase angle (θ):

$$\cos \theta = \frac{adjacent}{hypotenuse} = \frac{V_R}{V_A} = \frac{6.7}{10}$$
$$= 0.67$$
$$= inverse \ cosine \ 0.67$$
$$= 48°$$

*Slight difference is due to rounding off of numbers as they are calculated.

Figure 2-34. Series *RL* circuit example: (a) circuit; (b) procedure for finding circuit values; (c) circuit triangles.

currents in the branches of a parallel circuit.

The voltage of a parallel ac circuit is the same across each branch. The currents through the branches of a parallel ac circuit are shown by a right triangle called a *current triangle*. The current through the capacitor (I_C) is shown leading the current through the resistor (I_R) by 90°. The current through the inductor (I_L) is shown lagging I_R by 90°. I_L and I_C are 180° out of phase, and they are subtract-

ed to find the total reactive current (I_T). Since I_T, I_R, and I_X form a right triangle, the total current may be found by using the formula:

$$I_T = \sqrt{I_R{}^2 + I_X{}^2}$$

This method is used to find currents in parallel *RL*, *RC*, or *RLC* circuits.

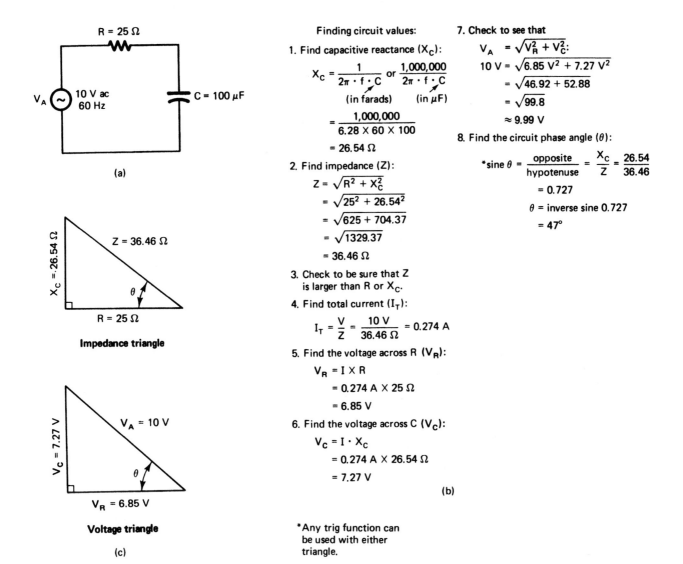

Finding circuit values:

1. Find capacitive reactance (X_C):

$$X_C = \frac{1}{2\pi \cdot f \cdot C} \text{ or } \frac{1,000,000}{2\pi \cdot f \cdot C}$$

(in farads)   (in μF)

$$= \frac{1,000,000}{6.28 \times 60 \times 100}$$

$$= 26.54 \ \Omega$$

2. Find impedance (Z):

$$Z = \sqrt{R^2 + X_C^2}$$

$$= \sqrt{25^2 + 26.54^2}$$

$$= \sqrt{625 + 704.37}$$

$$= \sqrt{1329.37}$$

$$= 36.46 \ \Omega$$

3. Check to be sure that Z is larger than R or X_C.

4. Find total current (I_T):

$$I_T = \frac{V}{Z} = \frac{10 \ V}{36.46 \ \Omega} = 0.274 \ A$$

5. Find the voltage across R (V_R):

$$V_R = I \times R$$

$$= 0.274 \ A \times 25 \ \Omega$$

$$= 6.85 \ V$$

6. Find the voltage across C (V_C):

$$V_C = I \cdot X_C$$

$$= 0.274 \ A \times 26.54 \ \Omega$$

$$= 7.27 \ V$$

7. Check to see that

$$V_A = \sqrt{V_R^2 + V_C^2}:$$

$$10 \ V = \sqrt{6.85 \ V^2 + 7.27 \ V^2}$$

$$= \sqrt{46.92 + 52.88}$$

$$= \sqrt{99.8}$$

$$\approx 9.99 \ V$$

8. Find the circuit phase angle (θ):

$$*sine \ \theta = \frac{opposite}{hypotenuse} = \frac{X_C}{Z} = \frac{26.54}{36.46}$$

$$= 0.727$$

$$\theta = \text{inverse sine } 0.727$$

$$= 47°$$

*Any trig function can be used with either triangle.

Figure 2-35. Series RC circuit: (a) circuit; (b) procedure for finding circuit values; (c) circuit triangles.

When components are connected in parallel, finding impedance is more difficult, since an impedance triangle cannot be used. A method that can be used to find impedance is to use an admittance triangle. The following quantities are plotted on the triangle: (1) admittance, $Y = 1/Z$; (2) conductance, $G = 1/R$; (3) inductive susceptance, $B_L = 1/X_L$; and (4) capacitive susceptance, $B_C = 1/X_C$. These quantities are the inverses of each type of opposition to ac. Since total impedance (Z) is the smallest quantity in a parallel ac circuit, its reciprocal ($1/Z$) becomes the largest quantity on the admittance triangle (just as $1/2$ is larger than $1/4$). The values are in siemens or mhos (ohms spelled backward).

Parallel ac circuits are similar in several ways to series ac circuits. Study the examples of Figure 2-37 through Figure 2-39.

Power in AC Circuits

Power values in ac circuits are found by using power triangles, as shown in Figure 2-40. The power delivered to an ac circuit is called *apparent power* and is equal to voltage times current. The unit of measurement is the volt-ampere (V_A) or kilovolt-ampere (WA). Meters are used to measure apparent power in ac circuits. Apparent power is the voltage applied to a circuit multiplied by the total current flow. The power converted to another form of energy by the load is measured with a wattmeter. This actual power converted is called *true power*. The ratio of true power converted in a circuit to apparent power delivered to an ac circuit is called the *power factor* (PF), which is found by using the formula

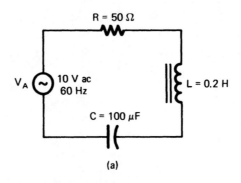

(a)

Finding circuit values:

1. Find inductive reactance (X_L):

   $X_L = 2\pi \cdot f \cdot L$

   $= 6.28 \times 60 \times 0.2$

   $= 75.36\ \Omega$

2. Find capacitive reactance (X_C):

   $X_C = \dfrac{1{,}000{,}000}{2\pi \cdot f \cdot C\ (\mu F)}$

   $= \dfrac{1{,}000{,}000}{6.28 \times 60 \times 100}$

   $= 26.54\ \Omega$

3. Find total reactance (X_T):

   $X_T = X_L - X_C$

   $= 75.36\ \Omega - 26.54\ \Omega$

   $= 48.82\ \Omega$

4. Find impedance (Z):

   $Z = \sqrt{R^2 + X_T^2}$

   $= \sqrt{50^2 + 48.82^2}$

   $= \sqrt{2500 + 2383.4}$

   $= \sqrt{4{,}883.4}$

   $= 69.88\ \Omega$

5. Check to be sure that Z is larger than X_T or R.

6. Find total current (I_T):

   $I_T = \dfrac{V}{Z} = \dfrac{10\ V}{69.88\ \Omega} = 0.143\ A$

7. Find voltage across R (V_R):

   $V_R = I \times R$

   $= 0.143\ A \times 50\ \Omega$

   $= 7.15\ V$

8. Find voltage across L (V_L):

   $V_L = I \cdot X_L$

   $= 0.143\ A \times 75.36\ \Omega$

   $= 10.78\ V$

9. Find voltage across C (V_C):

   $V_C = I \cdot X_C$

   $= 0.143\ A \times 26.54\ \Omega$

   $= 3.8\ V$

10. Find total reactive voltage (V_X):

    $V_X = V_L - V_C$

    $= 10.78\ V - 3.8\ V$

    $= 6.98\ V$

11. Check to see that

    $V_A = \sqrt{V_R^2 + V_X^2}$

    $= \sqrt{7.15\ V^2 + 6.98\ V^2}$

    $10\ V = \sqrt{51.12 + 48.72}$

    $= \sqrt{99.84}$

    $\approx 9.99\ V$

12. Find circuit phase angle (θ):

    $\text{tangent } \theta = \dfrac{\text{opposite}}{\text{adjacent}} = \dfrac{X_T}{R} = \dfrac{48.82}{50}$

    $= 0.976$

    $= \text{inverse tangent } 0.976$

    $= 44°$

(b)

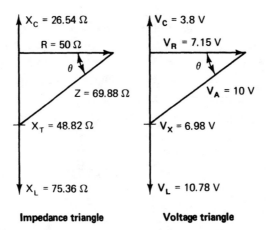

Impedance triangle   Voltage triangle

(c)

Figure 2-36. Series RLC circuit example: (a) circuit; (b) procedure for finding circuit values; (c) circuit triangles.

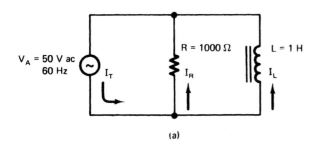

(a)

Finding circuit values:

1. Find inductive reactance (X_L):

$$X_L = 2\pi \cdot f \cdot L$$
$$= 6.28 \times 60 \times 1$$
$$= 376.8\ \Omega$$

2. Find current through $R(I_R)$:

$$I_R = \frac{V}{R} = \frac{50\ V}{1000\ \Omega} = 0.05\ A$$

3. Find current through $L\ (I_L)$:

$$I_L = \frac{V}{X_L} = \frac{50\ V}{376.8\ \Omega} = 0.133\ A$$

4. Find total current (I_T):

$$I_T = \sqrt{I_R^2 + I_L^2}$$
$$= \sqrt{0.05^2 + 0.133^2}$$
$$= \sqrt{0.0025 + 0.0177}$$
$$= \sqrt{0.0202}$$
$$= 0.142\ A$$

5. Check to see that I_T is larger than I_R or I_L.

6. Find impedance (Z):

$$Z = \frac{V}{I_T} = \frac{50\ V}{0.142\ A}$$
$$= 352.1\ \Omega$$

7. Check to see that Z is less than R or X_L.

8. Find circuit phase angle (θ):

$$\text{sine } \theta = \frac{\text{opposite}}{\text{hypotenuse}}$$
$$= \frac{I_L}{I_T} = \frac{0.133}{0.142}$$
$$= 0.937$$
$$\theta = \text{inverse sine } 0.937$$
$$= 70°$$

(b)

$$PF = \frac{\text{true power (W)}}{\text{apparent power } (V_A)}$$

or

$$\%PF = \frac{W}{VA} \times 100$$

The term W is the true power in watts and VA is the apparent power in volt-amperes. The highest power factor of an ac circuit is 1.0, or 100%. This value is called *unity power factor.*

The phase angle (θ) of an ac circuit determines the power factor. Remember that the phase angle is the angular separation between voltage applied to an ac circuit and current flow through the circuit. Also, more inductive or capacitive reactance causes a larger phase angle. In a purely inductive or capacitive circuit, a 90° phase angle causes a power factor of 0°. The power factor varies according to the values of resistance and reactance in a circuit.

There are two types of power that affect power conversion in ac circuits. Power conversion in the resistive part of the circuit is called *active power,* or *true power.* True power is measured in watts. The other type of power is caused by inductive or capacitive loads. It is 90° out of phase with the true power and is called reactive power. It

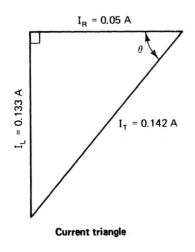

Current triangle

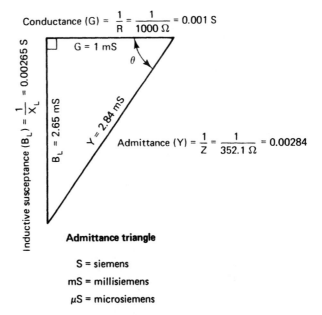

Conductance $(G) = \dfrac{1}{R} = \dfrac{1}{1000\ \Omega} = 0.001\ S$

$G = 1\ mS$

Admittance $(Y) = \dfrac{1}{Z} = \dfrac{1}{352.1\ \Omega} = 0.00284$

Admittance triangle

S = siemens
mS = millisiemens
μS = microsiemens

Figure 2-37. Parallel *RL* circuit example: (a) circuit; (b) procedure for finding circuit values; (c) circuit triangles.

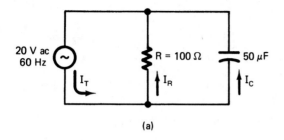

(a)

Finding circuit values:

1. Find capacitive reactance (X_C):

$$X_C = \frac{1,000,000}{2\pi \cdot f \cdot C\,(\mu F)}$$

$$= \frac{1,000,000}{6.28 \times 60 \times 50}$$

$$= 53\ \Omega$$

2. Find current through R (I_R):

$$I_R = \frac{V}{R} = \frac{20\ V}{100\ \Omega} = 0.2\ A$$

3. Find current through C (I_C):

$$I_C = \frac{1}{X_C} = \frac{20\ V}{53\ \Omega} = 0.377\ A$$

4. Find total current (I_T):

$$I_T = \sqrt{I_R^2 + I_C^2}$$

$$= \sqrt{0.2^2 + 0.377^2}$$

$$= \sqrt{0.04 + 0.142}$$

$$= \sqrt{0.182}$$

$$= 0.427\ A$$

5. Check to see that I_T is larger than I_R or I_C.

6. Find impedance (Z):

$$Z = \frac{V}{I_T} = \frac{20\ V}{0.427\ A} = 46.84\ \Omega$$

7. Check to see that Z is less than R or X_C.

8. Find circuit phase angle (θ):

$$\text{cosine } \theta = \frac{\text{adjacent}}{\text{hypotenuse}} = \frac{G}{Y} = \frac{10}{21.3}$$

$$= 0.469$$

$$\theta = \text{inverse cosine } 0.469$$

$$= 62°$$

(b)

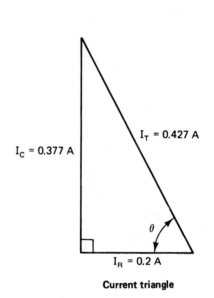

Current triangle

$I_C = 0.377\ A$ $I_T = 0.427\ A$

θ

$I_R = 0.2\ A$

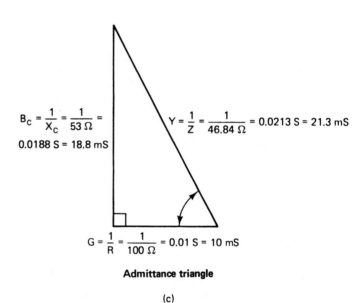

$B_C = \dfrac{1}{X_C} = \dfrac{1}{53\ \Omega} = 0.0188\ S = 18.8\ mS$

$Y = \dfrac{1}{Z} = \dfrac{1}{46.84\ \Omega} = 0.0213\ S = 21.3\ mS$

$G = \dfrac{1}{R} = \dfrac{1}{100\ \Omega} = 0.01\ S = 10\ mS$

Admittance triangle

(c)

Figure 2-38. Parallel *RC* circuit examples: (a) circuit; (b) procedure for finding circuit values; (c) circuit triangles.

is a type of "unused" power. Reactive power is measured in volt-amperes reactive (VAR).

In the power triangles of Figure 2-40, true power (watts) is marked as a horizontal line. Reactive power (VAR) is drawn at a 90° angle from the true power. Volt-amperes or apparent power (V_A) is the longest side (hy-potenuse) of the right triangle. This right triangle is similar to the impedance triangle and the voltage triangle for series ac circuits and the current triangle and admittance triangle for parallel ac circuits. Each type of right triangle has a horizontal line used to mark the resistive part of the circuit. The vertical lines are used to mark the reactive

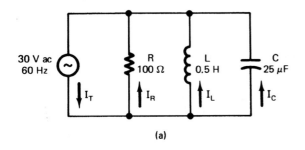

(a)

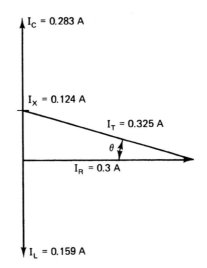

$I_C = 0.283$ A

$I_X = 0.124$ A

$I_T = 0.325$ A

θ

$I_R = 0.3$ A

$I_L = 0.159$ A

Current triangle

Finding circuit values:

1. Find inductive reactance (X_L):

$X_L = 2\pi \cdot f \cdot L$

$= 6.28 \times 60 \times 0.5$

$= 188.4\ \Omega$

2. Find capacitive reactance (X_C):

$X_C = \dfrac{1,000,000}{6.28 \times 60 \times 25}$

$= 106\ \Omega$

3. Find current through R (I_R):

$I_R = \dfrac{V}{R} = \dfrac{30\ V}{100\ \Omega} = 0.3$ A

4. Find current through L (I_L):

$I_L = \dfrac{V}{X_L} = \dfrac{30\ V}{188.4\ \Omega} = 0.159$ A

5. Find current through C (I_C):

$I_C = \dfrac{V}{X_C} = \dfrac{30\ V}{106\ \Omega} = 0.283$ A

6. Find total reactive current (I_X):

$I_X = I_C - I_L$

$= 0.283 - 0.159$

$= 0.124$

7. Find total current (I_T):

$I_T = \sqrt{I_R^2 + I_X^2}$

$= \sqrt{0.3^2 + 0.124^2}$

$= \sqrt{0.9 + 0.0154}$

$= \sqrt{0.1054}$

$= 0.325$ A

8. Check to see that I_T is larger than I_R or I_X.

9. Find impedance (Z):

$Z = \dfrac{V}{I_T} = \dfrac{30\ V}{0.325\ A} = 92.3\ \Omega$

10. Check to see that Z is less than R or X_L or X_C.

11. Find circuit phase angle (θ):

$\text{tangent } \theta = \dfrac{\text{opposite}}{\text{adjacent}} = \dfrac{I_X}{I_R} = \dfrac{0.124}{0.3}$

$= 0.413$

$\theta = \text{inverse tangent } 0.413$

$= 22°$

(b)

$B_C = \dfrac{1}{X_C} = \dfrac{1}{106\ \Omega} = 0.0094\ S = 9.4$ mS

$B_X = B_C - B_L = 9.4 - 5.3 = 4.1$ mS

$Y = \dfrac{1}{Z} = \dfrac{1}{92.3\ \Omega} = 0.0108\ S = 10.8$ mS

4.1 µs

θ

$G = \dfrac{1}{R} = \dfrac{1}{100\ \Omega} = 0.1\ S = 10$ mS

$B_L = \dfrac{1}{X_L} = \dfrac{1}{188.4\ \Omega} = 0.053\ S = 5.3$ mS

Admittance triangle

(c)

Figure 2-39. Parallel *RLC* circuit example: (a) circuit; (b) procedure for finding circuit values; (c) circuit triangles.

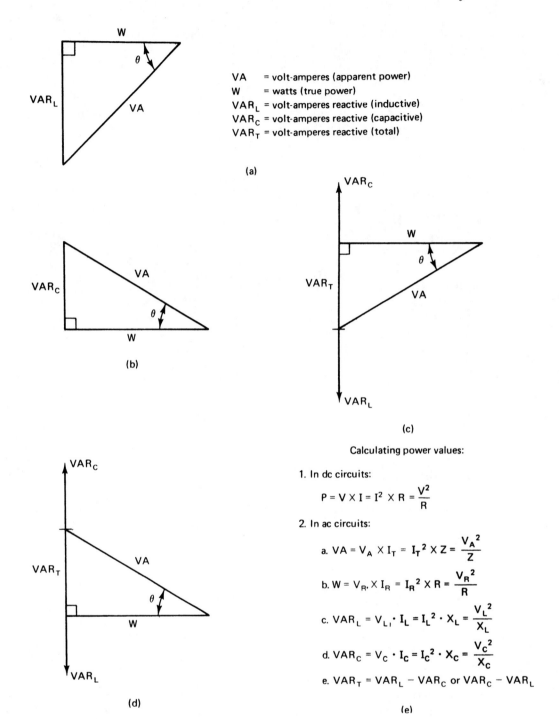

VA   = volt-amperes (apparent power)
W   = watts (true power)
VAR_L = volt-amperes reactive (inductive)
VAR_C = volt-amperes reactive (capacitive)
VAR_T = volt-amperes reactive (total)

(a)

(b)

(c)

(d)

Calculating power values:

1. In dc circuits:

$$P = V \times I = I^2 \times R = \frac{V^2}{R}$$

2. In ac circuits:

a. $VA = V_A \times I_T = I_T^2 \times Z = \dfrac{V_A^2}{Z}$

b. $W = V_R \times I_R = I_R^2 \times R = \dfrac{V_R^2}{R}$

c. $VAR_L = V_L \cdot I_L = I_L^2 \cdot X_L = \dfrac{V_L^2}{X_L}$

d. $VAR_C = V_C \cdot I_C = I_C^2 \cdot X_C = \dfrac{V_C^2}{X_C}$

e. $VAR_T = VAR_L - VAR_C$ or $VAR_C - VAR_L$

(e)

Figure 2-40. Power triangles for ac circuits: (a) *RL* circuits; (b) *RC* circuits; (c) *RLC* circuit with X_C larger than X_L; (d) *RLC* circuit with X_C larger than X_L (e) formulas for calculating power values.

part of the circuit. The longest side of the triangle (hypotenuse) is used to mark total values of the circuit. The size of the hypotenuse depends on the amount of resistance and reactance in the circuit. Vector diagrams and right triangles are extremely important for understanding ac circuits.

TRANSFORMERS

Transformers are used either to increase or to decrease ac voltage. Transformers are made by using two separate sets of wire windings, which are wound on a metal core and are called the *primary* and the *secondary* windings. A transformer that increases voltage is called a step-up transformer and one that decreases voltage is called a step-down transformer.

Transformer Operation

Transformers do not operate with dc voltage applied, since dc voltage does not change in value. Notice in Figure 2-41 that ac voltage is applied to the primary winding of the transformer. There is no connection of the primary and secondary windings. The transfer of energy from the primary to the secondary winding is due to magnetic coupling or mutual inductance. The transformer relies on electromagnetism to operate.

The primary and secondary windings of transformers are wound around a laminated iron core. The iron core is used to transfer the magnetic energy from the primary winding to the secondary winding. Remember that ac voltage is applied to the primary winding, and the secondary winding is connected to an electrical load.

There are many different types and sizes of transformers. However, the same basic principles of operation apply to all transformers. The operation of a transformer relies on the expanding and collapsing of the magnetic field around the primary winding. When current flows through a conductor, a magnetic field is developed around the conductor. When ac voltage is applied to the primary winding, it causes a constantly changing magnetic field around the winding. During times of increasing ac voltage, the magnetic field around the primary winding expands. After the peak value of the ac cycle is reached, the voltage decreases toward zero. When the ac voltage decreases, the magnetic field around the primary winding collapses. The collapsing magnetic field is transferred to the secondary winding.

Types of Transformers

There are three basic transformer types, air-core, iron-core, and powered-metal-core transformers. The core material used to construct transformers depends on the application of the transformers. Radios and televisions use air-core and powdered-metal-core transformers. Iron-core transformers are sometimes called power transformers. They are often used in the power supplies of electrical equipment and at substations that distribute electrical power.

Two other classifications of transformers are *step-up transformers* and *step-down transformers*. In step-up transformers, more turns of wire are used for the secondary windings than for the primary windings. In step-down transformers, more turns of wire are on the primary windings than on the secondary windings. An example of a step-up transformer is shown in Figure 2-42). A step-up transformer has a higher voltage across the secondary winding than the input voltage applied to the primary winding. The step up is caused by the turns ratio of the primary and secondary windings. The turns ratio is the ratio of primary turns of wire (N_p) to secondary turns of wire (N_s). If there are 100 turns of wire used for the primary winding and 200 turns for the secondary winding, the turns ratio is 1:2. The secondary voltage is two times as large as the primary voltage. If 10 V ac is applied to the primary, 20 V ac is developed across the load connected across the secondary winding. The voltage ratio (V_p/V_s) is the same as the turns ratio.

Figure 2-42(b) shows an example of a step-down transformer. The primary winding has 500 turns, and the secondary has 100 turns of wire. The turns ratio (N_p/N_s) is 5:1. This means that the primary voltage (V_p) is five times as great as the secondary voltage (V_s). If 25 V is applied to the primary, 5 V is developed across the secondary. The voltage formula shown in Figure 2-42 shows the relationship between the primary and secondary voltages and the turns ratio of the primary and secondary windings.

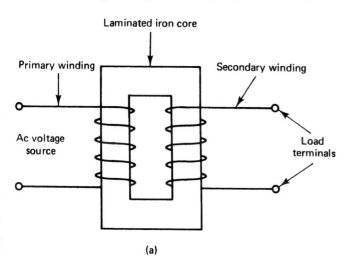

(a)

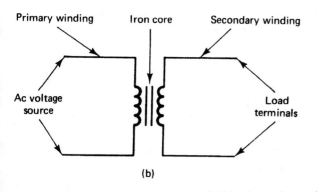

(b)

Figure 2-41. Transformer: (a) pictorial; (b) schematic symbol.

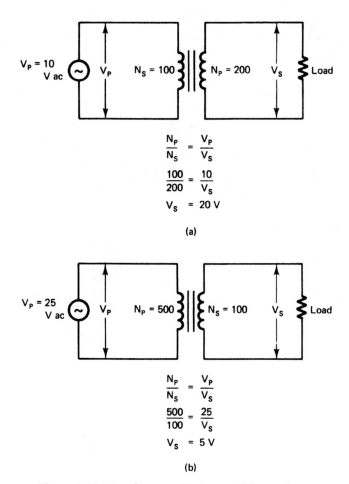

$$\frac{N_P}{N_S} = \frac{V_P}{V_S}$$

$$\frac{100}{200} = \frac{10}{V_S}$$

$$V_S = 20 \text{ V}$$

(a)

$$\frac{N_P}{N_S} = \frac{V_P}{V_S}$$

$$\frac{500}{100} = \frac{25}{V_S}$$

$$V_S = 5 \text{ V}$$

(b)

Figure 2-42. Transformers: (a) step-up; (b) step-down.

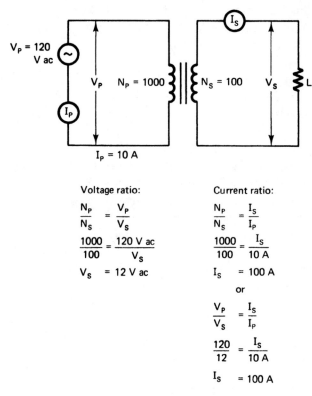

Voltage ratio:

$$\frac{N_P}{N_S} = \frac{V_P}{V_S}$$

$$\frac{1000}{100} = \frac{120 \text{ V ac}}{V_S}$$

$$V_S = 12 \text{ V ac}$$

Current ratio:

$$\frac{N_P}{N_S} = \frac{I_S}{I_P}$$

$$\frac{1000}{100} = \frac{I_S}{10 \text{ A}}$$

$$I_S = 100 \text{ A}$$

or

$$\frac{V_P}{V_S} = \frac{I_S}{I_P}$$

$$\frac{120}{12} = \frac{I_S}{10 \text{ A}}$$

$$I_S = 100 \text{ A}$$

Figure 2-43. Transformer current ratio.

Transformers are power-control devices. Power is equal to voltage (*V*) multiplied by current (*I*). In a transformer, primary power (Pr) is equal to secondary power (P8). Therefore, primary voltage (V_p) times primary current (In) is equal to secondary voltage (V8) times secondary current (Is):

$$P_p = P_s$$

or

$$V_p \times I_p = V_s \times I_s$$

The current ratio of transformers is an inverse ratio: $N_p/N_s = I_s/I_p$. An application of the current ratio is shown in Figure 2-43. If voltage is increased, current is decreased in the same proportion. The secondary voltage is decreased by 10 times and secondary current is increased by 10 times in the example shown. This is a step-down transformer, since voltage decreases.

Some transformers have more than one secondary winding. These are called multiple secondary transformers. Some are step-up and some are step-down windings, depending on the application. There are many other spe-cial types of transformers. *Auto transformers* have only one winding. They may be either step-up or step-down trans-formers, depending on how they are connected. Variable autotransformers are used in power supplies to provide variable ac voltages.

Current transformers are used to measure large ac currents. A conductor is placed inside the hole in the center. A current-carrying conductor produces a magnetic field. This magnetic field is transferred to the winding of the secondary of the current transformer. The high current is reduced by the transformer so that it can be measured.

Many transformers are used to control three-phase voltages. Large three-phase transformers are used at electrical power distribution substations to step down voltages. Another type of transformer is called an *isolation transformer*. When making tests with any equipment that plugs into an ac power outlet, an isolation transformer should be used. Its purpose is to isolate the power source of the test equipment from the equipment being tested. An isolation transformer is plugged into a power outlet. The equipment being tested is plugged into the secondary winding of the isolation transformer. There is no connection between the primary and secondary windings of the transformer. This "isolates" the hot and neutral wires of the two pieces of equipment. The possibility of electrical hot-to-ground shorts is eliminated by using an isolation transformer.

Transformer Efficiency

Transformers are very efficient electrical devices. A typical efficiency rating for a transformer is around 98%. Efficiency of electrical equipment is found by using the formula

$$\text{Efficiency } (\%) = \frac{P_{out}}{P_{in}} \times 100$$

where P_{out} is the secondary power ($V_s \times I_s$) in watts and P_{in} is the primary power ($V_p \times I_p$) in watts.

FREQUENCY-SENSITIVE AC CIRCUITS

Some types of ac circuits are designed to respond to ac frequencies. Circuits that are used to pass some frequencies and block others are called frequency-sensitive circuits. Two types of frequency-sensitive circuits are (1) filter circuits and (2) resonant circuits. Each type of circuit uses reactive devices to respond to different ac frequencies. These circuits have *frequency response curves*. Frequency is graphed on the horizontal axis and voltage output on the vertical axis. Sample frequency response curves for each type of filter and resonant circuit are shown in the examples that follow.

Filter Circuits

The three types of filter circuits are shown in Figure 2-44. Filter circuits are used to separate one range of frequencies from another. Low-pass filters pass low ac frequencies and block higher frequencies. High-pass filters pass high frequencies and block lower frequencies. Band-pass filters pass a midrange of frequencies and block lower and higher frequencies. All filter circuits have resistance and capacitance or inductance.

Figure 2-45 shows the circuits used for low-pass, high-pass, and band-pass filters and their frequency response curves. Many low-pass filters are series RC circuits, as shown in Figure 2-45(a). Output voltage (V_{out}) is taken across a capacitor. As frequency increases, capacitive reactance (X_C) decreases, since

$$X_C = \frac{1}{2\pi fC}$$

The voltage drop across the output is equal to I times X_C, so as frequency increase, X_C decreases and voltage output

decreases.

Series RL circuits may also be used as low-pass filters. As frequency increases, inductive reactance (X_L) increases, since $X_L = 2\pi fL$. Any increase in X_L reduces the circuit's current. The voltage output taken across the resistor is equal to $I \times R$. So when I decreases, V_{out}, also decreases. As frequency increases, X_L increases, I decreases, and V_{out}, decreases.

Figure 2-45(b) shows two types of high-pass filters. The series RC circuit is a common type. The voltage output (V_{out}) is taken across the resistor (R). As frequency increases, X decreases. A decrease in X_C causes current flow to increase. The voltage output across the resistor (V_{out}) is equal to $I \times R$, so as I increases, V_{out}, increases. As frequency increases, X_C decreases, I increases, and V_{out}, increases.

A series RL circuit may also be used as a high-pass filter. The value V_{out} is taken across the inductor. As frequency increases, X_L increases. V_{out}, is equal to $I \times X_L$, so as X_L increases, V_{out}, also increases. In this circuit, as frequency increases, X_L increases and V_{out}, increases.

The band-pass filter of Figure 2-45(c) is a combination of low-pass and high-pass filter sections. It is designed to pass a midrange of frequencies and block low and high frequencies. R_1 and C_1 form a low-pass filter,

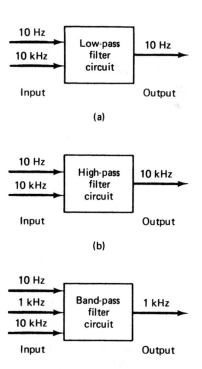

Figure 2-44. Three types of filter circuits: (a) low-pass filter—passes low frequencies and blocks high frequencies; (b) high-pass filter—passes high frequencies and blocks low frequencies; (c) band-pass—passes a midrange of frequencies and blocks high and low frequencies.

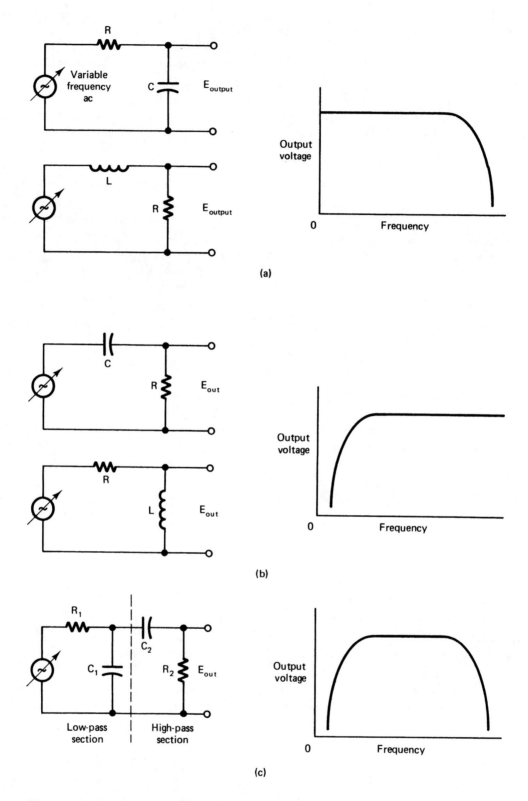

Figure 2-45. Circuits used to filter ac frequencies and their response curves: (a) low-pass filters; (b) high-pass filters; (c) band-pass filter.

and R_2 and C_2 form a high-pass filter. The range of frequencies to be passed is determined by calculating the values of resistance and capacitance.

Resonant Circuits

Resonant circuits are designed to pass a range of frequencies and block others. They have resistance, inductance, and capacitance. Figure 2-46 shows the two types of resonant circuits, series resonant and parallel resonant circuits, and their frequency response curves.

Series Resonant Circuits

Series resonant circuits are a series arrangement of inductance, capacitance, and resistance. A series resonant circuit offers a small amount of opposition to some ac frequencies and much more opposition to other frequencies. They are important for selecting or rejecting frequencies.

The voltage across inductors and capacitors in ac series circuits are in direct opposition to each other (180° out of phase) and tend to cancel each other out. The frequency applied to a series resonant circuit affects inductive reactance and capacitive reactance. At a specific input frequency, X_L will equal X_C. The voltages across the inductor and capacitor are then equal. The total reactive voltage (V_X) is 0V at this frequency. The opposition offered by the inductor and the capacitor cancel each other at this frequency. The total reactance (X_T) of the circuit ($X_L - X_C$) is zero. The impedance (Z) of the circuit is then equal to the resistance (R).

The frequency at which $X_L = X$ is called *resonant frequency*. *To* determine the resonant frequency (f_r) of the circuit, use the formula

$$f_r = \frac{1}{2\pi\sqrt{LC}}$$

In the formula, L is in henries, C is in farads, and f_r is in hertz. As either inductance or capacitance increases, resonant frequency decreases. When the resonant frequency is applied to a circuit, a condition called *resonance* exists. Resonance for a series circuit causes the following:

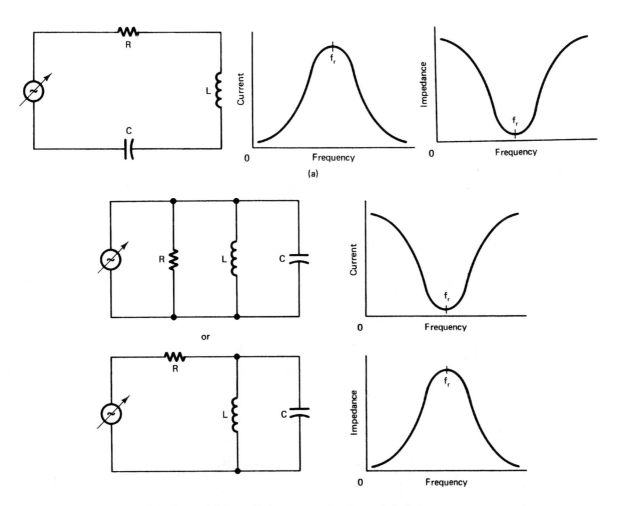

Figure 2-46. (a) Series and (b) parallel resonant circuits and their frequency response curves.

1. $X_L = X_C$.
2. $X_T = 0$.
3. $V_L = V_C$.
4. Total reactive voltage (V_X) is equal to zero.
5. $Z = R$.
6. Total current (I_T) is maximum.
7. Phase angle (θ) is 0°.

The ratio of reactance (X_L or X_C) to resistance (R) at resonant frequency is called *quality factor* (Q). This ratio is used to determine the range of frequencies or *bandwidth* (BW) a resonant circuit will pass. A sample resonant circuit problem is shown in Figure 2-47. The frequency range that a resonant circuit will pass (BW) is found by using the procedure of steps 5 and 6 in Figure 2-47.

The cutoff points are at about 70% of the maximum output voltage. These are called the low-frequency cutoff (f_{LC}) and high-frequency cutoff (f_{hc}). The bandwidth of a resonant circuit is determined by Q, and Q is determined by the ratio of X_L and X_C to R. Resistance determines bandwidth, for the most part. This effect is summarized as follows:

1. When R is increased, Q decreases, since $Q\ X_L/R$.

2. When Q decreases, BW increases, since $BW = f_r/Q$.

3. When R is increased, BW increases.

Two curves are shown in Figure 2-48 that show the effect of resistance on bandwidth. The curve of Figure 2-48(b) has high *selectivity*. This means that a resonant circuit with this response curve would select a small range of frequencies. This is very important for radio and television tuning circuits.

Parallel Resonant Circuits

Parallel resonant circuits are similar to series resonant circuits. Their electrical characteristics are somewhat different, but they accomplish the same purpose. Another name for parallel resonant circuits is *tank circuit*. A tank circuit is a parallel combination of L and C used to select or reject ac frequencies.

With the resonant frequency applied to a parallel resonant circuit, the following occurs:

1. $X_L = X_C$.
2. $X_T = 0$.
3. $I_L = I_C$.
4. $I_X = 0$, so the circuit current is minimum.
5. $Z = R$ and is maximum.
6. Phase angle (θ) = 0°.

Finding circuit values:

1. Find resonant frequency (f_r):

$$f_r = \frac{1}{2\pi \times \sqrt{L \times C}} = \frac{1}{6.28 \times \sqrt{(10 \times 10^{-3}) \times (0.05 \times 10^{-6})}} =$$

$$= \frac{1}{6.28 \times \sqrt{0.5 \times 10^{-9}}} = \frac{1}{6.28 \times (2.23 \times 10^{-5})} = \frac{1}{1.4 \times 10^{-4}} = 7121 \text{ Hz}$$

2. Find X_L^* and X_C at resonant frequency:

$X_L = 2\pi \cdot f \cdot L = 6.28 \times 7121 \times (10 \times 10^{-3}) = 447\ \Omega$

*Easier to calculate.

3. Find quality factor (Q):

$$Q = \frac{X_L}{R} = \frac{447\ \Omega}{100\ \Omega} = 4.47$$

4. Find bandwidth (BW):

$$BW = \frac{f_r}{Q} = \frac{7121 \text{ Hz}}{4.47} = 1593 \text{ Hz}$$

5. Find low-frequency cutoff (f_{lc}):

$f_{lc} = f_r - \frac{1}{2}BW = 7121 \text{ Hz} - 797 \text{ Hz} = 6324 \text{ Hz}$

6. Find high-frequency cutoff (f_{hc})

$f_{hc} = f_r + \frac{1}{2}BW = 7121 + 797 = 7918 \text{ Hz}$

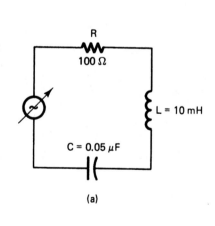

(a)

(b)

Figure 2-47. Sample resonant circuit problem: (a) circuit; (b) procedure to find circuit values.

The calculations used for parallel resonant circuits are similar to those for series circuits. There is one exception. The quality factor (Q) is found by using this formula for parallel circuits: $Q = R/X_L$. A parallel resonant circuit problem is shown in Figure 2-49.

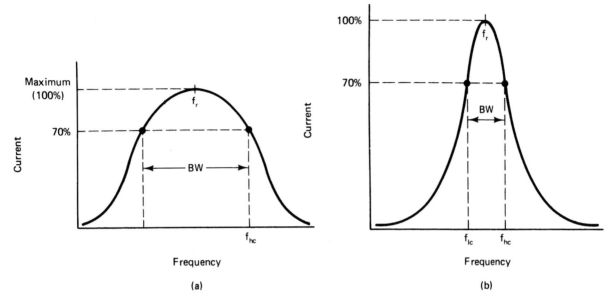

Figure 2-48. Effect of resistance on bandwidth of a series resonant circuit: (a) high-resistance, low selectivity; (b) low-resistance, high selectivity.

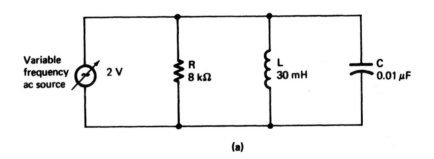

(a)

Finding circuit values:

1. Find resonant frequency (f_r):

$$f_r = \frac{1}{2\pi\sqrt{L \times C}} = \frac{1}{6.28 \times \sqrt{(30 \times 10^{-3}) \times (0.01 \times 10^{-6})}}$$

$$= \frac{1}{1.0877 \times 10^{-4}} = 9194 \text{ Hz}$$

2. Find X_L and X_C at resonant frequency:

$$X_L = 2\pi \cdot f \cdot L = 6.28 \times 9194 \times (30 \times 10^{-3}) = 1732 \ \Omega$$

3. Find quality factor (Q):

$$Q = \frac{R}{X_L} = \frac{8000 \ \Omega}{1732 \ \Omega} = 4.62$$

4. Find bandwidth (BW):

$$BW = \frac{f_r}{Q} = \frac{9194 \text{ Hz}}{4.62} = 1990 \text{ Hz}$$

5. Find low cutoff frequency (f_{lc}):

$$f_{lc} = f_r - 1/2 \ BW = 9194 \text{ Hz} - 995 \text{ Hz} = 8199 \text{ Hz}$$

6. Find high cutoff frequency (f_{hc}):

$$f_{hc} = f_r + 1/2 \ BW = 9194 \text{ Hz} + 995 \text{ Hz} = 10{,}189 \text{ Hz}$$

7. Find I_R at resonant frequency:

$$I_R = \frac{V_A}{R} = \frac{2 \text{ V}}{8 \text{ k}\Omega} = 0.25 \text{ mA}$$

8. Find I_L and I_C at resonant frequency:

$$I_L = \frac{V_A}{X_L} = \frac{2 \text{ V}}{1732 \ \Omega} = 1.15 \text{ mA}$$

(b)

Figure 2-49. Parallel resonant circuit: (a) circuit; (b) procedure for finding circuit values.

Chapter 3

Semiconductor Devices

INTRODUCTION

Nearly any study of semiconductor devices begins with an investigation of atomic theory. This is purposely done to familiarize the reader with a number of ideas related to semiconductor operation. Chapter 1 examined electrons, protons, and neutrons. Each of these parts had a specific role in the construction of an atom. Elements are classified according to the number of particles they possess. No two elements have the same physical structure.

OBJECTIVES

Upon completion of this chapter, you will be able to:

1. Explain how semiconductor devices functionally operate.

2. Distinguish between conductors, semiconductors, and insulators.

3. Describe the crystal structure of representative semiconductor diodes and amplifying devices.

4. Describe the operation of devices in a circuit.

5. Be familiar with semiconductor device packaging and symbol representations.

6. Recognize the functional operation of diodes and amplifying semiconductor devices.

7. Describe how to test semiconductor devices and evaluate their status.

8. Describe forward and reverse bias characteristics of diodes.

9. Explain voltage-current characteristics of semiconductor devices.

10. Describe the operation of zener diodes, light-emitting diodes (LEDs), photovoltaic cells, photodiodes, and varactor diodes.

11. Recognize NPN and PNP transistors.

12. Explain transistor biasing.

13. Describe transistor Alpha and Beta characteristics.

14. Explain the operation of JFET, MOSFET, and UJTs.

SEMICONDUCTOR THEORY

Chapter 1 pointed out that the electrons of a particular atom are not an equal distance from the nucleus. They rotate in well-defined orbits. Each shell or orbit can hold only a certain number of electrons. Outer-shell electrons have a great deal to do with the electrical conductivity of an atom. Uncontrolled movement of these electrons causes certain atoms to be good electrical conductors. A material composed of this type of atom has many free electrons. An insulating material, by comparison, has a relatively small number of free electrons. Insulators tend to hold outer-shell electrons firmly in place. Very little electrical conductivity takes place in an insulating material.

Midway between conductors and insulators is a third classification of atoms known as semiconductors. Silicon and germanium are the most common semiconductor elements. Also semiconductor compounds such as copper oxide, cadmium-sulfide, and gallium arsenide are frequently used. Semiconductor materials are generally classified as type IVB elements. This type of atom has four valence electrons. If they were to give up four valence electrons, stability could be achieved. If it could gain four electrons, stability could also be achieved.

Stability of an atom is an important concept in the status of semiconductor materials. Atoms cannot possess more than 8 electrons in the valence band. When exactly 8 electrons appear in the valence band, an atom is considered to be stable. The valence electrons in a stable atom become tightly bound together. Atoms of this type are excellent insulators. They do not have any free electrons for electric conduction. Gases such as neon, argon, krypton, and xenon are examples of stabilized elements. They will not mix with other materials because of this condition. A stabilized gas is generally called inert.

Atoms having fewer than 8 valence electrons are unstable. They have a tendency to seek electrons from neighboring atoms to become stable. Atoms with 5, 6, or 7 valence electrons tend to borrow electrons from other atoms to seek stability. Atoms with one, two, or three valence electrons try to release these electrons to other atoms.

ATOM COMBINATIONS

Through their valence electrons, different atoms frequently combine to form stabilized molecules or compounds. The process of linking or interconnection is generally called *bonding*. Several kinds of bonding occur in atom combinations. *Ionic bonding, covalent bonding,* and *metallic bonding* are of particular importance in semiconductor theory.

When atoms bond together to form molecules, each atom is seeking stability. A stabilized condition occurs when the valence band contains 8 electrons. In ionic bonding, the valence electrons of one atom join together with those of a different atom to become stable. When an atom has more than 4 valence electrons it is seeking additional electrons. This type of atom is often called an *acceptor.* Atoms containing less than 4 valence electrons try to rid themselves of these electrons. These atoms are called *donors.* Donor and acceptor atoms frequently combine together in ionic bonding. When this occurs the combination becomes stabilized.

Table salt is a common example of ionic bonding. Figure 3-1 shows an example of independent atoms and

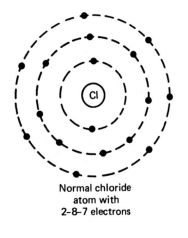

Normal chloride
atom with
2–8–7 electrons

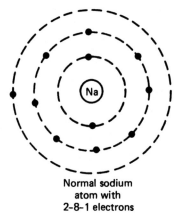

Normal sodium
atom with
2–8–1 electrons

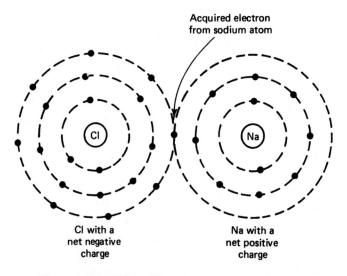

Acquired electron
from sodium atom

Cl with a
net negative
charge

Na with a
net positive
charge

Figure 3-1. Ionic bonding of sodium chloride atoms.

ionic bonding. The sodium (Na) atom donates its 1 valence electron to a chlorine (Cl) atom with 7 valence electrons. Upon acquiring the extra electron, Cl immediately becomes overbalanced negatively. This causes the atom to become a *negative ion*. At the same time, the sodium atom loses its valence electron. The sodium atom then becomes a *positive ion*. Since unlike charges attract, the Na and Cl are bonded together by an electrostatic force.

Copper oxide (Cu_2O) is also an example of ionic bonding. Two atoms of pure copper (Cu) combine with one atom of oxygen (O). Copper is a good electrical conductor. It has 1 valence electron in its atomic structure. Oxygen has 6 valence electrons. The combination of two copper atoms and one oxygen atom forms a stable copper oxide compound. The stability of this compound makes it a good insulating material.

Covalent bonding takes place when the valence electrons of neighboring atoms are shared with other atoms. When an atom contains four valence electrons, it may share one electron with four neighboring atoms. There is a unique difference in covalent bonding and ionic bonding. No ions are formed in covalent bonding. A *covalent force* is established between the two connecting electrons. These electrons alternately shift orbits between the atoms. This covalent force bonds the individual atoms together.

An illustration of covalent bonding is shown in Figure 3-2(a). A single silicon atom is shown independently in Figure 3-2(b). A simplified version of the silicon atom is used in the covalent bonding structure. Only the nucleus and valence electrons of each atom are shown in this arrangement. Individual atoms are bonded together by electron pairs. Five atoms are needed to complete the bonding function. The bonding process extends out in all directions. This links each atom together in a *lattice* network. A crystal structure is formed by the lattice network.

Metallic bonding occurs in good electrical conductors. Copper, for example, has one electron in its valence band. This electron has a tendency to wander around the material between different atoms. Upon leaving one atom, it immediately enters the orbit of another atom. The process is repeated on a continuous basis. When an electron leaves an atom, the atom becomes a positive ion. The process takes place in a random manner. This means that one electron is always associated with an atom but is not in one particular orbit. As a result, all atoms tend to share all the valence electrons.

In metallic bonding there is a type of electrostatic force between positive ions and electrons. In a sense, electrons float around in a cloud that covers the positive ions. This floating cloud bonds the electrons randomly to the ions. Figure 3-3 shows an example of the metallic bonding of copper.

CONDUCTION IN SOLID MATERIALS

Solid materials are primarily used in electrical devices to achieve electron conduction. These materials can be divided into conductors, semiconductors, and insulators. For discussion purposes, assume that we have a cubic centimeter block of these three materials. If the resistance is measured between opposite faces of each block, it shows some very unique differences. The insulator cube measures several million ohms. The conductor cube, by comparison, measures only $0.000001\,\mathcal{W}$. The semiconductor cube measures approximately $600\,\mathcal{W}$. In effect, this

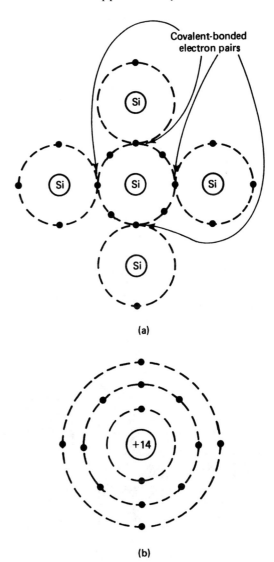

Figure 3-2. (a) Covalent-bonded silicon atoms; (b) a single silicon atom (Si).

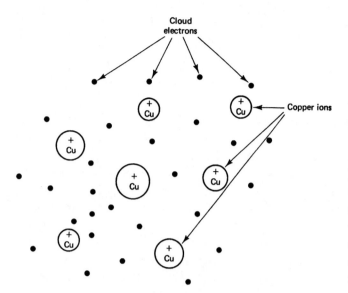

Figure 3-3. Metallic bonding of copper.

shows a rather wide range of resistance differences in the three materials. A comparison of this type shows only the general boundaries of the respective materials.

Scientifically, conductors, semiconductors, and insulators are evaluated by *energy-level diagrams*. This takes into account the amount of energy needed to cause an electron to leave its valence band and go into conduction. The diagram is a composite of all atoms within the material.

Energy-level diagrams of insulators, semiconductors, and conductors are shown in Figure 3-4. The valence band is located at the bottom, and the conduction band is at the top. There are other energy levels in the structure of

the atom. Each electron shell has a discrete energy level. In electronics we are interested primarily in the valence and conduction bands.

The valence band shows the energy levels of the valence electrons of each atom. A certain amount of energy must be added to the valence electrons to cause them to go into conduction. The *forbidden gap*, when it exists, separates the valence and conduction bands. A certain amount of energy is needed to cross the forbidden gap. If it is insufficient, electrons are not released for conduction. They will remain in the valence band. The width of the forbidden gap indicates conduction status of a particular material. In atomic theory the width of the gap is expressed in *electron volts* (eV). An electron volt is defined as the amount of energy gained or lost when an electron is subjected to a potential difference of 1 V. The atoms of each element have a distinct energy-level value that permits conduction.

Insulating materials have a very wide forbidden gap. It takes a very large amount of energy to cause an insulator to go into conduction. Thyrite is a material commonly used as a lightning arrester. At normal voltages it is an ideal insulator. When it is subjected to high voltage, electrons cross the forbidden gap and go into conduction. This shunts the high voltage to ground, thus protecting a device from lightning. When good insulators are operated at high temperatures, the increased heat energy causes valence electrons to go into conduction.

The forbidden gap of a semiconductor is much smaller than that of an insulator. Silicon, for example, needs to gain 0.7 eV of energy to go into conduction. In many cases, the addition of heat energy at room temperature may be sufficient to cause conduction in a semiconductor. This particular characteristic is extremely important in solid-state electronic devices.

In a conductor, the conduction band and valence band overlap one another. In a sense, the valence band and conduction band are one and the same. This means that valence band electrons are readily released to become free electrons. Some electrical conduction takes place within this material through the application of heat at normal room temperature.

SEMICONDUCTOR MATERIALS

Materials such as silicon, germanium, and carbon are found naturally in crystalline form. Instead of being in a random mass, these atoms are arranged in an orderly manner. A crystal of silicon or germanium forms a definite geometrical pattern. Figure 3-5 shows a simplification of the

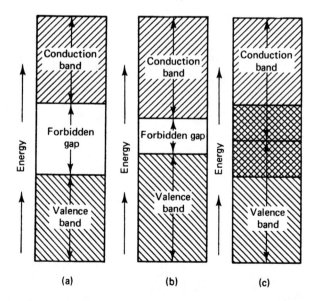

Figure 3-4. Energy-level diagrams for (a) insulators, (b) semiconductors and (c) conductors.

silicon (Si) crystal. Germanium (Ge) would have a similar type of crystal formation. Note that the individual atoms are covalently bonded together. This display shows a two-dimensional layout of these atoms. The drawing is also amplified by displaying only the nucleus and valence electrons of each atom. The geometrical pattern of silicon causes its crystal to be in the shape of a cube.

Crystals of silicon and germanium can be manufactured by melting the natural elements. The process is somewhat complex and rather expensive to achieve. Manufactured crystals must be made extremely pure to be usable in semiconductor devices. A very pure semiconductor crystal is called an *intrinsic* material. Germanium is considered to be intrinsic when only one part of impurity exists in 108 parts of germanium. Silicon is intrinsic when the impurity ratio is 1:10's. In more sophisticated solid-state devices the ratio may be even higher.

An intrinsic crystal of silicon would appear as the structure in Figure 3-5. Covalent bonding changes the electrical conductivity of the material, causing each group of atoms to become stable. Thus only a limited number of free electrons are available for conduction. Silicon and germanium crystals respond to some extent as an insulator.

Intrinsic silicon at absolute zero (–273°C) is considered to be a perfect insulator. The valence electrons of each atom are firmly bound together in perfect covalent bonds. No free electrons are available for conduction. In actual circuit operation, the absolute zero condition is not very meaningful. It cannot be readily attained. Any temperature above -273°C causes silicon to become somewhat conductive. The insulating quality of a semiconductor material is therefore dependent on its operating temperature.

At room temperature, such as 25°C, silicon atoms receive enough energy from heat to break their bonding. A number of free electrons become available for conduction. At room temperature, intrinsic silicon becomes somewhat conductive.

An important solid-state event occurs when intrinsic silicon goes into conduction. For every electron that is freed from its covalent bonding a void spot is created. This spot, which is normally called a *hole*, represents an electron deficiency. A hole in a covalent bonding group occurs when an electron is released. An increase in the temperature of a piece of silicon causes the number of free electrons and holes to increase.

It should be remembered that when a neutral atom loses an electron, it acquires a positive charge. The atom then becomes a positive ion. Since an atom acquires a hole at the same time as the positive charge, the hole bears a positive charge. It should be noted, however, that the crystal remains electrically neutral. For every free electron there is an equivalent hole. These two balance out the overall charge of the crystal.

When a valence electron leaves it covalent bond to become a free electron, a hole appears in its place. This hole can then attract a different electron from a nearby bonded group. Upon leaving its bonded group, this electron creates a new hole. The original hole is then filled and becomes void. Each electron that leaves its bonding to fill a hole creates a new hole in its original group. In a sense, this means that electrons move in one direction and holes in the opposite direction. Since electron movement is considered to be an electric current, holes are also representative of current flow.

Electrons are called negative current carriers and holes are called positive current carriers. These current carriers move in opposite directions in a semiconductor material. When voltage is applied, holes move toward the negative side of the source. Electron current flows toward the positive side of the source.

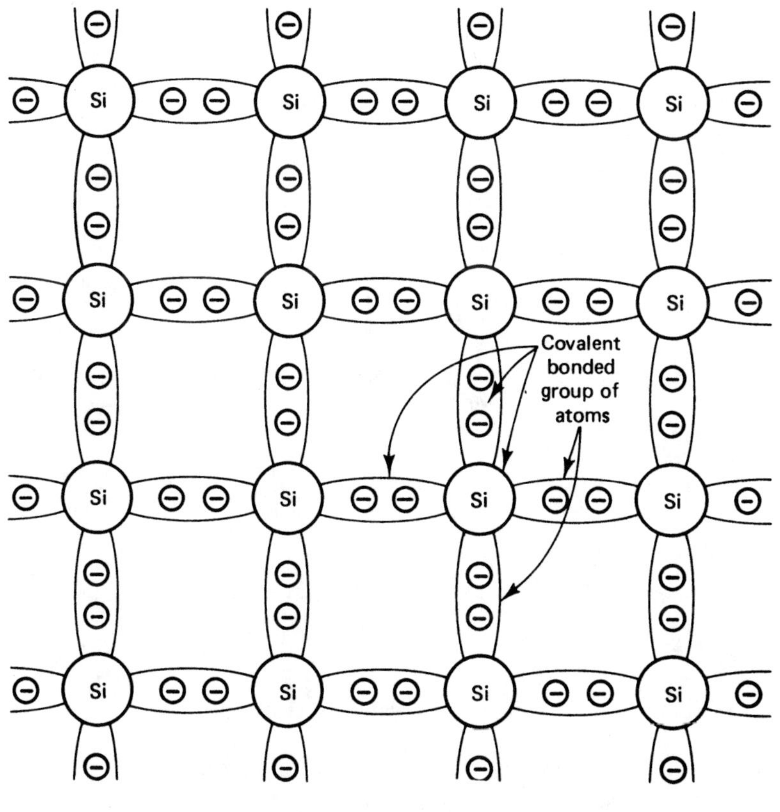

Figure 3-5 Intrinsic crystal of silicon.

This condition occurs only in a semiconductor material. In a conductor, such as copper, there is no covalent bonding. Current carriers are restricted to only electrons.

Figure 3-6 shows how an intrinsic piece of silicon will respond at room temperature when voltage is applied. Note that the free electrons move toward the positive terminal of the battery. Electrons leaving the semiconductor flow into the copper connecting wire. Each electron that leaves the material creates a hole in its place. The holes appear to be moving by jumping between covalent bonded groups. Holes are attracted by the negative terminal connection. For each electron that flows out of the material a new electron enters at the negative connection point. In effect, the new electron fills a hole at this point. The process of hole flow and electron flow through the material is continuous as long as energy is supplied. Current flow is the resulting carrier movement. An intrinsic semiconductor has an equal number of current carriers moving in each direction. The resulting current flow of a semiconductor is limited primarily to the applied voltage and the operating temperature of the material.

Pure silicon or germanium in its intrinsic state is rarely used as a semiconductor. Usable semiconductors must have controlled amounts of impurities added to them. The added impurities change the conductor capabilities of a semiconductor. The process of adding an impurity to an intrinsic material is called *doping*. The impurity is called a *dopant*. Doping a semiconductor causes it to be an extrinsic material. Extrinsic semiconductors are the operational basis of nearly all solid-state devices.

N-Type Material

When an impurity is added to silicon or germanium without altering the crystal structure, an N-type material is produced. Atoms of arsenic (As) and antimony (Sb) have five electrons in their valence band (see the periodic element table in Figure 1-

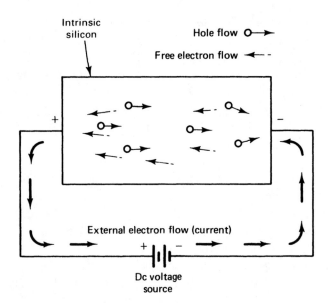

Figure 3-6. Current carrier movement in silicon.

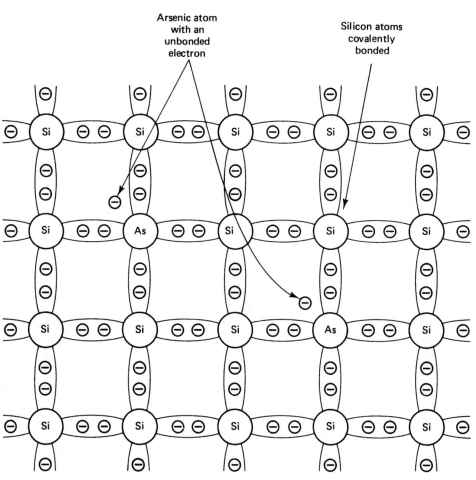

Figure 3-7. Extrinsic silicon crystal structure.

9). Adding either impurity to silicon must not alter the crystal structure or bonding process. Each impurity atom has an extra electron that does not take part in a covalent bonding group. These electrons are loosely held together by their parent atoms. Figure 3-7 shows how the crystal is altered with the addition of an impurity atom.

When arsenic is added to pure silicon, the crystal becomes an N-type material. It has extra electrons or negative (N) charges that do not take part in the covalent bonding process. Impurities that add electrons to a crystal are generally called *donor* atoms. An N-type material has more extra or free electrons than an intrinsic piece of material. A piece of N material is not negative charged. Its atoms are all electrically neutral. There are, however, several extra electrons that do not take part in the covalent bonding process. These electrons are free to move about through the crystal structure.

An extrinsic silicon crystal of the N type will go into conduction with only 0.005eV of energy applied. An intrinsic crystal requires 0.7eV to move electrons from the valence band into the conduction band. Essentially, this means that an N material is a fairly good electrical conduction. In this type of crystal, electrons are considered to be the *majority current carriers.* Holes are the minority current carriers. The amount of donor material added to silicon determines the number of majority current carriers in its structure.

The number of electrons in a piece of N-type silicon is a million or more times greater than the electron-hole pairs of a piece of intrinsic silicon. At room temperature, there is a decided difference in the electrical conductivity of this material. Extrinsic silicon becomes a rather good electrical conduction. There is a larger number of current carriers to take part in conduction. Current flow is achieved primarily by electrons in this material. Figure 3-8 shows how the current carriers respond in a piece of N material. There is a larger number of electrons indicated than holes. This shows that electrons are the majority current carriers and holes are the minority carriers.

If the voltage source of Figure 3-8 is reversed, the current flow would reverse its direction. Essentially, this means that N-type silicon will conduct equally as well in either direction. The flow of current carriers is simply reversed. This is an important consideration in the operation of a device that employs N material in its construction.

P-Type Material

P-type material is formed when indium (In) or gallium (Ga) is added to intrinsic silicon. Indium or gallium are type III elements according to the periodic table of Figure 1-9. These elements are often called *acceptors.* They are readily seeking a fourth electron. In effect, this type of dopant material has three valence electrons. Each

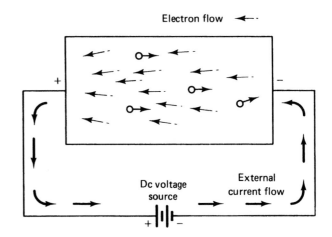

Figure 3-8. Current carriers in an N-type

covalent bond that is formed with an indium atom will have an electron deficiency or hole. This represents a positive charge area in the covalent bonding structure. A positive charge area crystal or P material describes this type of structure. Each hole in the P material can be filled with an electron. Electrons from neighboring covalent bonded groups require very little energy to move in and fill a hole.

The ratio of doping material to silicon is typically in the range of 1 to 10^6. This means that P material will have a million times more holes than the heat-generated electron-hole pairs of pure silicon. At room temperature, there is a very decided difference in the electrical conductivity of this material. Figure 3-9 shows how the crystal structure of silicon is altered when doped with an acceptor element—in this case, indium.

A piece of P material is not positively charged. Its atoms are primarily all electrically neutral. There are, however, holes in the covalent structure of many atom groups. When an electron moves in and fills a hole, the hole becomes void. A new hole is created in the bonded group where the electron left. Hole movement in effect is the result of electron movement.

A P material will go into conduction with only 0.05 eV of energy applied. Intrinsic silicon, by comparison, requires 0.7 eV for conduction. Extrinsic silicon is therefore considered to be a rather good electrical conduction. In this material, holes are the majority carriers and electrons are the minority carriers. The amount of acceptor material added to silicon determines the majority carrier content of the crystal.

Figure 3-10 shows how a P-type crystal will respond when connected to a voltage source. Note that there is a larger number of holes than electrons. With voltage applied, electrons are attracted to the positive battery terminal. Holes move, in a sense, toward the negative battery terminal. An electron is picked up at this point. The electron immediately fills a hole. The hole then becomes void. At the same time, an electron is pulled from the material by the positive battery terminal. Holes therefore move toward the negative terminal due to electrons shifting between different bonded groups. With energy applied, hole flow is continuous.

JUNCTION DIODES

The term junction *diode* is often used to describe a crystal structure made of P and N materials. A diode is generally described as a two-terminal device. One terminal is attached to P material and the other to N material. The common connecting point where these materials are joined is called a junction. A junction diode permits current carriers to flow readily in one direction and blocks the flow of current in the opposite direction.

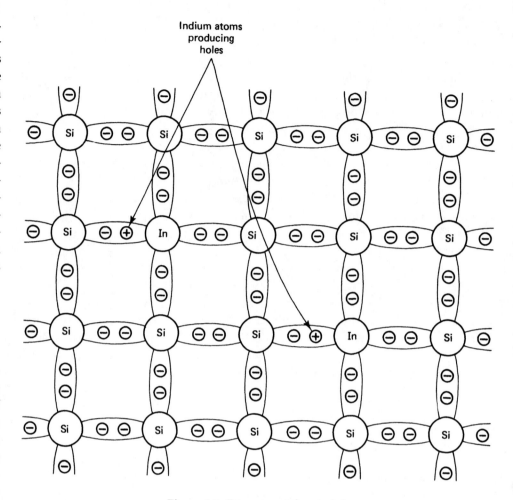

Figure 3-9. P-type crystal material.

Figure 3-11 shows the crystal structure of a junction diode. Notice the location of the P and N materials with respect to the junction. The crystal structure is continuous from one end to the other. The junction serves only as a dividing line that marks the ending of one material and the beginning of the other. This type of structure permits electrons to move readily through the entire structure.

Figure 3-12 shows two pieces of semiconductor material before they are formed into a P-N junction. As indicated, each piece of material as majority and minority current carriers. The number of carrier symbols shown in each material indicates the minority or majority function. Electrons are the majority carriers in the N material and are the minority carriers in the P material. Holes are the majority carriers of the P material and are in the minority in the N material. Both holes and electrons have freedom to move about in their respective materials.

Depletion Zone

When a junction diode is first formed, there is a unique interaction between current carriers. Electrons

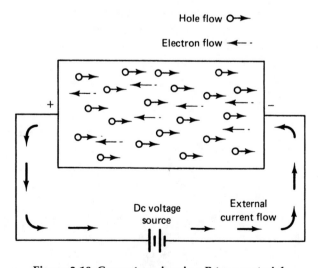

Figure 3-10. Current carriers in a P-type material.

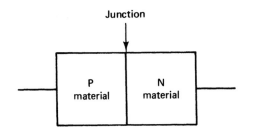

Figure 3-11. Crystal structure of a junction diode.

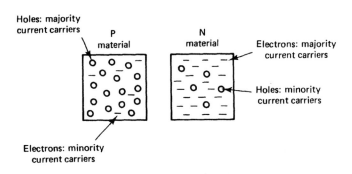

Figure 3-12. Semiconductor

the N side of the junction to take on a net positive charge. These respective charge creations tend to drive remaining electrons and holes away from the junction. This action makes it somewhat difficult for additional charge carriers to diffuse across the junction. The end result is a charge buildup or *barrier potential* appearing across the junction.

Figure 3-14 shows the resulting barrier potential as a small battery connected across the P-N junction. Note the

from the N material move readily across the junction to fill holes in the P material. This action is commonly called *diffusion*. Diffusion is the result of high concentration of carriers in one material and a lower concentration in the other. Only those current carriers near the junction take part in the diffusion process.

The diffusion of current carriers across the junction of a diode causes a change in its structure. Electrons leaving the N material cause positive ions to be generated in their place. Upon entering the P material to fill holes, these same electrons create negative ions. The area of each side of the junction then contains a large number of positive and negative ions. The number of holes and electrons in this area becomes depleted. The term *depletion zone* is used to describe this area. It represents an area that is void of majority current carriers. All P-N junctions develop a depletion zone when they are formed. Figure 3-13 shows the depletion zone of a function diode.

Barrier Potential

Before N and P materials are joined together at a common junction they are considered to be electrically neutral. After being joined, however, diffusion takes place immediately. Electrons crossing the junction to fill holes causes negative ions to appear in the P material. This action causes the area near the junction to take on a negative charge. Similarly, electrons leaving the N material causes it to produce positive ions. This, in turn, causes

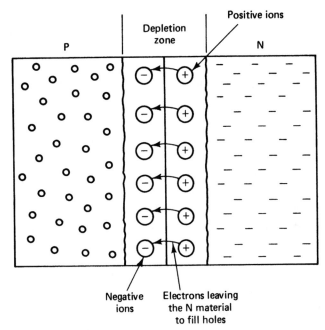

Figure 3-13. Depletion-zone formation.

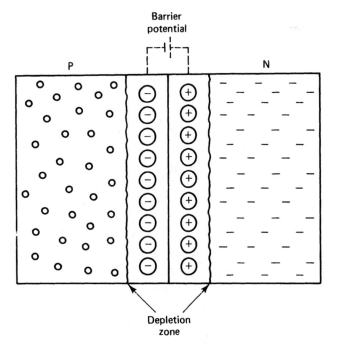

Figure 3-14. Barrier potential of a diode.

polarity of this potential with respect to P and N material. This voltage will exist even when the crystal is not connected to an outside source of energy. For germanium the barrier potential is approximately 0.3 V, and 0.7 V for silicon. These voltage values cannot, however, be measured directly. They appear only across the space charge region of the junction. The barrier potential of a P-N junction must be overcome by an outside voltage source in order to produce current conduction.

JUNCTION BIASING

When an external source of energy is applied to a P-N junction it is called a *bias voltage* or simply biasing. This voltage either adds to or reduces the barrier potential of the junction. Reducing the barrier potential causes current carriers to return to the depletion zone. *Forward biasing* is used to describe this condition. Adding external voltage of the same polarity to the barrier potential causes an increase in the width of the depletion zone. This action hinders current carriers from entering the depletion zone. The term reverse biasing is used to describe this condition.

Reverse Biasing

When a battery is connected across a P-N junction as in Figure 3-15 it causes reverse biasing. Note that the negative terminal of the battery is connected to the P material arid the positive to the N material. Connection in this way causes the battery polarity to oppose the material polarity of the diode. Since unlike charges attract, the majority charge carriers of each material are pulled away from the junction. Reverse biasing of a diode normally causes it to be nonconductive.

Figure 3-16 shows how the majority current carriers are rearranged in a reverse-biasing diode. As shown, electrons of the N material are pulled toward the positive battery terminal. Each electron that moves or leaves the diode causes a positive ion to appear in its place. This causes a corresponding increase in the width of the depletion zone on the N side of the junction.

The P side of the diode has a reaction very similar to the N side. In this case, a number of electrons leave the negative battery terminal and enter the P material. These

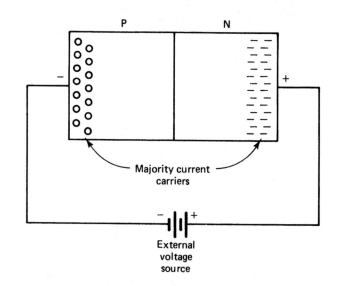

Figure 3-15. Reverse-biased diode.

electrons immediately move in and fill a number of holes. Each filled hole then becomes a negative ion. These ions are then repelled by the negative battery terminal and driven toward the junction. As a result of this, the width of the depletion zone increases on the P side of the junction.

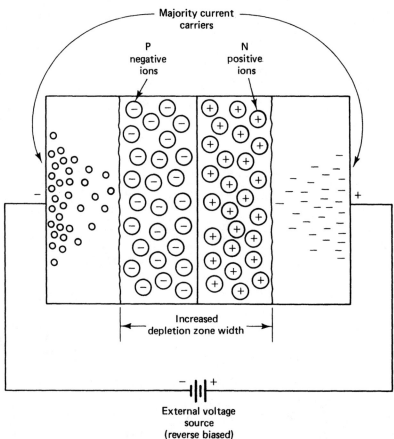

Figure 3-16. Reversed-biased P-N junction.

The overall depletion zone width of a reverse-biased diode is directly dependent on the value of the supply voltage. With a wide depletion zone a diode cannot effectively support current flow. The charge buildup across the junction will increase until the barrier voltage equals the external bias voltage. When this occurs a diode is effectively a nonconductor.

Leakage Current

When a diode is reverse biased, its depletion zone increases in width. Normally, we would expect this condition to restrict current carrier formation near the junction. The depletion zone in effect represents an area that is primarily void of majority current carriers. Because of this, the depletion zone will respond as an insulator. Ideally, current carriers do not pass through an insulator. In a reverse-biased diode some current actually flows through the depletion zone. This is called *leakage current*. Leakage current is dependent on minority current carriers.

Remember that minority carriers are electrons in the P material and holes in the N material. Figure 3-17 shows how these carriers respond when a diode is reverse biased. It should be noted that the minority carriers of each material are pushed through the depletion zone to the

junction. This action causes a very small leakage current to occur. Normally, leakage current is so small that it is often considered negligible.

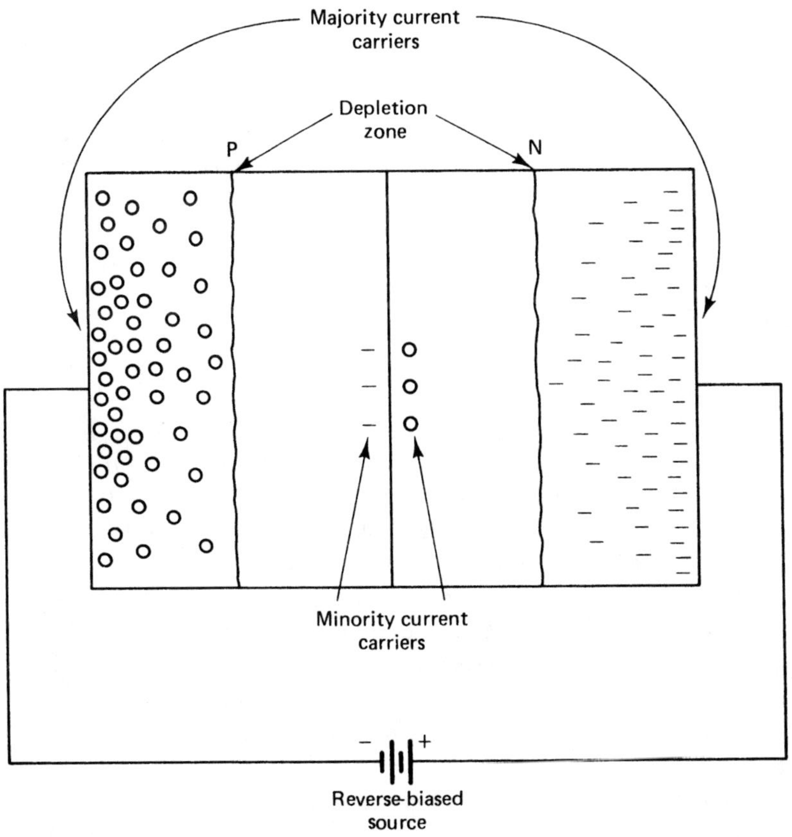

The minority current carrier content of a semiconductor is primarily dependent on temperature. At normal room temperatures of 25°C or 78°F there is a rather limited number of minority carriers present in a semiconductor. However, when the surrounding temperature rises it causes considerable increase in minority carrier production. This causes a corresponding increase in leakage current.

Leakage current occurs to some extent in all reverse-biased diodes. In germanium diodes, leakage current is only a few microamperes. Silicon diodes normally have a lower minority carrier content. This in effect means less leakage current. Typical leakage current values for silicon are a few nanoamperes. The construction material of a diode is an important consideration to remember. Germanium is much more sensitive to temperature than is silicon. Germanium therefore has a higher level of leakage current. This factor is largely responsible for the widespread use of silicon in modern semiconductor devices.

Forward Biasing

When voltage is applied to a diode as in Figure 3-18, it causes forward biasing. In this example, the positive battery terminal is connected to the P material and the negative terminal to the N material. This voltage repels the majority current carriers of each material. A large number of holes and electrons therefore appear at the junction. On the N side of the junction, electrons move in to neutralize the positive ions in the depletion zone. In the P material, electrons are pulled from negative ions which causes them to become neutral again. This action means that forward biasing causes the depletion zone to collapse and the barrier potential to be removed. The P-N junction will therefore support a continuous current flow when it is forward biased.

Figure 3-19 shows how the current carriers of a forward-biased diode respond. Since the diode is connected to an external voltage source, it has a constant supply of electrons. Large arrows are used in the diagram to show the direction of current flow outside the diode. Inside the diode, smaller arrows show the movement of majority current carriers. Remember that electron flow and current refer to the same thing.

Figure 3-17. Minority current carriers.

Starting at the negative battery terminal, assume now that electrons flow through a wire to the N material. Upon entering this material, they flow immediately to the junction. At the same time, an equal number of electrons are removed from the P material and are returned through a resistor to the positive battery terminal. This action generates new holes and causes them to move toward the junction. When these holes and electrons reach the junction they combine and effectively disappear. At the same time new holes and electrons appear at the outer ends of the diode. These majority carriers are generated on a continuous basis. This action continues as long as the external voltage source is applied.

It is important to realize that electrons flow through the entire diode when it is forward biased. In the N material this is quite obvious. In the P material, however, holes are the moving current carriers. Remember that hole movement in one direction must be initiated by electron movement in the opposite direction. Therefore, the combined flow of holes and electrons through a diode equals the total current flow.

The current-limiting resistor of Figure 3-19 is essential in a forward-biasing diode circuit. This resistor is needed to keep the current flow at a safe operating level. The maximum current (I) rating of a diode is representative of this value. As long as diode current does not exceed this value it can operate satisfactorily. Resistors are used to limit current flow to a reasonable operating level.

DIODE CHARACTERISTICS

Now that we have seen how a junction diode operates, it is time to examine some of its electrical characteristics. This takes into account such things as voltage, current, and temperature. Since these characteristics vary a great deal in an operating circuit, it is best to look at them graphically. This makes it possible to see how the device will respond under different operating conditions.

Forward Characteristic

When a diode is connected in the forward bias direction it conducts forward current I_F. The value of I_F is directly dependant on the amount of forward voltage V_F. The relationship of forward voltage and forward current is called the ampere-volt, or IV characteristic of a diode. A typical diode forward IV characteristic is shown in Figure 3-20(a) with its test circuit in Figure 3-20(b). Note in particular that V_F is measured across the diode and that I_F is a measure of what flows through the diode. The value of the source voltage V_S does not necessarily compare in

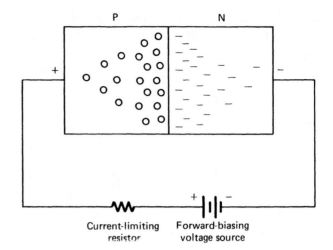

Figure 3-18. Forward-biased diode.

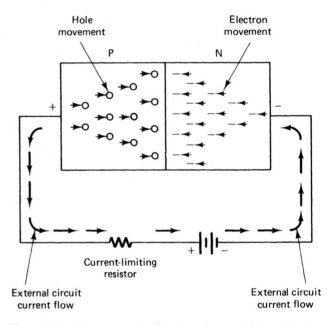

Figure 3-19. Current carrier flow in a forward-biased diode.

value with V_F.

When the forward voltage of the diode equals 0V, the I_F equals 0 mA. This value starts at the origin point (0) of the graph. If V_F is gradually increased in 0.1-V steps, I_F begins to increase. When the value of V_F is great enough to overcome the barrier potential of the P-N junction, a substantial increase in I_F occurs. The point at which this occurs is often called the knee voltage V_K. For germanium diodes V_K is approximately 0.3 V and 0.7 V for silicon.

If the value of I_F increases much beyond V_K, the forward current becomes quite large. This, in effect, causes heat to develop across the junction. Excessive junction heat can destroy a diode. To prevent this from happening a protective resistor is connected in series with the di-

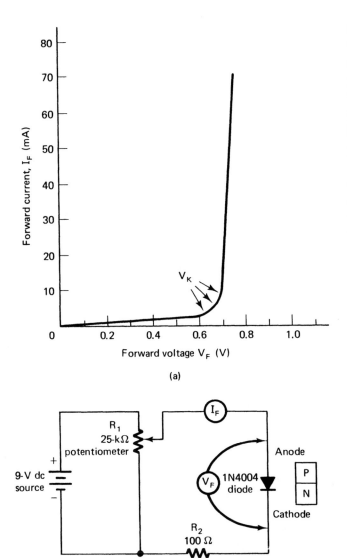

Figure 3-20. (a) IV characteristics of a diode; (b) diode characteristic test circuit.

ode. This resistor limits the I_F to its maximum rated value. Diodes are rarely operated in the forward direction without a current-limiting resistor.

Reverse Characteristic

When a diode is connected in the reverse bias direction it has an I_R—V_R characteristic. Figure 3-21(a) shows the reverse *IV* characteristic of a diode and its test circuit. This characteristic has different values of I_R and V_R. Reverse current is usually quite small. The vertical I_R line in this graph has current values graduated in microamperes. The number of minority current carriers that take part in I_R is quite small. In general, this means that I_R remains rather constant over a large part of V_R. Note also

that V_R is graduated in 100-V increments. Starting in zero when the reverse voltage of a diode is increased there is a very slight change in I_R At the voltage breakdown V_{BR} point current increases very rapidly. The voltage across the diode remains fairly constant at this time. This constant-voltage characteristic leads to a number of reverse bias diode applications. Normally, diodes are used in application where the V_{BR} is not reached.

The physical processes responsible for current conduction in a reverse-biased diode are called *zener breakdown* and *avalanche breakdown*. Zener breakdown takes place when electrons are pulled from their covalent bonds in a strong electric field. This occurs at a rather high value of $1R$ When large numbers of covalent bonds are broken at the same time there is a sudden increase in I_R

Avalanche breakdown is an energy-related condition of reverse biasing. At high values of V_R minority carriers gain a great deal of energy. This gain may be great enough to drive electrons out of their covalent bonding. This action creates new electron-hole pairs. These carriers then move across the junction and produce other ionizing collisions. Additional electrons are produced. The process continues to build up until an avalanche of current carriers takes place. When this occurs the process is irreversible.

Combined *IV* Characteristics

The forward and reverse *IV* characteristics of a diode are generally combined on a single characteristic curve. Figure 3-22 shows a rather standard method of displaying this curve. Forward and reverse bias voltage V_F and V_R are usually plotted on the horizontal axis of the graph. V_F extends to the right and V_R to the left. The point of origin or zero value is at the center of the horizontal line. Forward and reverse current values are shown vertically on the graph. I_F extends above the horizontal axis with I_R extending downward. The origin point serves as a zero indication for all four values. This means that combined V_F and I_F values are located in the upper right part of the graph and V_R and I_R in the lower left corner. Different scales are normally used to display forward and reverse values.

A rather interesting comparison of silicon and germanium characteristic curves is shown in Figure 3-23. Careful examination shows that germanium requires less forward voltage to go into conduction than silicon. This characteristic is a distinct advantage in low-voltage circuits. Note also that a germanium diode requires less voltage drop across it for different values of current. This means that germanium has a lower resistance to forward current flow. Germanium therefore appears to be a better

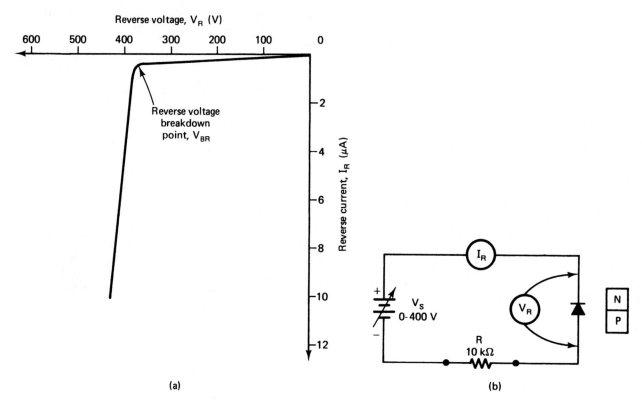

Figure 3-21. (a) Reverse IV characteristics of a diode; (b) reverse characteristic test circuit.

conductor than silicon. Silicon is more widely used, however, because of its low leakage current and cost.

The reverse bias characteristics of silicon and germanium diodes can also be compared in Figure 3-23. The I_R of a silicon diode is very small compared to that of a germanium diode. Reverse current is determined primarily by the minority current content of the material. This condition is influenced primarily by temperature. For germanium diodes, I_R doubles for each 10°C rise in temperature. In a silicon diode the change in I_R is practically negligible for the same rise in temperature. As a result of this, silicon diodes are preferred over germanium diodes in applications where large changes in temperature occur. Comparisons of this type are quite obvious through the study of characteristic curves.

DIODE SPECIFICATIONS

When selecting a diode for a specific application some consideration must be given to specifications. This type of information is usually made available through the manufacturer. Diode specifications usually include such things as absolute maximum ratings, typical operating conditions, mechanical data, lead identification, mounting procedures, and characteristic curves. Figure 3-24

shows a representative diode data sheet. Some of the important specifications are explained in the following:

1. *Maximum reverse voltage (V_{RM}):* The absolute maximum or peak reverse bias voltage that can be applied to a diode. This may also be called the peak inverse voltage (PIV) or peak reverse voltage (PRy).

2. *Reverse breakdown voltage (V_{BR}):* The minimum steady-state reverse voltage at which breakdown will occur.

3. *Maximum forward current (I_{FM}):* The absolute maximum repetitive forward current that can pass through a diode at 25°C (77°F). This is reduced for operation at higher temperatures.

4. *Maximum forward surge current (I_{FM}-surge):* The maximum current that can be tolerated for a short interval of time. This current value is much greater than IFM. This represents the increase in current that occurs when a circuit is first turned on.

5. *Maximum reverse current (I_R):* The absolute maximum reverse current that can be tolerated at device operating temperature.

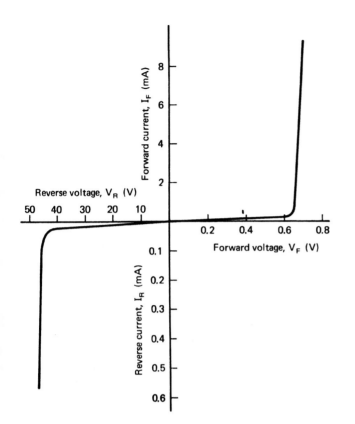

Figure 3-22. Diode current-voltage characteristics.

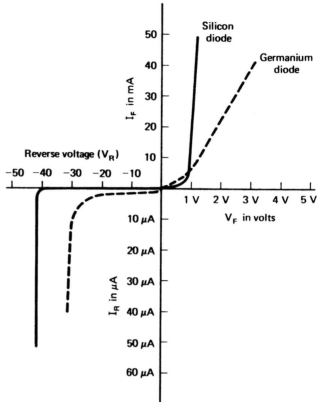

Figure 3-23. Silocon and germanium diode characteristics.

6. *Forward voltage* (V_F): Maximum forward voltage drop for a given forward current at device operating temperature.

7. *Power dissipation* (P_D): The maximum power that the device can safely absorb on a continuous basis in free air at 25°C (77°F).

8. *Reverse recovery time* (T_{rr}): The maximum time that it takes the device to switch from its on to its off state.

DIODE PACKAGING

Diodes are manufactured today in a wide range of case styles and packages. A person working with these devices must be familiar with some of the common methods of packaging. Element identification and lead marking techniques are essential for proper installation and testing. A diode improperly connected in a circuit may be damaged or cause damage to other parts. Figure 3-25 shows three general classes of diodes according to their current-handling capability. These represent some of the more popular package styles in use today.

Low-current diodes (Figure 3-25(a)) are the smallest of all packages. Body length rarely exceeds 0.3 cm in length. The cathode is usually denoted by a color band. Glass-packaged diodes often have two or three color bands to indicate specific number types. This group of diodes is generally capable of passing forward current values of approximately 100 mA. The peak reverse voltage rating rarely ever exceeds 100 V. Low reverse current values for these devices are typically 5 A at 25°C.

Medium-current diodes (Figure 3-25(b)) are slightly larger in size than the low-current devices. Body size is approximately 0.5 cm with larger connecting leads. Diodes in this group can pass forward current values up to 5 A. Peak reverse voltage ratings generally do not exceed 1000 V. The anode and cathode terminals may be identified by a diode symbol on the body of the device. A color band near one end of the case is also used to identify the cathode. Low- and medium-current diodes are usually mounted by soldering. Any heat generated during operation is carried away by air or lead conduction.

High-current, or power diodes are the largest of all diode types (see Figure 3-25(c)). These devices normally generate a great deal of heat. Air convection of heat is generally not adequate for most installations. These devices are

TYPES 1N4001 THROUGH 1N4007
DIFFUSED-JUNCTION SILICON RECTIFIERS

***electrical characteristics at specified ambient† temperature**

PARAMETER		TEST CONDITIONS		MAX	UNIT
I_R	Static Reverse Current	V_R = Rated V_R,	T_A = 25°C	10	μa
		V_R = Rated V_R,	T_A = 100°C	50	μa
$I_{R(avg)}$	Average Reverse Current	V_{RM} = Rated V_{RM}, f = 60 cps,	I_O = 1 a, T_A = 75°C	30	μa
V_F	Static Forward Voltage	I_F = 1 a,	T_A = 25°C to 75°C	1.1	v
$V_{F(avg)}$	Average Forward Voltage	V_{RM} = Rated V_{RM}, f = 60 cps,	I_O = 1 a, T_A = 25°C to 75°C	0.8	v
V_{FM}	Peak Forward Voltage	V_{RM} = Rated V_{RM}, f = 60 cps,	I_O = 1 a, T_A = 25°C to 75°C	1.6	v

*Indicates JEDEC registered data.

THERMAL INFORMATION

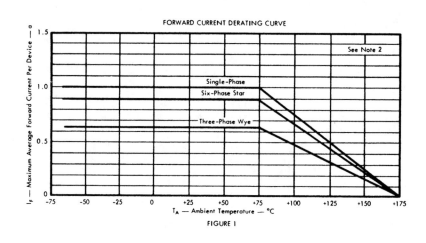

FORWARD CURRENT DERATING CURVE

FIGURE 1

†The ambient temperature is measured at a point 2 inches below the device. Natural air cooling shall be used.

NOTE 2: This rectifier is a lead-conduction-cooled device. At (or above) ambient temperatures of 75°C, the lead temperature ⅜ inch from case must be no higher than 5°C above the ambient temperature for these ratings to apply.

TEXAS INSTRUMENTS RESERVES THE RIGHT TO MAKE CHANGES AT ANY TIME IN ORDER TO IMPROVE DESIGN AND TO SUPPLY THE BEST PRODUCT POSSIBLE.

Figure 3-24. Data sheet for a silicon diode. (Courtesy of Texas Instruments.)

designed to be mounted on metal heat sinks. Heat is conducted away from the diode by metal. Diodes of this classification can pass hundreds of amperes of forward current. Peak reverse voltage ratings are in the 1000 V range. Numerous packaging types and styles are used to house power diodes. As a rule, the rating of the device and method of installation usually dictate its package type.

ZENER DIODES

The characteristics of a regular junction diode show that it is designed primarily for operation in the forward direction. Forward biasing will cause a large I_F with a rather small value of V_F. Reverse biasing will generally not cause current conduction until higher values of re-

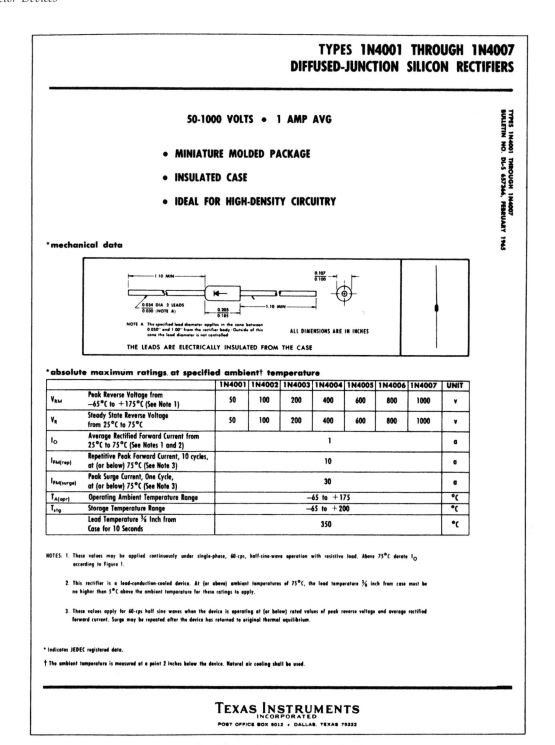

TYPES 1N4001 THROUGH 1N4007
DIFFUSED-JUNCTION SILICON RECTIFIERS

TYPES 1N4001 THROUGH 1N4007
BULLETIN NO. DL-S 6572A4, FEBRUARY 1965

50-1000 VOLTS • 1 AMP AVG

- **MINIATURE MOLDED PACKAGE**
- **INSULATED CASE**
- **IDEAL FOR HIGH-DENSITY CIRCUITRY**

*mechanical data

NOTE A The specified lead diameter applies in the zone between 0.050" and 1.00" from the rectifier body · Outside of this zone the lead diameter is not controlled

ALL DIMENSIONS ARE IN INCHES

THE LEADS ARE ELECTRICALLY INSULATED FROM THE CASE

*absolute maximum ratings at specified ambient† temperature

		1N4001	1N4002	1N4003	1N4004	1N4005	1N4006	1N4007	UNIT
V_{RM}	Peak Reverse Voltage from −65°C to +175°C (See Note 1)	50	100	200	400	600	800	1000	v
V_R	Steady State Reverse Voltage from 25°C to 75°C	50	100	200	400	600	800	1000	v
I_O	Average Rectified Forward Current from 25°C to 75°C (See Notes 1 and 2)	1							a
$I_{FM(rep)}$	Repetitive Peak Forward Current, 10 cycles, at (or below) 75°C (See Note 3)	10							a
$I_{FM(surge)}$	Peak Surge Current, One Cycle, at (or below) 75°C (See Note 3)	30							a
$T_{A(opr)}$	Operating Ambient Temperature Range	−65 to +175							°C
T_{stg}	Storage Temperature Range	−65 to +200							°C
	Lead Temperature ⅜ Inch from Case for 10 Seconds	350							°C

NOTES: 1. These values may be applied continuously under single-phase, 60-cps, half-sine-wave operation with resistive load. Above 75°C derate I_O according to Figure 1.

2. This rectifier is a lead-conduction-cooled device. At (or above) ambient temperatures of 75°C, the lead temperature ⅜ inch from case must be no higher than 5°C above the ambient temperature for these ratings to apply.

3. These values apply for 60-cps half sine waves when the device is operating at (or below) rated values of peak reverse voltage and average rectified forward current. Surge may be repeated after the device has returned to original thermal equilibrium.

* Indicates JEDEC registered data.

† The ambient temperature is measured at a point 2 inches below the device. Natural air cooling shall be used.

TEXAS INSTRUMENTS
INCORPORATED
POST OFFICE BOX 5012 • DALLAS, TEXAS 75222

Figure 3-24 (Continued)

verse voltage are reached. If V_R is great enough, however, breakdown will occur and cause a reverse current flow. Junction diodes are usually damaged when this occurs. Special diodes are now designed to operate in the reverse direction without being damaged. Zener diodes are manufactured specifically for this purpose.

Figure 3-26 shows a test circuit and the voltampere characteristic of a typical zener diode. In the forward bias direction this device behaves like an ordinary silicon diode. In the reverse bias direction, there is practically no reverse current flow until the breakdown voltage is reached. When this occurs there is a sharp increase in reverse current. Varying amounts of reverse current can pass through the diode without damaging it. The break-

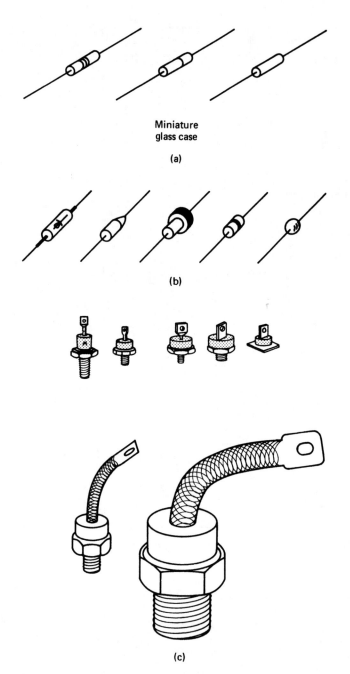

Figure 3-25. Diode packages: (a) low-power diodes; (b) medium-power diodes; (c) high-power diodes.

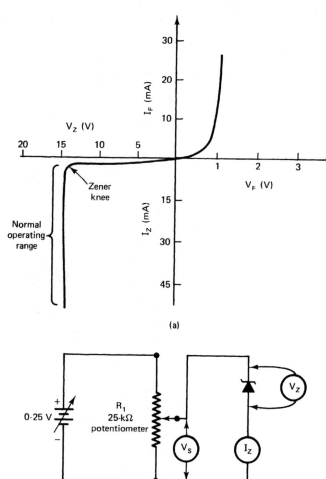

Figure 3-26. Zener diode: (a) IV characteristics; (b) test circuit.

$$\text{maximum } I_R \quad = \quad \frac{PD}{V_Z}$$

$$= \quad \frac{1\ W}{15\ V}$$

$$= \quad 0.066A \text{ or } 66\ mA$$

down voltage or zener voltage (V_Z) across the diode remains relatively constant. The maximum reverse current is limited, however, by the wattage rating of the diode.

Manufacturers rate zener diodes according to their V_Z value and the maximum power dissipation (P_D) at 25°C. This gives an indication of the maximum reverse current or I_R that a diode can safely conduct. A 1.0-W 15-V zener diode can, for example, conduct an I_R of 0.066 A, or 66 mA. This is determined by the basic power equation $P = IV$ by solving for I. Thus,

Zener diodes are available commercially in a wide range of breakdown voltages. V_Z values range from 1.4 to 200 V with accuracies between 1 and 20%. Each V_Z value is generally specified at minimum zener current (I_Z). Figure 3-26 shows that 1 mA of I_Z is needed to enter the knee of the I_R curve. After this value has been reached, the V_Z remains fairly constant over a wide range of I_Z. Power dissipation values are therefore used to indicate the safe operating range. Typical P_D ratings are from 150 mW to 50 W. When purchasing this kind of a device, one usually asks for a 1-W zener diode at 8.2 V_Z with an accuracy of 10%.

The symbol and crystal structure of a zener diode is shown in Figure 3-27. Notice that it is very similar to that of a regular diode. The cathode is, however, drawn with a bent line to distinguish it from a regular diode symbol. This is supposedly done to represent the letter Z. Zener diodes are normally used only in the reverse bias direction. This means that the anode must be connected to the negative side of the voltage source and the cathode to the positive.

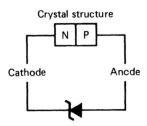

Figure 3-27. Zener diode symbol and crystal structure.

An important difference in zener diodes and regular silicon diodes is in the way they are used in circuits. Because of the constant V_Z characteristic of a zener diode, it is used primarily to regulate voltages. A large change in I_R will cause only a small change in V_Z. This means that a zener diode can be used as an alternate current path. The constant V_Z developed across the diode can then be applied to a load device. The load voltage thus remains at a constant value by altering the current flow through the zener diode.

Figure 3-28 shows a representative zener diode regulator circuit. Notice that the circuit contains a series resistor, zener diode, load device, and a dc power supply. The primary function of this circuit is to maintain the load voltage at a constant 15 V. With 25 V applied, current through R_S will cause a voltage drop of 10 V. The output voltage or V_Z across the diode is therefore 15 V. An increase in the supply voltage to 30 V would simply cause the zener to conduct more current. The voltage drop across R_S would therefore increase to 15 V. This in turn would cause V_Z and the load voltage to remain at 15 V. A decrease in supply voltage to 20 V would cause a reduction in total current. The voltage drop across R_S would now drop to only 5 V. The load voltage across the diode would continue to be 15 V. This means that the regulated output will remain at 15 V for a change in supply voltage from 20 to 30 V. Voltage regulation is an important function of many electronic circuits today.

LIGHT-EMITTING DIODES

The *light-emitting* diode (LED) is another important diode device. It contains a P-N junction that emits light when passing forward current. A schematic symbol and typical package of the LED are shown in Figure 3-29(a) and (b). The lens of an LED (Figure 3-29(c)) is transparent and focuses on the P-N junction.

When an LED is forward-biased, electrons cross the junction and combine with holes. This action causes electrons to fall out of conduction and return to the valence band. The energy possessed by each free electron is then released. Some of this appears as heat energy and the rest of it is given off as light energy. Special materials such as gallium arsenide (GaAs) and gallium phosphide (GaP) cause this to take place when used in a P-N junction. These materials are purposely used because they emit visible light energy.

Light-emitting diodes are relatively small, ruggedly constructed, and have a very long life expectancy. They can be switched on and off very quickly. Construction techniques permit this type of device to be made in various shapes and patterns. A very common application is in numeric displays indicating the numbers 0 through 9. A representative LED seven-segment display is shown in Figure 3-30. By selecting different combinations of segments, a desired number can be produced. Displays of this type are found in calculators, digital meters, clocks, and wristwatches.

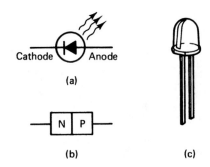

Figure 3-29. Light-emitting diode: (a) symbol; (b) crystal structure; (c) transparent lens.

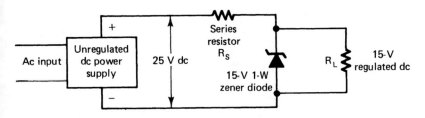

Figure 3-28. Zener diode regulator circuit.

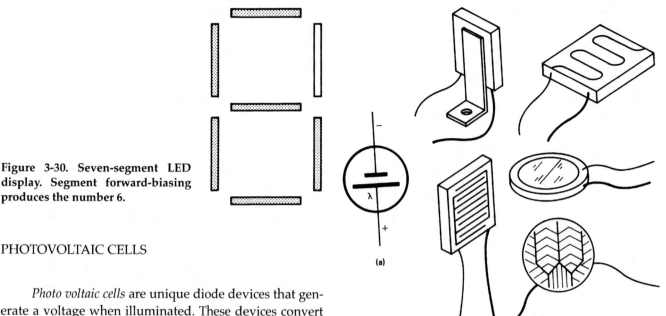

Figure 3-30. Seven-segment LED display. Segment forward-biasing produces the number 6.

Figure 3-31. Photovoltaic cells: (a) symbol; (b) representative cells.

PHOTOVOLTAIC CELLS

Photo voltaic cells are unique diode devices that generate a voltage when illuminated. These devices convert light energy directly into electrical energy. No outside source of electricity is needed to produce current. Figure 3-31 shows the schematic symbol and some representative photovoltaic cells.

Figure 3-32 shows the crystal structure of a junction photovoltaic cell. Basic construction is very similar to that of a junction diode. In operation with no light applied, the cell will not cause a current flow. Essentially, this means that the junction is not being properly biased. However, when light is applied, an interesting condition can be observed. Photocurrent (I_{PH}) will flow in the circuit because of the photovoltage (V_{PH}) that is developed across the junction. V_{PH} is due to the generation of new electron-hole pairs in the depletion region. Electrons move into the N material because of its normal positive ion concentration. Holes are likewise swept into the P material because of its negative-ion content. This action causes the N material to immediate-

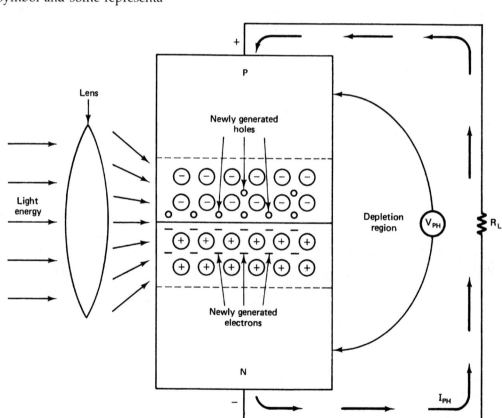

Figure 3-32. Photovoltaic cell operation.

ly take on a negative charge and the P material to take on a positive charge. The P-N junction therefore responds as a small voltage cell. Because of this, photovoltaic cells

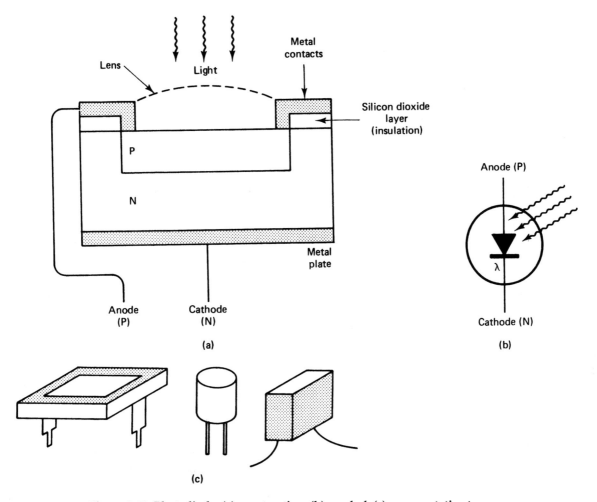

Figure 3-33. Photodiode: (a) construction; (b) symbol; (c) representative types.

are often called solar cells or sun batteries.

Photovoltaic cells are generally classified as direct energy power converters. They will supply electrical power to a load device when light is applied. The operational efficiency of this device is based on the developed electrical power output divided by the light power input. Ideally, a high level of efficiency is desirable. The most efficient cells manufactured today are in the range 10 to 15%. For this reason, photovoltaic cells are used in applications where high levels of light energy are available.

PHOTODIODES

A photodiode is a P-N junction device that will conduct current when energized by light energy. This device is designed for operation in the reverse bias direction. An external voltage source is required to reverse bias the device. Figure 3-33 shows a schematic symbol and some representative photodiodes. These devices have a very fast re-

sponse time. They are used in high-speed computer counting circuits and as movie projector sound track pickups.

In an earlier discussion, it was found that a very small reverse current flows when a diode is reverse biased. This leakage current is primarily caused by thermally generated electron-hole pairs in the depletion region. Current carriers of this type are swept across the junction by the reverse-biased electric field. An increase in junction temperature causes a corresponding increase electron-hole pair generation. This in turn causes an increase in leakage current. The same effect takes place when light energy is applied to a P-N junction. Light travels in small bundles of energy called photons. When photons are absorbed by the P-N junction, their energy is used to free new electron-hole pairs. Increasing the intensity of light causes a reverse-biased photodiode to become more conductive.

Characteristically, the reverse current of a photodiode increases with the intensity of light. A reverse current-light intensity graph and its test circuit are shown in Figure 3-34. The source voltage V_S is adjusted to a typical operating value. In this case –20 V. Note particularly that an increase in illumination level causes a linear rise in reverse current. As shown, the value of I_R is limited to only a few microamperes. As a rule, its current must be amplified before it can be made useful. Photodiodes are generally used to drive transistor amplifiers. Phototransistors are now available that have a photodiode and a transistor combined in a single device.

VARACTOR DIODES

In an ordinary junction diode, the depletion region is an area which separates the P and N material on each side of the junction. This area develops when the junction

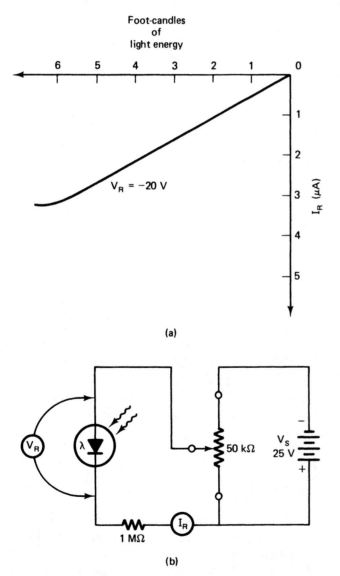

(a)

(b)

Figure 3-34. Characteristic (a) reverse current light intensity graph and (b) its test circuit.

is initially formed. It represents a unique part of the diode that is essentially void of current carriers. In this regard, the depletion region serves as a dielectric medium or insulator. When a diode has bias voltage applied, its depletion region will change in width. In a sense, this means that a diode responds as a voltage-variable capacitor. By definition, a capacitor is two or more conductors separated by an insulating material. The P and N materials, being semiconductors, are separated by a depletion region insulator. Devices designed to respond to the capacitance effect are called varactors, varicap diodes, or voltage-variable capacitors.

A varactor diode is a specially manufactured P-N junction with a variable concentration of impurities in its P-N materials. In a conventional diode doping impurities are usually distributed equally throughout the material. Varactors have a very light dose of impurities near the junction. Moving away from the junction the impurity level increases. This type of construction produces a much steeper voltage-capacitance relationship. Figure 3-35 shows a comparison of the capacitance between a conventional silicon diode and a varactor diode. It can be seen that an ordinary diode possesses only a small value of internal capacitance. In general this capacitance is too small to be of practical value.

Varactor diodes are normally operated in the reverse bias direction. With an increase in reverse biasing the depletion region increases its width. This means less resulting capacitance. A decrease in reverse bias voltage causes a corresponding increase in capacitance. In effect, the capacitance of a diode varies inversely with its bias voltage. This relationship, however, does not change linearly. Notice the nonlinear area between 0 and 0.2 V_R of the varactor diode of Figure 3-35. Its normal range of operation is between zero and the reverse breakdown voltage.

A number of schematic symbols are used to represent the varactor diode. Figure 3-36 shows three rather popular methods of representation. The symbol on the left has a small capacitor inside the circle surrounding the diode. This symbol tends to be used more frequently than the other two. Industrial electronic circuits and specialized engineering schematics seem to use the other two symbols.

Varactor diodes are used primarily in resonant circuits where some level of tuning or frequency control is desired. Figure 3-37 shows a representative varactor tuned *LC* circuit. The resonant frequency of this circuit is determined by the inductor L and the series combination of C_D and C_1. C_1 normally has a larger capacitance value than C_D. C_1 is also used to prevent the dc bias voltage of

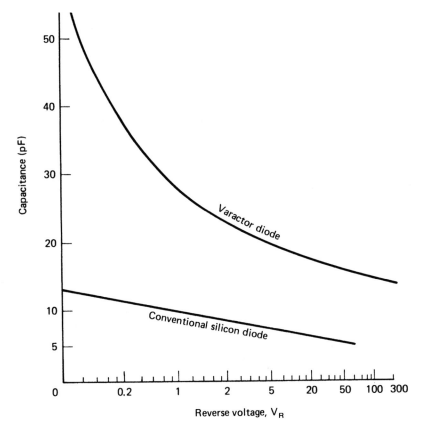

Figure 3-35. Diode capacitance values, $T_A = 25''$ □.

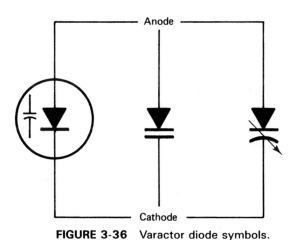

FIGURE 3-36 Varactor diode symbols.

Figure 3-36. Varactor diode symbols.

C_D from going into the inductor. R_1, R_2, and R_3 serve as the bias voltage control network for C_D. Adjusting R_2 will change the amount of reverse bias voltage to C_D. Moving the wiper arm of R_2 *up causes more reverse bias voltage to C_D. This increases the capacitance of C_D.* Moving the wiper arm down reduces the bias voltage, which causes a reduction in capacitance. Used in this manner the varactor becomes a tuning component for resonant frequency. Other

applications of the varactor diode are automatic frequency control AFC in FM radio and television receivers.

BIPOLAR TRANSISTORS

When the word *transistor* is used, it usually has some kind of letter designations to identify its type. NPN and PNP are two types of bipolar transistors. The letters are used to denote the polarity of the semiconductor material used in its construction. Material polarity is determined by the type of impurity added to silicon or germanium when it is manufactured.

A PNP transistor has a thin layer of N material placed between two pieces of P material. The schematic symbol of the PNP transistor and its crystal structure are shown in Figure 3-38. Leads attached to each piece of material are identified as the emitter, base, and collector. In the symbol, when the arrowhead of the emitter lead "Points iN" toward the base it indicates that the device is of the PNP type. Lead identification is extremely important when working with transistors.

An NPN transistor has a thin layer of P material placed between two pieces of N material. The schematic symbol, crystal structure, and lead designations of the transistor are shown in Figure 3-39. The schematic symbol of this device is the same as the PNP except the pointing

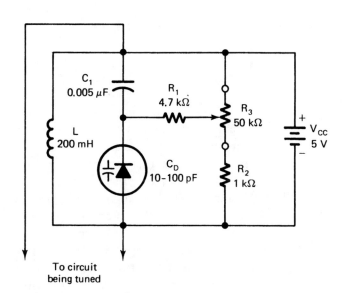

Figure 3-37. Varactor LC circuit.

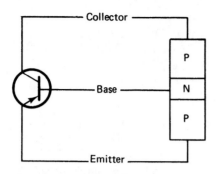

Figure 3-38. PNP transistor symbol and crystal structure.

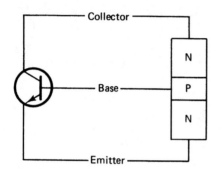

Figure 3-39. NPN transistor symbol and crystal structure.

direction of the arrowhead. An NPN transistor has the arrow "Not Pointing" toward the base. The polarity of the source voltage applied to each element is determined by the material used in its construction.

PNP and NPN transistors are considered to be bipolar devices. The term bipolar refers to conduction by holes and electrons. Electrons are the majority current carriers of the N material. In the P material, holes are the majority current carriers. Current conduction in the transistor is based on the movement of both current carriers. The term *bipolar* denotes this condition of conduction.

TRANSISTOR BIASING

For a transistor to function properly, it must have electrical energy applied to all its electrodes. In semiconductor circuits the source voltage is called bias voltage or simply bias. Bipolar transistors must have both junctions biased in order to function. In an earlier chapter we discussed diode biasing. Forward biasing of a P-N junction takes place when the polarity of the source voltage is positive to the P material and negative to the N material. This condition causes a current flow. The depletion zone of the device is reduced and majority current carriers are driven toward the junction. One junction of a transistor must be forward biased when it operates.

Reverse biasing of a P-N junction is achieved when the polarity of the source is negative to the P material and positive to the N material. This condition does not ordinarily permit current flow. The width of the depletion zone is increased by this action and minority current carriers move toward the junction. In normal circuit operation reverse bias current is extremely small. One junction of a transistor must be reverse biased in order for the device to be operational.

NPN Transistor Biasing

Consider now the biasing of the NPN transistor in Figure 3-40. In this diagram an external voltage source has been applied only to the emitter-base regions of the transistor. In an actual circuit, this source is called the emitter-base voltage or V_{BE}. Note that the polarity of V_{BE} causes the emitter-base junction to be forward biased. This condition causes majority current carriers to be forced together at the emitter-base junction. The resulting current flow in this case would be quite large. In our example, note that the same amount of current flows into the emitter and out of the base. Emitter current (I_E) and base current (I_R) are used to denote these values. Under normal circuit conditions a V_{BE} source would not be used independently.

Figure 3-41 shows an external voltage source connected across the base-collector junction of an NPN transistor. The base-collector voltage or V is used to reverse bias the base-collector junction. In this case there is no indication of base current I_R or collector current I_F. In an actual reverse-biased junction there could be a very minute amount of current. This would be supported primarily by minority current carriers. As a general rule, this is called *leakage current*. In a silicon transistor leakage current is usually considered to be negligible. In an actual circuit V_{CB} is not normally applied to a transistor without the V_{BE} voltage source.

For a transistor to function properly the emitter-base junction must be forward biased and the base-collector junction reverse biased. Both junctions must have bias voltage applied at the same time. In some circuits this voltage may be achieved by separate V_{BE} and V_{CB} sources. Other circuits may utilize a single battery with specially connected bias resistors. In either case, the transistor responds quite differently when all its terminals are biased.

Consider now the action of a properly biased NPN transistor. Figure 3-42 shows separate V_{BE} and V_{CB} sources connected to the transistor. V_{BE} provides forward bias for the emitter-base junction while V_{CB} reverse biases the collector junction. Connected in this manner, the two junctions do not respond as independent diodes.

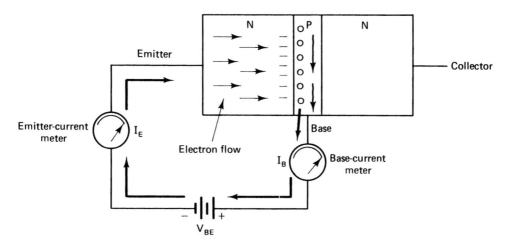

Figure 3-40. Emitter-base biasing.

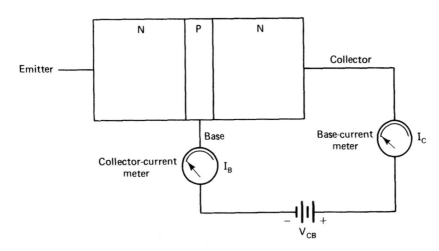

Figure 3-41. Base-collector biasing.

Figure 3-43 shows how current carriers pass through a properly biased NPN transistor. The emitter-base junction being forward biased causes a large amount of 'E to move into the emitter-base junction. Upon arriving at the junction, a large number of the electrons do not effectively combine with holes in the base. The base is usually made very thin (0.0025 cm or 0.001 in.) and it is lightly doped. This means that the majority current carriers of the emitter exceed the majority carriers of the base. Most of the electrons that cross the junction do not combine with holes. They are, however, immediately influenced by the positive V_{CB} voltage applied to the collector. A very high percentage of the original emitter current enters the collector. Typically, 95 to 99% of I_E flows into the collector junction. It then becomes collector current. After passing through the collector region, I_C returns to V_{CB}, V_{BE}, and ultimately to I_E. The current flow inside the transistor is indicated by a large arrow. Outside the transistor current flow is indicated by small arrows.

The difference between the amount of I_E and I_C of Figure 3-43 is base current. Essentially, base current is due to the combining of a small number of electrons and holes in the emitter-base junction. In a typical circuit, I_B is approximately 1 to 5% of I_E. With the base region being very narrow, it cannot support a large number of current carriers. A small amount of base current is needed however, to make the transistor operational. Note the direction of I_B and its flow path in the diagram.

The relationship of I_E, I_B, and I_C in a transistor can be expressed by the equation

emitter current = base current + collector current

or

$$I_E = I_B + I_C$$

This shows that the emitter current is equal to the sum of I_B and I_C. The largest current flow in a transistor takes place in the emitter. I_C is slightly less than I_E.

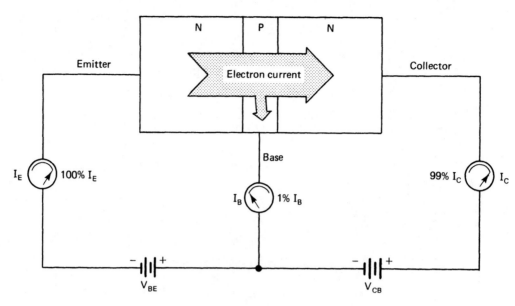

Figure 3-42. NPN transistor biasing.

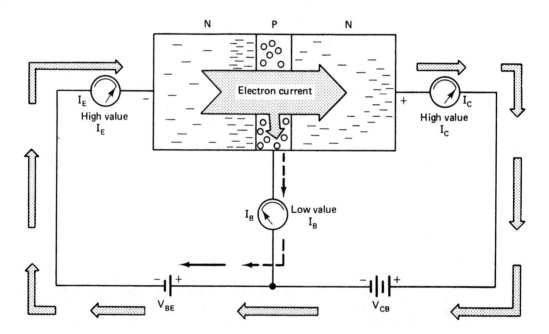

Figure 3-43. Current carriers passing through an NPN transistor.

This means that the current flow through a reverse-biased collector-base junction is nearly equal to that of the forward-biased emitter-base junction. The difference in I_E and I_C is I_B.

If the base of a transistor were not made extremely thin, it would not respond as just described. The thin base region makes it possible for large amounts of I_E to pass through the base and into the collector region. A thicker base would cause more I_E to combine with holes in the base. This would reduce the operational effectiveness of the transistor.

The amount of base current that flows in a transistor is very small but extremely important. Suppose, for example, that the base lead of the transistor in Figure 3-43 is momentarily disconnected. With this element open it should be obvious that there will be no base current. Closer examination will also show that V_{BE} and V_{CB} are now connected in series. This means that their voltages are added together. As a result of this action, the collector becomes more positive and the emitter more negative.

One would immediately think that this condition would cause the transistor to conduct very heavily. It does not, however, permit any conduction at all. In effect, this means that base current has a direct influence on emitter and collector current. The base-emitter junction of a transistor must be forward biased in order for it to produce collector current.

Transistor Beta

A small change in transistor base current is capable of producing a very large change in collector current. This, in effect, shows that a transistor is capable of achieving current gain. The term *beta* is commonly used as an expression of current gain. Beta can be determined by dividing the collector current by the base current. This relationship is expressed by the equation

$$\text{Beta} = \frac{\text{collector current}}{\text{base current}} \quad \text{or} \quad \beta = \frac{I_C}{I_B}$$

For most transistors, beta is usually quite large. In a previous circuit, we described I_C as being 95 to 99% of I_E, with I_B being from 1 to 5%. For this example, the beta of a transistor with this ratio would be

$$\beta = \frac{95\%}{5\%} = 19 \quad \text{or} \quad \beta = \frac{99\%}{1\%} = 99$$

In both examples note that beta is expressed as a pure number. The percentage signs in the numerator and denominator of the equation cancel each other. If actual current values in milliamperes or amperes were used, they would also cancel. Beta is simply expressed as a whole-number value.

Transistors are manufactured today with a very wide range of beta capabilities. Power transistors, which respond to large current values, usually have a rather low beta. Typical values are in the range of 20. Small-signal transistors may be capable of beta in the range of 400. A good average beta for all transistors is in the range of 100. Beta is a very important transistor characteristic. It is not very meaningful unless it is applied to a specific transistor and its circuit components.

PNP Transistor Biasing

Biasing of a PNP transistor is very similar to that of the NPN transistor. The emitter-base junction must, for example, still be forward biased and the collector-base reverse biased. The polarity of the bias voltage is, however, reversed for each transistor. The majority current carriers of a PNP transistor are holes instead of electrons as in the NPN device. Except for these two differences, operation is primarily the same.

Figure 3-44 shows how current carriers pass through a properly biased PNP transistor. The emitter-base junction being forward biased causes a large number of holes to move through the junction. Upon arriving in the base region, a very small number of these holes combine with electrons. This is representative of the base current. Ninety-five to 99% of the holes move

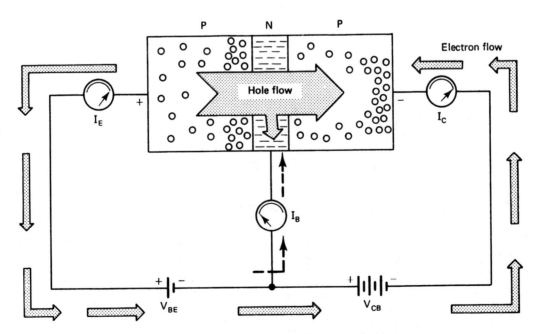

Figure 3-44. Current carriers of a PNP transistor.

through the base region and enter the collector junction. These holes do not find electrons to combine with in the thin base region. They are, however, immediately influenced by the negative V_{CB} voltage of the collector. This action causes hole current to flow through the reverse-biased base-collector junction. The end result of this is I_C passing through the collector region. It then leaves the collector and flows into V_{CB}, V_{BE}, and returns to the emitter. Hole current flow is indicated by the large arrow inside the transistor. Outside the transistor, current flow is achieved by electrons. Small arrows in the diagram show current outside the transistor.

It is rather interesting to compare the transistors of Figures 3-43 and 3-44. Specific attention should be directed to the polarity of V_{BE} and V_{CB}. This shows the primary difference in device biasing. Note also the type of majority current flow through each transistor. This is represented by the large arrow of each diagram. It is different for each device. This means that the two transistor types are not directly interchangeable. Substituting one for the other would necessitate a complete reversal of bias voltage.

PNP transistors are not as widely used today as is the NPN type. Electron current flow of the NPN device has much better mobility than the hole flow of a PNP type. This means that electrons have a tendency to move more quickly through the crystal material than do holes. Because of this characteristic, NPN transistors tend to respond better at high frequencies. In general, this means that the NPN device has a wider range of applications.

TRANSISTOR CHARACTERISTICS

When a transistor is manufactured it has characteristic data that distinguish it from other devices. These data are extremely important in predicting how a transistor will perform in a circuit. One very common way of showing this information is with the use of graphs. A collector family of characteristic curves is widely used to show transistor data.

A collector family of characteristic curves for an NPN transistor is shown in Figure 3-45. The vertical part of this graph shows different collector current values in milliamperes. The horizontal part shows the collector-emitter voltage. Individual

lines of the graph represent different values of base current. The zero base-current line if normally omitted from the graph. This condition shows when the transistor is not conducting.

A family of collector curves tells a great deal about the operation of a transistor. Take for example, point A on the 60-μA base-current line. If the transistor has 60 μA of I_B and a V_{CE} of 6 V, there is a collector current of 3 mA. This is determined by projecting a line to the left of the intersection of 60 μA and 6 V. Any combination of I_C, I_B, and V_{CE} can be quickly determined from the display of these curves.

If the values of I_C and I_B can be determined from a family of curves, they can be used to predict other values. The emitter current (I_E) of a transistor can be determined from the values of I_C and I_B. Remember that $I_E = I_B + I_C$. For point A, the I_E is 60 μA + 3 mA, or 3.06 mA.

The current gain or beta of a transistor can also be determined from a family of collector curves. Since beta is I_C/I_B, it possible to determine this condition of operation. For point A on the curves, I_C is 3 mA and I_B is 60 μA. The current gain of a transistor operating at this point would

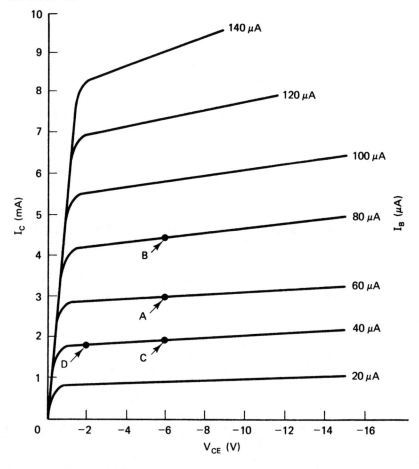

Figure 3-45. Collector family of characteristic curves for an NPN transistor.

be determined by the equation

$$\beta = \frac{I_C}{I_B} = \frac{3\text{ mA}}{60\text{ }\mu\text{A}} = \frac{0.003}{0.00006} = 50$$

This condition is called the direct-current beta. It shows how the transistor will respond when dc voltage is applied. Dc beta is an important consideration when a transistor is operated as a direct current amplifier.

In many applications a transistor must amplify ac signals. Dc voltage values are applied to the device to make it operational. An ac signal voltage applied to the base would then cause a change in the value of I_B. In Figure 3-45, assume that the input voltage causes I_B to change between points B and C. Point A is considered to be the operating point. The change in I_B is from 80 μA to 40 μA. This is normally called ΔI_B. The Greek letter delta (Δ) is used to denote a changing value in I_B. A corresponding change in I_C is determined by projecting a line to the left of each I_B value. ΔI_C is therefore 4.5 mA to 2 mA.

The ac beta of a transistor is determined by dividing ΔI_C by ΔI_B. For points B-C this is

$$\text{Ac beta} = \frac{\Delta I_C}{\Delta I_B} = \frac{4.5\text{ mA to 2 mA}}{80\text{ }\mu\text{A to 40 }\mu\text{A}} = \frac{2.5\text{ mA}}{40\text{ }\mu\text{A}}$$

$$= \frac{0.0025}{0.00004} = 6.25$$

The ac beta in this case is somewhat different from that of the dc beta. As a general rule, ac beta is larger than dc beta. To amplify an ac voltage, a transistor must operate over a range of different values. This usually accounts for the difference in beta values.

The same family of collector curves can also be used to show the effectiveness of a transistor in controlling collector current. Note points D and C on the 40-μA curve. A change in V_{CE} from 2 to 6 V occurs between these two points. Projecting points D and C to the left indicates a change in I_C from 1.8 to 2 mA. A 4-V change in V_{CE}, therefore, causes only a 0.2-mA change in I_C. Using the ac beta for comparison, a 40-μA change in I_B causes a ΔI_C of 2.5 mA. This indicates that I_C is more effectively controlled by I_B changes than by V_{CE} changes.

Operation Regions

A collector family of characteristic curves is also used to show the desirable operating regions of a transistor. When there is operation in the center area it is called the active region. This area is located anywhere between the two shaded areas of Figure 3-46. Amplification is achieved when the transistor operates in this region.

A second possible region of operation is called cutoff. On the family of curves, this is where $I_C = 0$ mA. When a transistor is cut off, no I_B and I_C flows. Any current flow that occurs in this region is due to leakage current. As a general rule, a transistor operating in the cutoff region responds as a circuit with an open switch. When a transistor is cut off there is infinite resistance between the emitter and collector.

The shaded area on the left side of the family of curves is called the saturation region. This area of operation is where maximum I_C flows. A transistor operating in this region responds as a closed switch. A transistor is considered to be fully conductive in the saturation region. The resistance between emitter and collector is extremely small when a transistor is saturated.

Figure 3-47 shows how a transistor responds in the three regions of operation. In part (a) the transistor is operating in the active area. The value of V_C is somewhat less than that of the source voltage V_{CC}. Collector current through the transistor causes a voltage drop across R_C. The base current of a transistor operating in this region would be some moderate value between zero and saturation. Transistor amplifiers usually operate in this area.

Figure 3-47(b) shows how a transistor responds when it is in the cutoff region. With no current flow through the transistor, there is no voltage drop across R_C. V_C of this circuit is therefore the same as the source voltage. A transistor is cut off when no I_B occurs.

Figure 3-47(c) shows how a transistor operates in the saturation region. In this condition of operation the transistor responds as a closed switch. V_C is approximately equal to 0 V. Heavy I_C values cause nearly all V_{CC} to appear across R_C. In a sense, V_C is considered to be at ground. Current passing through the transistor is limited by the value of V_{CC} and R_C. A large value of I_B is needed to cause a transistor to saturate.

Characteristic Curves

Special circuits are used to obtain the data for a family of characteristic curves. Figure 3-48 shows a test circuit that can be used to find the data points of a family of collector curves. Three meters are used to monitor this information. Base current and collector current are observed by meters connected in series with the base and collector. Collector-emitter voltage is measured with a voltmeter connected across these two terminals. V_{CE} is adjusted to different values by a variable power supply.

The data points for a single curve are developed by first adjusting I_B to a constant value. A representative val-

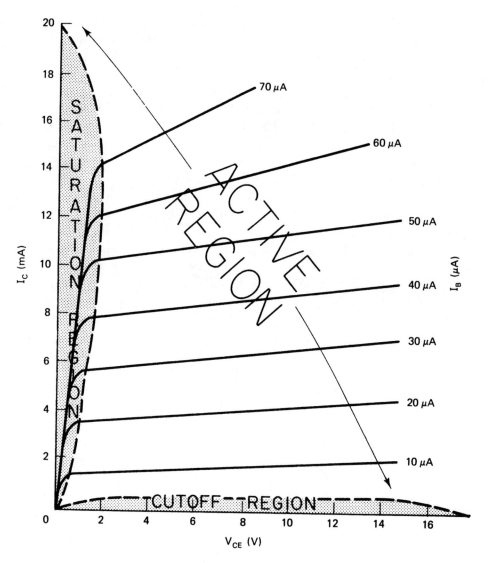

Figure 3-46. Operating regions of a transistor.

ue for a small-signal transistor is 10 μA. Resistor R_B is used to adjust the value of I_B. The V_{CE} is then adjusted through its range starting at 0 V. As V_{CE} is increased in 0.1-V steps up to 1 V, the I_C, increases very quickly. Corresponding I_C and V_{CE} values are plotted on the graph. V_{CE} should then be increased in 1-V steps while monitoring I_C. The resulting collector current usually levels off to a fairly constant value. The corresponding I_C and V_{CE} values are then added to the first data points. All points are then connected together by a continuous line. The completed curve would be labeled 10 μA of I_B.

To develop the second curve, V_{CE} is first returned to zero. I_B is then increased to a new value. A common value for a small-signal transistor would be 20 μA. The second curve is then plotted by recording the I_C values for each V_{CE} value.

To obtain a complete family of curves, the process would be repeated for several other I_B values. A representative family of curves may have 8 to 10 different I_B values. The step values of V_{CE} and the increments of I_B and I_C will change with different transistors. Small-signal transistors may use 10-μA I_B steps with I_C values ranging from 0 to 20 mA. Large-signal transistors may use 1-mA I_B steps with I_C values going to 100 mA. Each transistor may require different values of I_B, I_C, and V_{CE} to obtain a suitable curve.

TRANSISTOR PACKAGING

Once a transistor is constructed, the entire crystal assembly must be placed in an enclosure or package. The unit is then sealed to protect it from dust, moisture, or outside contaminates. Electrical connections are made

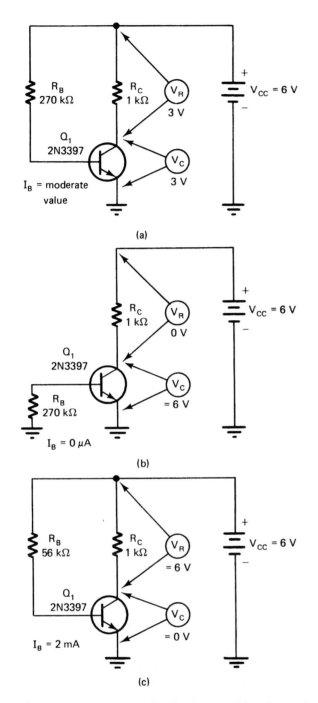

Figure 3-47. Transistor operation by regions: (a) active region; (b) cutoff region; (c) saturation region.

to each element through leads attached to the package. When metal packages are used, the housing also serves as a heat sink. Heat is carried away from the crystal by the metal package. A common practice in this kind of unit is to attach the collector to the metal case. This helps the collector to dissipate heat. Packages of this type should not be permitted to touch other circuit components when it is in operation.

Ceramic and epoxy packaged transistors are also available. As a rule, this type of packaging is somewhat cheaper to produce. These devices are usually not as rugged as the metal-packaged devices. Most small-signal transistors are housed in epoxy packages. The outside case of this device is insulated from all transistor elements. It is specifically designed for printed-circuit-board construction and close placement with other circuit components.

Several transistor package styles are shown in Figure 3-49. Part(a) shows devices designed for small-signal applications. The letter designation "TO" stands for transistor outline. These same packages are also used to house other solid-state devices. Part (b) shows some of the packages used to house large-signal or power transistors. Heavy metal cases of this type are used to dissipate heat generated by the transistor. The collector is usually attached to the metal case. Part (c) shows a few of the common epoxy packages. The letter designation "SP" stands for semiconductor package. This type of package is used to house a number of other semiconductor devices.

With the wide range of package styles available today, it is difficult to have a standard method of identifying transistor leads. Each package generally has a unique lead location arrangement. It is usually a good idea to consult the manufacturer's specifications for proper lead identification.

UNIPOLAR TRANSISTORS

A transistor that has only one P-N junction in its construction is called a *unipolar* device. Unlike the bipolar transistor, a unipolar transistor conducts current through a single piece of semiconductor material. The current carriers of this device have only one polarity. The term unipolar is used to describe this characteristic.

Field-effect transistors and unijunction transistors are unipolar devices. These transistors are very similar in construction but differ a great deal in operation and function. Field-effect transistors are normally used to achieve amplification. Unijunction transistors are used as waveshaping devices. Both devices play an important role in solid-state electronics.

JUNCTION FIELD-EFFECT TRANSISTORS

The junction field-effect transistor (JFET) is a three-element electronic device. Its operation is based on the

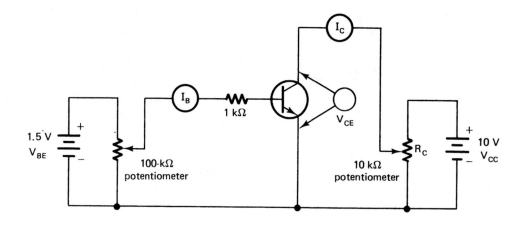

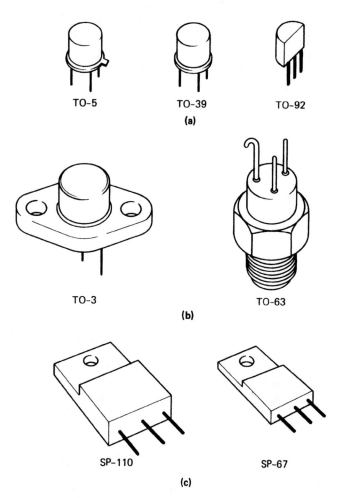

Figure 3-49. Transistor packages: (a) small-signal devices; (b) large-scale devices; (c) epoxy devices.

conduction of current carriers through a single piece of semiconductor material. This piece of material is called the *channel*. An additional piece of semiconductor material is diffused into the channel. This element is called the *gate*. The entire unit is built on a third piece of semicon-

ductor known as the *substrate*. The assembled device is housed in a package. The physical appearance of a JFET in its housing is very similar to that of a bipolar transistor.

N-Channel JFETs

Figure 3-50 shows the crystal structure, element names, and schematic symbol of an N-channel JFET. Its construction has a thin channel of N material formed on the *P substrate*. The P material of the gate is then formed on top of the N channel. Lead wires are attached to each end of the *channel* and the gate. No connection is made to the substrate of this device.

The schematic symbol of an N-channel JFET is somewhat representative of its construction. The bar part of the symbol refers to the channel. The drain (D) and source (S) are attached to the channel. The gate (G) has an arrow. This shows that it forms a P-N junction. The arrowhead of an N-channel symbol "*Points iN*" toward the channel. This indicates that it is a P-N junction. The arrowhead or gate is P material and the channel N material. This part of the device responds as a junction diode.

The operation of a JFET is somewhat unusual when compared with that of a bipolar transistor. Figure 3-51 shows the source and drain leads of a JFET connected to a dc voltage source. In this case, maximum current will flow through the channel. The I_S and I_D meters will show the same amount of current. The value of V_{DD} and the internal resistance of the channel determines the amount of channel current flow. Typical source-drain resistance values of a JFET are several hundred ohms. Essentially this means that full conduction will take place in the channel even when the gate is open. A JFET is considered to be a normally on device.

Current carriers passing through the channel of a JFET are controlled by the amount of bias voltage applied

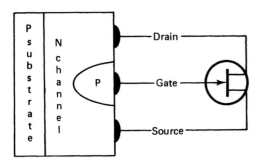

Figure 3-50. N-channel junction field-effect transistor.

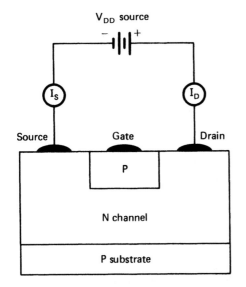

Figure 3-51. Source-drain conduction of a JFET.

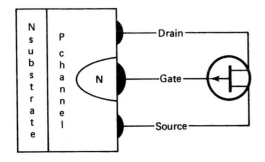

Figure 3-52. P-channel junction field-effect transistor.

to the gate. In normal circuit operation, the gate is reverse biased with respect to the source. Reverse biasing of the gate-source of any P-N junction will increase the size of its depletion region. In effect, this will restrict or deplete the number of majority carriers that can pass through the channel. This means that drain current (I_D) is controlled by the value of V_{GS}. If V_{GS} becomes great enough, no I_D will be permitted to flow through the channel. The voltage that causes this condition is called the cutoff voltage.

I_D can be controlled anywhere between full conduction and cutoff by a small change in gate voltage.

P-Channel JFETs

The crystal structure, element names, and schematic symbol of a P-channel JFET are shown in Figure 3-52. This device has a thin channel of P material formed on an N substrate. The N material of the gate is then formed on top of the P channel. Lead wires are attached to each end of the channel and to the gate. Other construction details are the same as those of the N-channel device.

The schematic symbol of a JFET is different only in the gate element. In a P-channel device the arrow is "Not Pointing" toward the channel. This means that the gate is N material and the channel is P material. In conventional operation, the gate is made positive with respect to the source. Varying values of reverse bias gate voltage will change the size of the P-N junction depletion zone. Current flow through the channel can be altered between cutoff and full conduction. P-channel and N-channel JFETs cannot be used in a circuit without a modification in the source voltage polarity.

JFET Characteristic Curves

The JFET is a very unique device compared with other solid-state components. A small change in gate voltage will, for example, cause a substantial change in drain current. The JFET is therefore classified as a voltage-sensitive device. By comparison, bipolar transistors are classified as current-sensitive devices. A JFET has a rather unusual set of characteristics compared with other solid-state devices.

A family of JFET characteristic curves is shown in Figure 3-53. The horizontal part of the graph shows the voltage appearing across the source-drain as V_{DS}. The vertical axis shows the drain current (I_D) in milliamperes. Individual curves of the graph show different values of gate voltage (V_{GS}). The cutoff voltage of this device is approximately –7 V. The control range of the gate is from 0 to –6 V.

A drain family of characteristic curves tells a great deal about the operation of a JFET. Refer to point *A* on the –3 V_{GS} curve. If the device has –3 V applied to its gate and a V_{DS} of 6 V, there is approximately 4 mA of I_D flowing through the channel. This is determined by projecting a line to the left of the intersection of –3 V_{GS} and 6 V_{DS}. Any combination of I_D, V_{GS}, and V_{DS} can be determined from the family of curves.

Developing JFET Characteristic Curves

A special circuit is used to develop the data for a

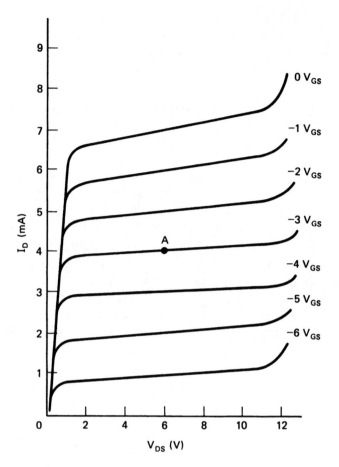

Figure 3-53. Family of drain curves.

Drain current is measured by a milliampere meter connected in series with the drain. V_{DS} is measured with a voltmeter connected across the source and drain. V_{GS} and V_{DS} are adjusted to different values while monitoring I_D. Two variable dc power supplies are used in the test circuit.

The data points of a single curve are developed by first adjusting V_{GS} to 0 V. V_{DS} is then adjusted through its range starting at 0 V. Normally, V_{DS} is increased in 0.1-V steps up to 1 V. I_D increases very quickly during this operational time. V_{DS} is then increased in 1-V steps recording the I_D values. Corresponding I_D and V_{GS} values are then plotted on a graph as the 0 V_{GS} curve.

To develop the second curve, V_{DS} must be returned to zero. V_{GS} is then adjusted to a new value. 0.5 V would be a suitable value for most JFETs. V_{DS} is again adjusted through its range while monitoring I_D. Data for the second curve are then recorded on the graph.

To obtain a complete family of drain curves, the process would be repeated for several other V_{GS} values. A typical family of curves may have 8 to 10 different values. The step values of V_{DS}, V_{GS}, and I_D will obviously change with different devices. Full conduction is usually determined first. This gives an approximation of representative I_D and V_{DS} values. Normally, I_D will level off to a rather constant value when V_{DS} is increased. V_{DS} can be increased in value to a point where I_D starts a slight increase. Generally, this indicates the beginning of the breakdown region. JFETs are usually destroyed if conduction occurs in this area of operation.

Maximum V_{DS} values for a specific device are available from the manufacturer. It is a good practice to avoid operation in or near the breakdown region of the device.

drain family of characteristic curves. Figure 3-54 shows a circuit that is used to find the data points of a characteristic curve for an N-channel JFET. Three meters are used to monitor this data. Gate voltage is monitored by a V_{GS} meter connected between the gate-source leads.

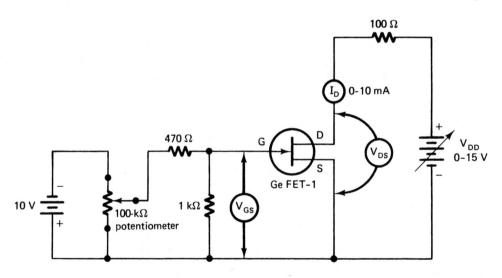

Figure 3-54. Drain curve test circuit for N-channel JFET.

MOS FIELD-EFFECT TRANSISTORS

Metal-oxide semiconductor field-effect transistors (MOSFETs) are a rather new addition to the unipolar transistor family. A distinguishing feature of this device is its gate construction. The gate is, for example, completely insulated from the channel. Voltage applied to the gate will cause it to develop an electrostatic charge. No current is permitted to flow in the gate area of the device. The gate is simply an area of the device coated with metal. Silicon dioxide is used as an insulating material between the gate and the channel. Construction of a MOSFET is rather easy to accomplish.

The channel of a MOSFET has a number of differences in its construction. One type of MOSFET responds to the depletion of current carriers in its channel. Depletion-type MOSFETs have an interconnecting channel built on a common substrate. Direct connection is made between the source and drain of this device. Channel construction is very similar to that of the JFET. A second type of MOSFET is called an enhancement type of device. E-MOSFETs do not have an interconnecting channel. Current carriers pulled from the substrate form an induced channel. Gate voltage controls the size of the induced channel. Channel development is a direct function of the gate voltage. E- and D-type MOSFETs both have unique operating characteristics.

Enhancement-type MOSFET construction, element names, and schematic symbols are shown in Figure 3-55. Note in the crystal structure that no channel appears between the source and drain. The gate is also designed to cover the entire span between the source and drain. The gate is actually a thin layer of metal oxide. Conductivity of this device is controlled by the voltage polarity of the gate. A P-N junction is not formed by the gate, source, and drain.

An E-MOSFET is considered to be a normally off device. This means that without an applied gate voltage, there is no conduction of I_D. When the gate of an N-channel device is made positive, electrons are pulled from the substrate into the source-drain region. This action causes the induced channel to be developed. With the channel complete, conduction occurs between the source and drain. In effect, drain current is aided or enhanced by the gate voltage. Without V_G, the channel is not properly developed and no I_D will flow.

A family of characteristic curves for an E-type MOSFET is shown in Figure 3-56. This particular set of curves is representative of an N-channel device. A P-channel device would have the polarity of the gate voltage reversed. Note than an increase of gate voltage causes

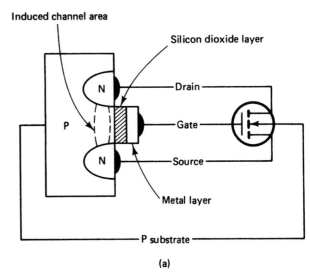

(a)

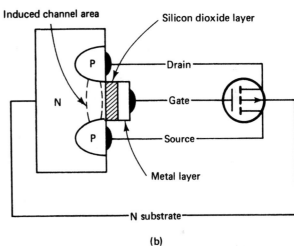

(b)

Figure 3-55. Enhancement-mode MOSFET: (a) N channel; (b) P channel.

a corresponding increase in I_D. The input impedance of this device is extremely high regardless of the crystal material used or the polarity of the gate voltage.

Depletion-type MOSFET construction, element names, and schematic symbols are shown in Figure 3-57. This N-channel device has a thin channel of N material formed on a P substrate. Source and gate connections are made directly to the channel. A thin layer of silicon dioxide (SiO_2) insulation covers the channel. The gate is a metal-plated area formed on the silicon dioxide layer. The entire unit is then built on a substrate of P material. The arrow of the schematic symbol refers to the material of the substrate. When it "*Points iN*" this shows that the substrate is P material and the channel is N.

The construction of P- and N-channel devices is essentially the same. The crystal material of the channel and substrate is the only difference. Current carriers are holes in the P-channel device, while electrons are carriers in the N-channel unit. The schematic symbol differs only in the direction of the substrate arrow. It does "Not Point" toward the substrate in a P-channel device. This means that the substrate is N material and the channel is P material.

The operation of a D-MOSFET is very similar to that of a JFET. With voltage applied to the source and drain, current carriers pass through the channel. The gate does not need to be energized in order to produce conduction. In this regard, a D-MOSFET is considered to be a normally on device. Control of I_D is, however, determined by the polarity of gate voltage. When the polarity of V_G is the same as the channel, current carrier depletion occurs. A reverse in polarity causes I_D to increase due to the enhancement of the current carriers. In a sense, the D-MOSFET should be classified as a depletion-enhancement device. Its conduction normally responds to both conditions of operation.

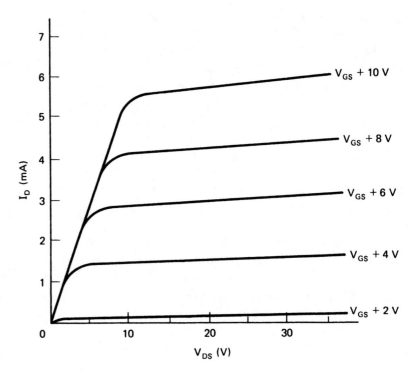

Figure 3-56. Family of drain curves for an E-type MOSFET (N-channel type).

A family of characteristic curves for an N-channel D-MOSFET is shown in Figure 3-58. The horizontal part of the graph shows the source-drain voltage as V_{DS}. The vertical axis shows the drain current in milliamperes. Individual curves of the graph show different values of gate voltage. Note that zero V_G is near the center of the curve. This means that the gate can swing above or below zero. For an N-channel device, negative gate voltage reduces I_D. This voltage, in effect, pulls holes from the substrate into the initial channel area. Electrons normally in the channel move in to fill the holes. This action depletes the number of electrons in the initial channel. An increase in negative gate voltage causes a corresponding decrease in I_D.

When V_G swings positive, it causes the number of current carriers in the initial channel to be increased. A positive gate voltage attracts electrons from the P substrate. This action increases the width of the initial channel. As a result, more flows. Making V_G more positive increases I_D. This means that the channel is aided or enhanced by a positive gate voltage. A D-MOSFET therefore responds to both the depletion and enhancement of its current carriers. Selection of an operating point is easy to accomplish when this device is used as an amplifier.

In a P-channel D-MOSFET, operation is very similar to that of the N-channel device. In the P-channel de-

vice, holes are the current carriers. Current conduction is normally on when V_G is zero. A positive gate voltage will reduce I_D. Electrons are drawn out of the substrate to fill holes in the channel. This causes a reduction in channel current. When the gate voltage swings negative the channel current carriers are enhanced. Holes pulled into the initial channel cause an increase in current carriers. Current carriers are enhanced or depleted according to the polarity of V_G.

UNIJUNCTION TRANSISTORS

The unijunction transistor (UJT) is often described as a voltage-controlled diode. This device is not considered to be an amplifying device. It is, however, classified as a unipolar transistor. The UJT and an N-channel JFET are often confused because of the similarities in their schematic symbols and crystal construction. The operation and function of each device is entirely different.

Figure 3-59 shows the crystal construction, schematic symbol, and element names of a UJT. Note that the UJT is a three-terminal, single-junction device. The crystal is an N-type bar of silicon with contacts at each end. The end connections are called base 1 (B_1) and base 2 (B_2). A small, heavily doped P region is alloyed to one side of the silicon bar. This serves as the emitter (E). A P-

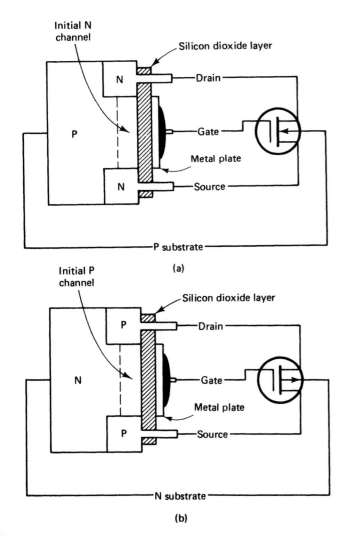

Figure 3-57. D-type MOSFET: (a) N channel; (b) P channel.

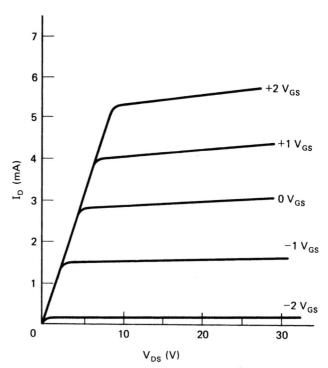

Figure 3-58. Family of drain curves for a D-type N-channel MOSFET.

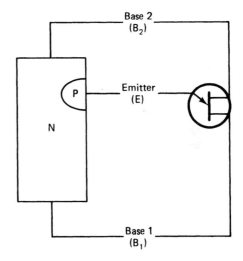

Figure 3-59. Unijunction transistor.

N junction is formed by the emitter and the silicon bar. The arrow of the symbol "*Points in N*," which means that the emitter is P material and the silicon bar is N material. The arrow of the symbol is slanted, which distinguishes it from the N-channel JFET.

An interbase resistance (R_{BB}) exists between B_1 and B_2. Typically, R_{BB} is between 4 and 10 k Ω. This value can be easily measured with an ohmmeter. The resistance of the silicon bar is represented by R_{BB}. This resistance can be divided into two values. R_{B1} is between the emitter and B_1. R_{B2} is between the emitter and B_2. Normally, R_{B2} is somewhat less than R_{B1}. The emitter is usually closer to B_2 than B_1. When a UJT is made operational, the value of R_{B1} will change with different emitter voltages.

In circuit applications B_1 is usually placed at circuit ground or the source voltage negative. The emitter then serves as the input to the device. B_2 provides circuit output. A change in E–B_1 voltage will cause a change R_{BB}. The output current of the device will increase when E–B_1 turns on. A UJT is normally used as a trigger device. It is used to control some other solid-state device. No amplification is achieved by a UJT.

A characteristic curve for a typical UJT is shown in Figure 3-60. The vertical part of this graph shows the emitter voltage as VE. An increase in V causes the curve to rise vertically. The horizontal part of the graph shows the emitter current (I_E). Note that the curve has *peak voltage* (V_P) and *valley voltage* (V_V) points. An increase in I_E causes V_P to rise until it reaches the peak point. A further increase in I_E will cause the emitter voltage to drop to the valley point. This condition is called the negative resistance region. A device that has a negative resistance characteristic is capable of regeneration or oscillation.

In operation, when the emitter voltage reaches the peak voltage point, $E–B_1$ becomes forward biased. This condition draws holes and electrons to the P-N junction. Holes are injected into the B_1 region. This action causes B_1 to be more conductive. As a result, the $E–B_1$ region become low resistant. A sudden drop in R_{B1} will cause a corresponding increase in B_1-B_2 current. This change in current can be used to trigger or turn on other devices. The trigger voltage of a UJT is a predictable value. Knowing this value will permit the device to be used as a control element.

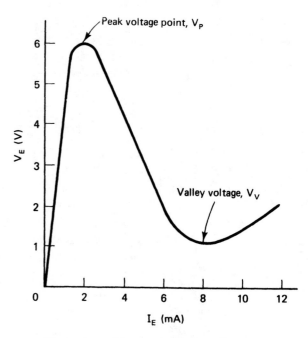

Figure 3-60. Characteristic curve for a UJT.

Chapter 4

Electronic Circuits

INTRODUCTION

All electronic systems require certain voltage and current values for operation. The energy source of the system is primarily responsible for this function. As a general rule, the source is more commonly called a *power supply*. In some systems, the power supply may be a simple battery or a single dry cell. Portable radios, DVD players, and calculators are examples of systems that employ this type of supply. Television receivers, computer terminals, stereo amplifier systems, and electronic instruments usually derive their energy from the ac power line. This part of the system is called an *electronic power supply*. Electronic devices are used in the power supply to develop the required output energy.

Electronic systems must perform a variety of basic functions in order to accomplish a particular operation. An understanding of these functions is essential. In this unit we investigate the *amplification* function. Amplification is achieved by devices that produce a change in signal amplitude. Transistors and integrated circuits are typical amplifying devices.

In this chapter we look at *amplification* in general with no particular amplifying device considered. Second, we discuss how a particular type of device achieves amplification. Circuit operation, biasing methods, circuit configurations, classes of amplification, and operational conditions are also considered.

Many electronic systems employ circuits that convert the dc energy of a power supply into a useful form of ac. Oscillators, generators, and electronic clocks are typical circuits. In a radio receiver, for example, dc is converted into high-frequency ac to achieve signal tuning. Television receivers also have an oscillator in their tuner. In addition to the tuner, oscillators are used to produce horizontal and vertical sweep signals in television sets. These signals control the electron beam of the picture tube. In the same way, calculators and computers employ an electronic clock circuit. Timing pulses are produced by this circuit. Every radiofrequency (RF) transmitter employs an oscillator. This part of the system generates the signal that is radiated into space. Oscillators are extremely important in electronics.

An *oscillator* generally uses an amplifying device to aid in the generation of the ac output signal. Transistors and ICs can be used for this function. In this chapter we will also study solid-state oscillator circuits.

OBJECTIVES

Upon completion of this chapter you will be able to:

1. List basic types of amplifiers and explain the function of each.
2. Identify the three basic amplifier configurations and describe their characteristics.
3. Explain the basic biasing arrangements used with transistor circuits.
4. Determine if a transistor amplifier is operating in the class A, AB, B, or C mode.
5. Identify basic coupling techniques and explain advantages and disadvantages of each type.
6. Understand thermal stability of basic amplifier circuits.
7. Discuss the operation and characteristics of a common-emitter amplifier circuit.
8. Identify different types of power amplifiers.
9. Define common mode rejection ratio, input resistance, output resistance, offset voltage, offset current, bias current, slew rate, and other op-amp characteristics.
10. Analyze operational amplifier circuits.
11. Explain basic inverting and noninverting operational amplifier circuits.
12. Calculate the input and output resistances of an operational amplifier circuit.
13. Explain the characteristic difference among half-wave, full-wave, and bridge rectifier circuits.
14. Explain the effect of a filter capacitor on the output voltage and ripple voltage of a rectifier circuit.

15. Describe the characteristics of capacitor, resistor-capacitor, and pi filters.
16. Explain zener regulator circuits.
17. Understand simple transistor and op-amp regulator circuits.
18. Explain the operation of series and shunt transistor regulators.
19. Calculate percent ripple of a power supply circuit.
20. Calculate percent regulation for a power supply.
21. Explain the operation of half-wave, full-wave, and bridge rectifier circuits.
22. Draw typical output waveforms before and after adding a filter for each type of rectifier circuit.
23. List the classes of feedback oscillators.
24. Calculate the frequency of common *LC* oscillators.
25. Identify and explain the operation of oscillator circuits.
26. Recognize crystal oscillator circuits and explain their frequency selective characteristics.
27. Calculate the frequency of oscillators using their *RC* component values.
28. Explain the sine wave output and damped oscillation of an *LC* tank circuit.
29. Describe the operation of astable, monostable, and bistable multivibrator circuits.
30. Investigate the 555 timer integrated circuit and its applications.

POWER SUPPLY CIRCUITS

The energy source of an electronic power supply is alternating current. In most systems this is supplied by the ac power line. Most small electronic systems use 120-V, single-phase, 60-Hz ac as their energy source. This energy is readily available in homes, buildings, and industrial facilities.

All electronic power supplies have a number of functions that must be performed in order to operate. Some of these functions are achieved by all power supplies. Others are somewhat optional and are dependent primarily on the parts being supplied. A block diagram of a general-purpose *electronic power supply* is shown in Figure 4-1. Each block represents a specific power supply function.

Transformer

Electronic power supplies rarely operate today with ac obtained directly from the power line. A *transformer* is commonly used to step the line voltage up or down to a desired value. A schematic symbol and a simplification of a power supply transformer are shown in Figure 4-2.

The coil on the left of the symbol is called the *primary winding*. The ac *input* is applied to this winding. The coil on the right side is the *secondary winding*. The *output* of the transformer is developed by the secondary winding. The parallel lines near the center of the symbol are representative of the *core*. In a transformer of this type, the core is usually made of laminated soft steel. Both windings are placed on the same core. Power supply transformers may have more than one primary and secondary winding on the same core. This permits it to accommodate different line voltages and to develop alternate output values.

The *output voltage* of a power supply transformer may have the same polarity as the input voltage or it may be reversed. This depends on the winding direction of the coils. A dot is often used on the symbol to indicate the *polarity* of a winding. The dot is used to designate the beginning or the end of a winding. This designation allows consistency when connecting windings together.

The output voltage developed by a transformer is primarily dependent on the *turns ratio* of its windings. If the primary winding has 500 turns of wire and the secondary has 1000 turns, the transformer has a 1:2 turns ratio. This type of transformer will *step up* the input voltage by a factor of 2. With 120 V applied from the power line, the output voltage will be approximately 240 V. It is interesting to note that the current capability of a transformer

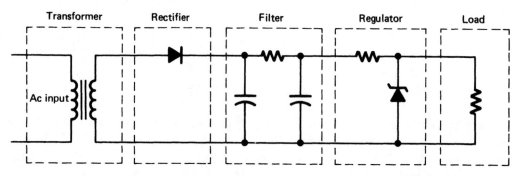

Figure 4-1. Functional block diagram of an electronic power supply.

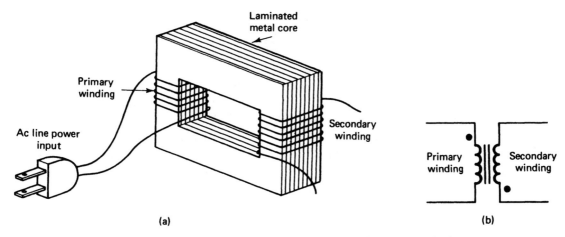

Figure 4-2. (a) Power supply transformer; (b) schematic symbol.

is the reverse of its turns ratio. A 1:2 turns ratio will step up the voltage and *step down* the current capability. One ampere of primary current is capable of producing only 0.5 A of secondary current. The resistance of the wire is largely responsible for this condition. A turns ratio of 1:2 will have twice as much resistance in the secondary as it has in the primary winding. An increase in resistance will, therefore, cause a decrease current. This assumes the wire size of the two windings is the same.

Power supplies for solid-state systems generally develop low voltage values. Transformers for this type of supply are designed to *step down* the line voltage. A turns ratio of 10:1 is very common. With 120 V applied to the input, the output will be approximately 12 V. The output current capabilities of this transformer are increased by a ratio of 1:10. This means that a step-down transformer has a low voltage output with a high current capacity. Supplies of this type are well suited for solid-state applications.

Rectification

The primary function of a power supply is to develop dc for the operation of electronic devices. Most power supplies are initially energized by ac. This energy must be changed into dc before it can be used. The process of changing ac into dc is called *rectification*. All electronic power supplies have a rectification function. Figure 4-3 shows a graphic representation of the rectification func-

tion. Notice that rectification can be either half-wave or full-wave. Half-wave rectification uses one alternation of the ac input, while full-wave uses both alternations.

Half-Wave Rectification

In half-wave rectification, only one alternation of the ac input appears in the output. The negative alternation of Figure 4-4(a) has been eliminated by the diode rectifier. The resulting output of this circuit is called *pulsating dc*. A half-wave rectifier can remove either the positive or negative alternation, depending on its polarity in the circuit. When the output has a positive alternation, the resulting dc is positive with respect to the common point or ground. A negative output occurs when the positive al-

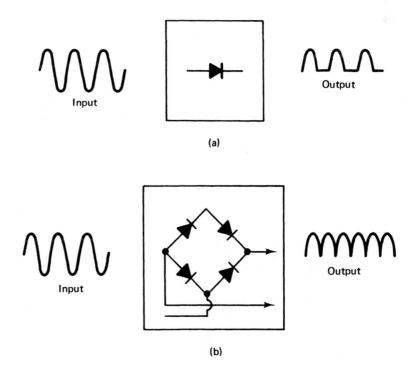

Figure 4-3. Rectifier: (a) half wave; (b) full wave.

ternation is removed. Figure 4-4(b) shows an example of a rectifier with a negative output.

The rectification function of a modern power supply is performed by *solid-state diodes.* This type of device has two electrodes, known as the *anode* and *cathode.* Operation is based on its ability to conduct easily in one direction but not in the other direction. *As* a result of this, only one alternation of the applied ac will appear in the output. This occurs only when the diode is forward-biased. The anode must be positive and the cathode negative. Reverse-biasing will not permit conduction through a diode.

A simplification of the rectification process is shown in Figure 4-5 Part (a) shows how the diode will respond during the *positive alternation.* This alternation causes the diode to be forward-biased. Note that the anode is positive and the cathode is negative. The resulting current flow is shown by arrows. The output of the rectifier appears across R_L. It is positive at the top of R_L and negative at the bottom.

Figure 4-6(b) shows how the diode will respond during the *negative alternation* This alternation causes the diode to be reverse-biased. The anode is now negative and the cathode is positive. No current will flow during this alternation. The output across R_L is zero for this alternation.

The dc output of a half-wave rectifier appears as a series of pulses across R_L. A dc voltmeter or ammeter would therefore indicate the average value of these pulses over a period of time. Figure 4-6(a) shows an example of the *pulsating dc output.* The average value of one pulse is 0.637 of the peak value. Since the second alternation of the output is zero, the composite average value must take into account the time of both alternations. The composite output is therefore 0.637 ÷ 2, or 0.318, of the peak value. Figure 4-6(b) shows a graphic example of this value.

The equivalent dc output of a half-wave rectifier without filtering can be calculated when the value of the ac input is known. The *peak value* of one ac alternation is determined by first multiplying the RMS value by 1.414. The composite dc output is then determined by multiplying the peak ac value by 0.318.

A simplified method of calculating the dc output of a half-wave rectifier is to combine different values, such as 1.414 × 0.637 ÷ 2 = 0.45. The composite dc output of a half-wave rectifier is therefore 45%, or 0.45, of the RMS input voltage. In an actual circuit we must also take into account the voltage drop across the diode. When a silicon diode is used, this value is 0.6 V. The dc output of an unfiltered half-wave rectifier is, therefore, 0.45 × RMS – 0.6 V. This procedure can be used to determine equivalent dc voltage of the half-wave rectifier.

The frequencies of the output pulses of a half-wave rectifier occur at the same rate as the applied ac input. With 60-Hz input, the pulse rate, or *ripple frequency,* of a half-wave rectifier is 60 Hz. Only one pulse appears in the output for each complete sine-wave input. Potentially, this means that only 45% of the ac input is transformed into a usable dc output. Half-wave rectification is therefore only 45% efficient. Due to its low efficiency rating, half-wave rectification is not widely used today.

Full-Wave Rectification

A full-wave rectifier responds as two half-wave rectifiers that conduct on opposite alternations of the input. Both alternations of the input are changed into pulsating dc

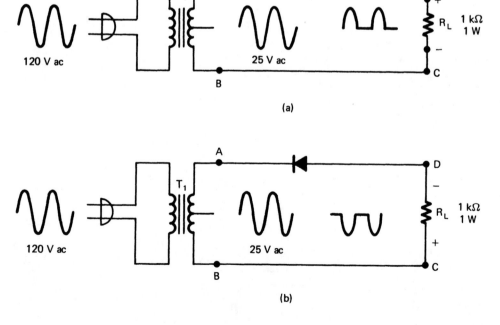

(a)

(b)

Figure 4-4. Half-wave power supplies: (a) positive output; (b) negative output.

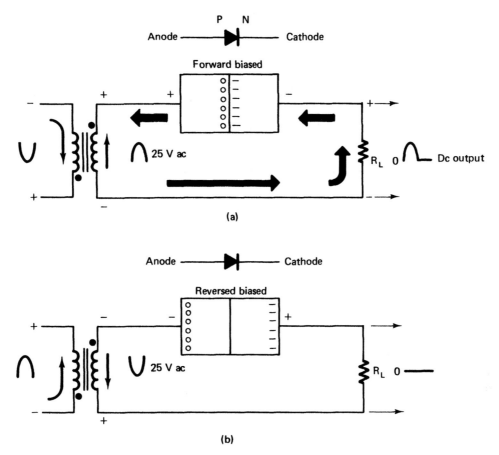

P N

Anode ————▷|———— Cathode

Forward biased

25 V ac

R_L 0 Dc output

(a)

Anode ————▶|———— Cathode

Reversed biased

25 V ac

R_L 0

(b)

Figure 4-5. Simplification of rectification: (a) positive alternation; (b) negative alternation.

output. Full-wave rectification can be achieved by two diodes and a *center-tapped transformer* or by four diodes in a *bridge circuit*. Both types of rectifiers are widely used in solid-state power supplied today.

A schematic diagram of a two-diode, full-wave rectifier is shown in Figure 4-7 Note that the anodes of each diode are connected to opposite ends of the transformer's secondary winding. The cathodes of the diodes are then connected together to form a common positive output. The *load* of the power supply (R_L) is connected between the common cathode point and the center-tap connection of the transformer. A complete path for current is formed by the transformer, two diodes, and the load resistor.

When ac is applied to the primary winding of the transformer, it steps the voltage down in the secondary winding. The *center tap* (CT) serves as the electrical neutral, or center, of the secondary winding. Half of the secondary voltage will appear between points CT and A and the other half, between CT and B. These two voltage values are equal and will always be 180° out of phase with respect to point CT.

Assume now that the 60-Hz input is slowed down in order to show how one complete cycle responds. In Figure 4-8 note the polarity of the secondary winding

voltage. For the first alternation, point A is positive and point B is negative with respect to the center tap. The second alternation causes point A to be negative and point B to be positive with respect to CT The center tap is always negative with respect to the positive end of the winding. This point then serves as the *negative output* of the power supply.

Conduction of a specific diode in a full-wave rectifier is based on the polarity of the applied voltage during each alternation. In Figure 4-8 this is made with respect to point CT. For the first alternation, point A is positive and B is negative. This polarity causes D_1 to be forward-biased and D_2 to be reverse-biased. Current flow for this alternation is shown by the solid arrows. Starting at CT, electrons flow through a conductor to R_L and D_1 and return to point A. This current flow causes a pulse of dc to appear across R_L for the first alternation.

For the second alternation, point A becomes negative, and point B is positive. This polarity forward-biases D_2 and reverse-biases D_1. Current flow for this alternation is shown by the dashed arrows. Starting at point CT, electrons flow through a conductor to R_L, flow through D_2, and return to point B. This causes a pulse of dc to appear across R_L for the negative alternation.

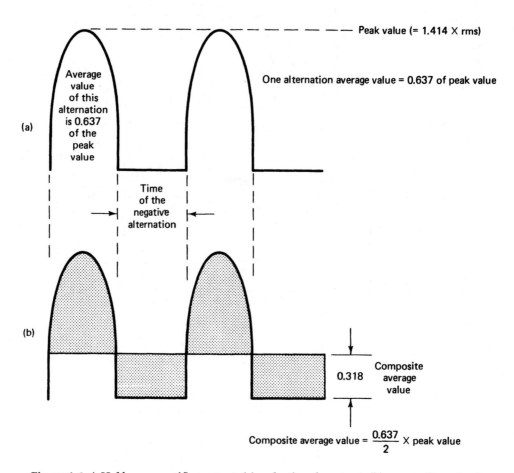

Figure 4-6. A Half-wave rectifier output: (a) pulsating dc output; (b) composite output.

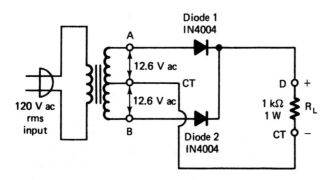

Figure 4-7. Two-diode full-wave rectifier.

It should be obvious at this point that current flow through R_L is in the same direction for each alternation of the input. This means that each alternation is transposed into a pulsating dc output. Full-wave rectification therefore changes the entire ac input into dc output.

The resulting dc output of a full-wave rectifier is 90%, or 0.90, of the ac voltage between the center tap and the outer ends of the transformer. This voltage value is determined by calculating the peak value of the RMS input voltage, then multiplying it by the average value. Since 1.414 × 0.637 equals 0.90, 90% of the RMS value is

the equivalent dc output. The output of a full-wave rectifier is 50% more efficient than that of an equivalent half-wave rectifier.

The actual dc output of a full-wave rectifier will be slightly less than 0.90 of the RMS input just described. Each diode, for example, reduces the dc voltage by 0.6 V. A more practical calculation is dc output = RMS × 0.90 − 0.6 V. In a low-voltage power supply, a reduction of the dc by 0.6 V may be quite significant.

The *ripple frequency* of a full-wave rectifier is somewhat different from that of a half-wave rectifier. Each alternation of the input, for example, produces a pulse of output current. Therefore, the ripple frequency is twice the alternation frequency. For a 60-Hz input the ripple frequency of a full-wave rectifier is 120 Hz. As a general rule, a 120-Hz ripple frequency is easier to filter than the 60-Hz frequency of a half-wave rectifier.

Full-Wave Bridge Rectifiers

A bridge rectifier requires four diodes to achieve full-wave rectification. In this configuration, two diodes will conduct during the positive alternation and two will conduct during the negative alternation. A center-tapped

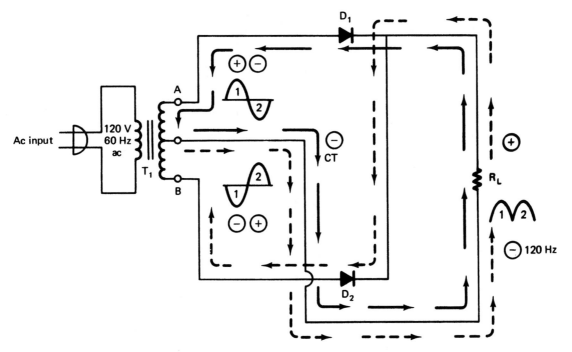

Figure 4-8. Conduction of a full-wave rectifier.

transformer is not required to achieve full-wave rectification with a bridge circuit.

The component parts of a full-wave bridge rectifier are shown in Figure 4-9. In this circuit, one side of the ac input is applied to the junction of diodes D_1 and D_2. The alternate side of the input is applied to diodes D_3 and D_4. The diodes at each input are reversed with respect to each other. Output of the bridge occurs at the other two junctions. The cathodes of D_1 and D_3 serve as the positive output. The *load* of the power supply is connected across the common anode and common cathode connection points.

When ac is applied to the primary winding of a power supply transformer, it can be stepped either up or down, depending on the desired dc output. In Figure 4-10 the 120-V input is stepped down to 25 V RMS. In normal operation, one alternation will cause the top of the transformer to be positive and the bottom to be negative. The next alternation will cause the bottom to be positive and the top to be negative. Opposite ends of the secondary winding will always be 180° out of phase with each other.

Assume now that the ac input causes point A to be positive and B to be negative for one alternation. The schematic diagram of a bridge in Figure 4-10 shows this condition of operation. With the indicated polarity, diode D_1 is forward-biased and D_2 is reverse-biased at the top junction.

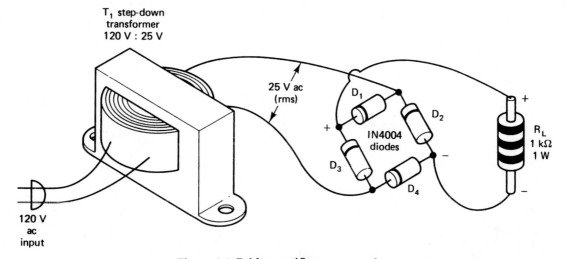

Figure 4-9. Bridge rectifier components.

At the bottom junction D_4 is forward-biased and D_3 is reverse-biased. When this occurs, electrons will flow from point B through D_4, up through R_L, and through D_1 and then return to point A. Solid arrows are used to show this current path. The load resistor sees one pulse of dc across it for this alternation.

With the next alternation, point A of the schematic diagram becomes negative, and point B becomes positive. When this occurs, the bottom diode junction becomes positive and the top junction goes negative. This condition forward-biases diodes D_2 and D_3 while reverse-biasing D_1 and D_4. The resulting current flow is indicated by the dashed arrows starting at point A. Electrons flow from A through D_2, up through R_L, and through D_3 and return to the transformer at point B. Dashed arrows are used to show this current path. The load resistor also sees a pulse of dc across it for this alternation.

The *output* of the bridge rectifier has current flow through R_L in the same direction for each alternation of the input. Ac is therefore changed into a pulsating dc output by conduction through a bridge network of diodes. The dc output, in this case, has a *ripple frequency* of 120 Hz. Each alternation produces a resulting output pulse. With 60 Hz applied, the output is 2 × 60, or 120, Hz.

The dc output voltage of a bridge circuit is slightly less than 90% of the RMS input. Each diode, for example, reduces the output by 0.6 V. With two diodes conducting during each alternation, the dc output voltage is reduced by 0.6 × 2, or 1.2 V. In the circuit of Figure 4-10, the dc output is 25 V × 0.9 − 1.2 V, or 21.3 V. The output of a bridge circuit is slightly less than that of an equivalent two-diode

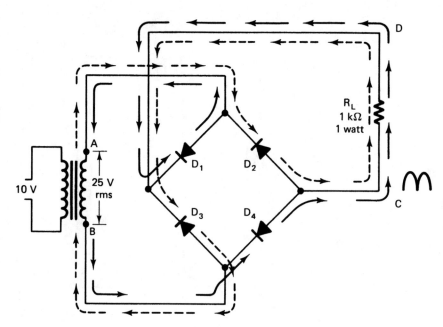

Figure 4-10. Bridge rectifier conduction.

full-wave rectifier. The voltage drop of the second diode accounts for this difference.

The four diodes of a bridge rectifier can be obtained in a single package. Figure 4-11 shows several different package types. As a general rule, there are two ac input terminals and two dc output terminals. These devices are generally rated according to their current-handling capability and peak reverse voltage rating. Typical current ratings are from 0.5 to 50 A in single-phase ac units. PRV ratings are 50, 200, 400, 800 and 1000 V. A bridge package usually takes less space than four single diodes. Nearly all bridge power supplies today employ the single-package assembly.

Dual Power Supplies

A new variation of the power supply is now being used in some systems. This supply has both nega-

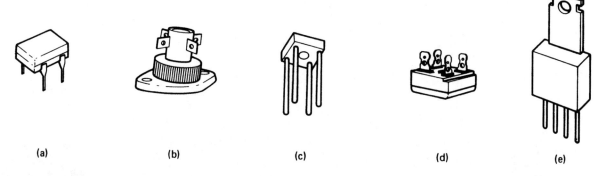

(a)   (b)   (c)   (d)   (e)

Figure 4-11. Packaged bridge rectifier assemblies: (a) dual-in-line package; (b) sink-mount package; (c) epoxy package; (d) tab-pack enclosure; (e) thermotab package.

tive and positive output with respect to ground. *Dual, or split, power supplies,* as they are known, have been developed as a voltage source for integrated circuits. The secondary winding of the input transformer is divided into two parts. The center tap, or neutral, serves as a common connection for the two outside windings. Each half of the winding has a complete full-wave rectifier.

A *dual power supply* with two full-wave rectifiers is shown in Figure 4-12. Notice that the output of this supply is + 10.7 and – 10.7 V.

Diodes D_1 and D_2 are rectifiers for the positive supply. They are connected to opposite ends of transformer T_1. Diodes D_3 and D_4 are rectifiers for the negative supply. They are connected in a reverse direction to the opposite ends of the transformer. The positive and negative outputs are with respect to the center tap of the transformer.

For one alternation assume that the top of the transformer is positive and the bottom is negative. Current flows out of point B through D_4, R_{L2}, R_{L1}, D_1, and returns to terminal A of the transformer. The bottom of R_{L2} becomes negative and the top of R_{L1} becomes positive. This current conduction path is made complete by forward biasing diodes D_4 and D_1. Diodes D_2 and D_3 are reverse biased by this transformer voltage. Solid arrows show the path of current flow for this alternation.

The next alternation makes the top of the transformer negative and the bottom positive. Current flows out of point A through diode D_3, R_{L2}, R_{L1}, D_2, and returns to terminal B of the transformer. The top of R_{L1} continues to be positive and the bottom of R_{L2} is negative. This alternation forward biases D_3 and D_2 while reverse biasing D_4 and D_1. The direction of current flow through R_{L2} and R_{L1} is the same as the first alternation. The resulting output voltage of the supply appears at the top and bottom of R_{L2} and R_{L1}. Dashed arrows show the direction of current flow for the second alternation.

Filtering

The output of a half- or full-wave rectifier is pulsating direct current. This type of output is generally not usable for most electronic circuits. A pure form of dc is usually required. The *filter* section of a power supply is designed to change pulsating dc into a pure form of dc. Filtering takes place between the output of the rectifier and the input to the load device. Power supplies discussed up to this point have not employed a filter circuit.

The pulsating dc output of a rectifier contains two components. One of these deals with the dc part of the output. This component is based on the combined average value of each pulse. The second part of the output refers to its ac component. Pulsating dc, for example, occurs at 60 Hz or 120 Hz, depending on the rectifier being employed. This part of the output has a definite ripple frequency. Ripple must be minimized before the output of a power supply can be used by most electronic devices.

Power supply filters fall into two general classes, ac-

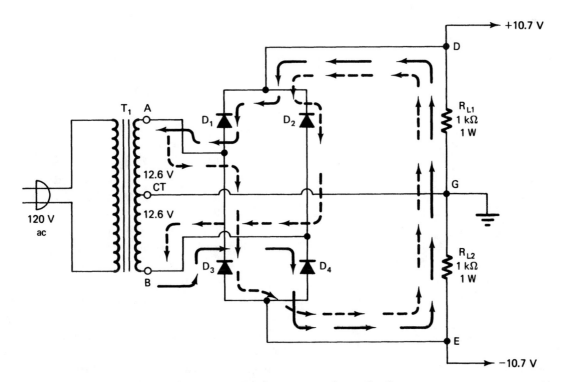

Figure 4-12. Dual power supply conduction.

cording to the types of components used in their input. If filtering is first achieved by a capacitor, it is classified as a *capacitor, or C-input, filter.* When a coil of wire or inductor is used as the first component, it is classified as an *inductive, or L-input filter.* C-input filters develop higher values of dc output voltage than do L filters. The output voltages of C filters usually drop in value when the load increases. L filters, by comparison, tend to keep the output voltage at a constant value. This is particularly important when large changes in the load occur. The output voltages of L filters are, however, somewhat lower than those of C filters. Figure 4-13 shows a graphic comparison of output voltage and load current for C and L filters.

C-input Filters

The ac component of a power supply can be effectively reduced by a *C-input* filter. A single capacitor is simply placed across the load resistor, as shown in Figure 4-14. For alternation 1 of the circuit (Figure 4-14(a)), the diode is forward-biased. Current flows according to the arrows of the diagram. C charges very quickly to the peak voltage value of the first pulse. At the same point in time, current is also supplied to R_L. The initial surge of current through a diode is usually quite large. This current is used to charge C and supply R_L at the same time. A large capacitor responds somewhat like low resistance when it is first being charged. Notice the amplitude of the I_D waveform during alternation 1.

When alternation 2 of the input occurs, the diode is reverse-biased. Figure 4-14(b) shows how the circuit responds for this alternation. Notice that there is no cur-

rent flow from the source through the diode. The charge acquired by C during the first alternation now finds an easy discharge path through R_L. The resulting discharge current flow is indicated by the arrows between C and R_L. In effect, R_L is now being supplied current even when the diode is not conducting. The voltage across R_L is therefore maintained at a much higher value. See the V_{RL} waveform for alternation 2 in Figure 4-14(c).

Discharge of C continues for the full time of alternation 2. Near the end of the alternation there is somewhat of a drop in the value of V_{RL}. This is due primarily to a depletion of capacitor charge current. At the end of this time, the next positive alternation occurs. The diode is again forward biased. The capacitor and R_L both receive current from the source at this time. With C still partially charged from the first alternation, less diode current is needed to recharge C. In Figure 4-14(c), note the amplitude change in the second I_D pulse. The process from this point on repeats alternations 1 and 2.

The effectiveness of a capacitor as a filter device is based on a number of factors. Three very important considerations are

1.   The size of the capacitor being used
2.   The value of the load resistor R_L
3.   The time duration of a given dc pulse

These three factors are related to one another by the formula

$$T = R \times C$$

where T is the time in seconds, R the resistance in ohms, and C the capacitance in farads. The product RC is an expression of the filter circuit's *time constant.* RC is a measure of how rapidly the voltage and current of the filter respond to changes in input voltage. A capacitor will charge to 63.2% of the applied voltage in one time constant. A discharging capacitor will live a 63.2% drop in its original value in one time constant. It takes five time constants to charge or discharge a capacitor fully.

The *filter capacitor* of Figure 4-14 charges very quickly during the first positive alternation. Essentially, there is very little resistance for the RC time constant during this period. The discharge of C is, however, through R_L. If R_L is small, C will discharge very quickly. A large value of R_L will cause C to discharge rather slowly. For good filtering action, C must not discharge very rapidly during the time of one alternation. When this occurs, there is very little change in the value of V_{RL}. A C-input filter works very well when the value of R_L is relatively large. If the value

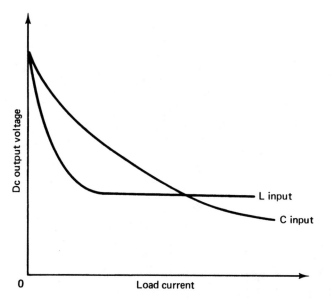

Figure 4-13. Load current versus output voltage comparison for L and C filters.

of R_L is small, as in a heavy load, more ripple will appear in the output.

A rather interesting comparison of filtering occurs between half-wave and full-wave rectifier power supplies. The time between reoccurring peaks is twice as long in a half-wave circuit as it is for full-wave. The capacitor of a half-wave circuit therefore has more time to discharge through R_L. The ripple of a half-wave filter will be much greater than that of a full-wave circuit. In general, it is easier to filter the output of a full-wave rectifier. Figure 4-15 compares half- and full-wave rectifier filtering with a single capacitor.

The *dc output voltage* of a filtered power supply is usually a great deal higher than that of an unfiltered power supply. In Figure 4-16, the waveforms show an obvious difference between outputs. In the filtered outputs, the capacitor charges to the peak value of the rectified output. The amount of discharge action that takes

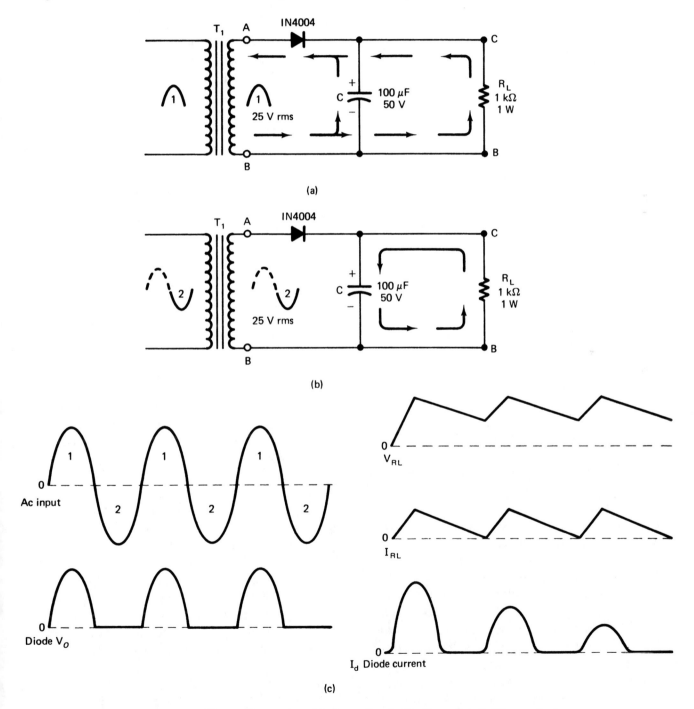

Figure 4- 14. C-input filter action: (a) alternation 1; (b) alternation 2; (C) waveforms.

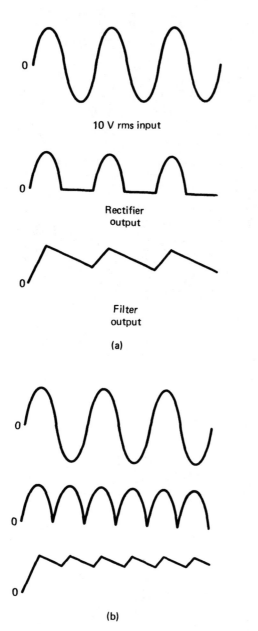

10 V rms input

Rectifier
output

Filter
output

(a)

(b)

Figure 4-15. (a) Half-wave and (b) full-wave filter comparisons.

place is based on the resistance of R_L. For a light load or high resistance R_L, the filtered output remains charged to the peak value. The peak value of the 10-V RMS is 14.14 V. For the unfiltered full-wave rectifier, the dc output is approximately 0.9 × RMS, or 9.0 V. Comparing 9 V to 14 V shows a rather decided difference in the two outputs. For the half-wave rectifier, the output difference is even greater. The unfiltered half-wave output is approximately 45% of 10 V RMS, or 4.5 V. The filtered output is 14 V less 20% for the added ripple, or approximately 11.2 V. The filtered output of a half-wave rectifier is nearly 2.5 times as much as that of the unfiltered

output. These values are only rough generalizations of power supply output with a light load.

Inductance Filtering

An inductor is a device that has the ability to store and release electrical energy. It does this by taking some of the applied current and changing it into a magnetic field. An increase in current through an inductor causes the magnetic field to expand. A decrease in current causes the field to collapse and release its stored energy.

The ability of an inductor to store and release energy can be used to achieve filtering. An increase in the current passing through an inductor will cause a corresponding increase in its magnetic field. Voltage induced into the inductor due to a change in its field will oppose a change in the current passing through it. A decrease in current flow causes a similar reaction. A drop in current will cause its magnetic field to collapse. This action also induces voltage into the inductor. The induced voltage in this case causes a continuation of the current flow. As a result of action, the added current tends to bring up its decreasing value. An inductor opposes any change in its current flow. Inductive filtering is well suited for power supplies with a large load current.

An inductive input filter is shown in Figure 4-16. The inductor is simply placed in series with the rectifier and the load. All the current being supplied to the load must pass through the inductor (L). The filtering action of L does not let a pronounced change in current take place. This prevents the output voltage from reaching an extreme peak or valley. In general, inductive filtering does not produce as high an output voltage as does capacitive filtering. Inductors tend to maintain the current at an average value. A larger load current can be drawn from an inductive filter without causing a decrease in output voltage.

When an inductor is used as the primary filtering element, it is commonly called a *choke*. The term choke refers to the ability of an inductor to reduce ripple voltage and current. In electronic power supplies, choke filters are rarely used as a single filtering element. A combination inductor-capacitor, or *LC*, filter is more widely used. This filter has a series inductor with a capacitor connected in parallel with the load. The inductor controls large changes in load current. The capacitor, which follows the inductor, is used to maintain the load voltage at a constant voltage value. The combined filtering action of the inductor and capacitor produces a rather pure value of load voltage.

Figure 4-17 shows a representative *LC* filter and its output.

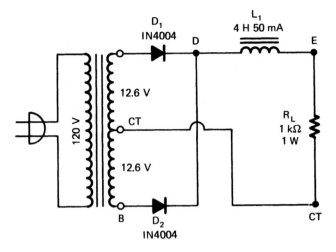

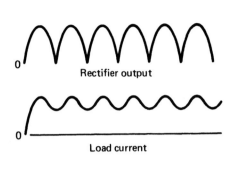

Figure 4-16. Inductive filtering.

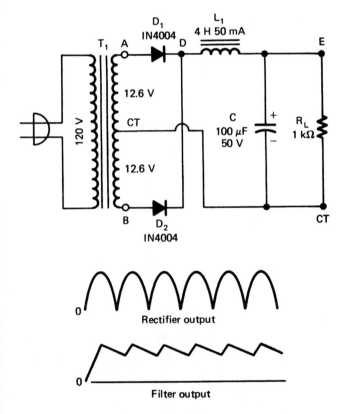

Figure 4-17. *LC* filter.

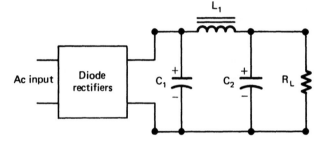

Figure 4-18. *CLC* (pi) filter.

PI Filters

When a capacitor is placed in front of the inductor of an *LC* filter it is called a *pi filter*. In effect, this circuit becomes a *CLC* filter. The two capacitors are in parallel with R_L and the inductor is in series. Component placement of this circuit in a schematic diagram resembles the capital Greek letter pi (π), which accounts for the name of the filter. See the schematic diagram of a pi filter in Figure 4-18.

The operation of a pi filter can best be understood by considering L_1 and C_2 as an *LC* filter. This part of the circuit acts upon the output voltage developed by the capacitor-input filter C_1. C_1 charges to the peak value of the rectifier input. It has a ripple content that is very similar to that of C-input filter of Figure 4-14. This voltage is then applied to C_2 through inductor L_1. C_1 charges C_2 through L_1, and C_2 then holds its charge for a time interval determined by the time constant of $C_2 - R_L$. As a result of this action, there is additional filtering by L_1 and C_2. The ripple content of this filter is much lower than that of a single C-input filter. There is, however, a slight reduction in the dc supply voltage to R_L due to the voltage drop across L_1.

A pi filter has a very low ripple content when used with a light load. An increase in load, however, tends to lower its output voltage. This condition tends to limit the number of applications of pi filters. Today we find them used to supply radio circuits, in stereo amplifiers, and in TV receivers. The input to this filter is nearly always supplied by a full-wave rectifier.

RC Filters

In applications where less filtering can be tolerated, an *RC* filter can be used in place of a pi filter. As shown in Figure 4-19, the inductor is replaced with a resistor. An inductor is rather expensive, quite large physically, and weighs a great deal more than a resistor. The performance of the *RC* filter is not quite as good as that of a pi filter. There is usually a reduction in dc output voltage and increased ripple.

In operating, C_1 charges to the peak value of the rectifier input. A drop in rectifier input voltage causes C_1 to discharge through R_1 and R_L. The voltage drop across R_1 lowers the dc output to some extent. C_2 charges to the peak value of the R_L voltage. The dc output of the filter is dependent on the load current. High values of load current cause more voltage drop across R_1. This in turn lowers the dc output. Low values of load current have less voltage drop across R_1. The output voltage therefore increases with a light load. In practice, *RC* filters are used in power supplies that have 100 mA or less of load current. The primary advantage of an *RC* filter is reduced cost.

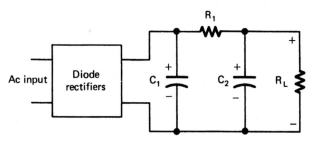

Figure 4-19. *RC* filter.

Voltage Regulation

The dc output of an unregulated power supply has a tendency to change value under normal operating conditions. Changes in ac input voltage and variations in the load are primarily responsible for these fluctuations. In some power supply applications, voltage changes do not represent a serious problem. In many electronic circuits, voltage changes may cause improper operation. When a stable dc voltage is required, power supplies must employ a voltage regulator. The block diagram of a power supply in Figure 4-20, shows where the *regulator* is located.

A number of voltage regulator circuits have been developed for use in power supplies. One very common method of regulation employs the *zener diode*. Figure 4-21 shows this type of regulator located between the filter and the load. The zener diode is connected in parallel or shunt with R_L. This regulator requires only a zener diode D_Z and a series resistor (R_S). Notice that D_Z is placed across the filter circuit in the reverse bias direction. Connected in this way, the diode will go into conduction only when it reaches the zener breakdown voltage V_Z. This voltage then remains constant for a large range of zener current I_Z. Regulation is achieved by altering the conduction of I_Z and load current I_L must all pass through the series resistor. This current value then determines the amount of voltage drop across R_S. Variations in current through R_S are used to keep the output voltage at a constant value.

A schematic diagram of a 9-V *regulated power supply* is shown in Figure 4-22. This circuit derives its input voltage from a 12.6-V transformer. *Rectification* is achieved by a self-contained bridge rectifier assembly. *Filtering* is accomplished by an *RC* filter. The series resistor of this circuit has two functions. It first couples capacitors C_1 and C_2 together in the filter circuit. Second, it serves as the series resistor for the regulator circuit. Diode D_Z is a 9-V, 1-W zener diode.

Operation of the regulated power supply is similar to that of the bridge circuit discussed earlier. Full-wave output from the rectifier is applied to C_1 of the filter circuit. C_1 then charges to the peak value of the RMS input less the voltage drop across two silicon diodes (1.2 V). This represents a value of 12.6 × 1.414 − 1.2, or 16.6 V dc input. R_S therefore has a voltage drop of 16.6 − 9 V, or 7.6 V. This represents a total current flow passing through R_S of 7.6 V / 100Ω = 0.076 A, or 76 mA. With the bleeder resistor R_S serving as a fixed load, there is 9 V / 10 kΩ = 0.9 mA, or 0.0009 A, of load current. The difference in I_{RS} and I_L is therefore 0.076 A − 0.0009 A = 0.0751 A, or 75.1 mA. This current must all pass through the zener diode when the circuit is in operation. Nine volts dc will then appear at the two output terminals of the power supply. This represents the *no-load (NL) condition* of operation.

When the power supply is connected to an external load, more current is required. Ideally, this current should

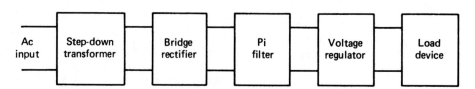

Figure 4-20. Zener diode regulator.

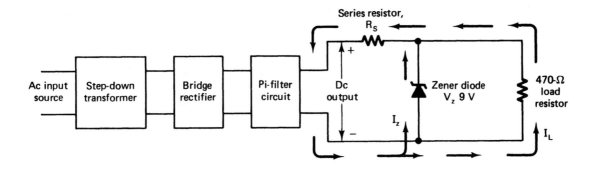

Figure 4-21. Zener diode regulator.

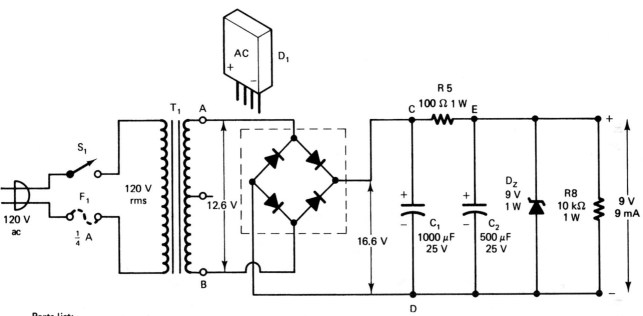

Parts list:

T_1 : 12.6-V 2-A Stancor P-8130, or Triad F-44X

D_1 : silicon bridge rectifier 2 A 50 V; General Instrument Co., KBF-005 or Motorola MDA-200

C_1 : 1000-μF 25-V electrolytic capacitor

C_2 : 500-μF 25-V electrolytic capacitor

R_5 : 100-Ω 1-W resistor

R_8 : 10-kΩ $\frac{1}{2}$-W resistor

D_Z : 9-V 1-W zener diode IN5346A or HEP Z-2513

S_1 : SPST toggle switch

F_1 : Fuse, $\frac{1}{4}$ A 250 V

Misc. : metal chassis, PC board, line cord, fuse holder, grommets, solder, cabinet

Figure 4-22. 9-V regulated power supply.

be available with 9 V of output. Assume now that the power supply is connected to a 270-Ω external load. The total resistance that the power supply sees at its output is 10 kΩ in parallel with 270Ω. This represents a load resistance of 263Ω. With 9 V applied, the total load current is 9 V/263 = 0.0342, or 34.2 mA. The total output current

available for the power supply is the current that passes through R_S. This was calculated to be 0.076 A, or 76 mA. With the 270–Ω load, I_L will be 34.2 mA with an I_Z of 41.8 mA. The output voltage therefore remains at 9 V with the increase in load.

With a slightly smaller external load, the power

supply goes out of its regulation range. Should this occur, no I_Z flows, and the value of I_L alone determines the current passing through R_5. An I_{R5} value in excess of 76 mA causes a greater voltage drop across R_S. This, in effect, causes the output voltage to be less than 9 V. All voltage regulator circuits of this type have a maximum load limitation. Exceeding this limit causes the output voltage to go out of regulation.

If the resistance of the external load is increased in value, it causes a reduction in load current. This condition causes a reverse in the operation of the regulator. With a larger R_L, there is less I_L. As a result of this, the zener diode must conduct more heavily. The maximum current-handling rating of D_Z determines this condition. An infinite load would, for example, demand no I_L D_Z would therefore conduct the full current passing through R_5. This value was calculated to be 76 mA for our circuit. For a 1-W, 9-V zener diode, the maximum current rating is 1 W/9 V = 0.111A, or 111 mA. In this case, the diode is capable of handling the maximum possible I_L that occurs for an infinite load. In the actual circuit the bleeder resistor R_S demands 0.9 mA of current when the load is infinite. The current through I_D is, therefore, not in excess of 75.1 mA, which is well below its maximum rating. In effect, the regulating range of this power supply is from an infinite value to approximately 120 Ω. The output voltage remains at 9 V over this entire range of load values.

A regulator must also be responsive to changes in input voltage. If the input voltage, for example, were to increase by 10%, it would cause the supply to receive 13.86 V instead of 12.6 V. The peak value of 13.86 V would then be 19.6 V – 1.2 V, or 18.4 V, instead of 16.6 V. The voltage drop across R_S would therefore increase from 7.6 V to 9.4 V. This, in turn, would cause an increase in total current through R_S of 94 mA. The current flow of the zener diode would increase to 94.0 – 0.9 = 93.1 mA with no external load. Since the zener is capable of handling up to 111 mA, it could respond to this change to maintain the output voltage at 9 V.

The response of a power supply to a decrease in input voltage must also be taken into account. A 10% decrease in input voltage would cause the supply to receive only 11.34 V instead of 12.6 V. The peak value of 11.34 V is 16 V. Capacitor C_1 would then charge to 16.0 – 1.2 V, or 14.8 V, where 1.2 V is the voltage drop across two silicon diodes in the bridge. The voltage drop across R_5 is now 5.8 V. With this reduced voltage, the total current drops to 5.8 V/100 = 0.058 A, or 58 mA. This value can certainly be handled by the zener diode. The maximum low-resistance value of the load increases somewhat. Under this condition the load may drop to only

160 before going out of regulation.

AMPLIFICATION PRINCIPLES

Amplification is achieved by an electronic device and its associated components. As a general rule, the amplifying device is placed in a circuit. The components of the circuit usually have about as much influence on amplification as the device itself. The circuit and the device must be supplied electrical energy for it to function. Typically, *amplifiers* are energized by direct current. This may be derived from a battery or a rectified ac power supply. The amplifier then processes a signal of some type when it is placed in operation. The signal may be either ac or dc, depending on the application.

REPRODUCTION AND AMPLIFICATION

The signal to be processed by an amplifier is first applied to the *input* part of the circuit. After being processed, the signal appears in the *output* circuitry. The output signal may be reproduced in its exact form, amplified, or both amplified and reproduced. The value and type of input signal, operating source energy, device characteristics, and circuit components all have some influence on the output signal.

Figure 4-23, illustrates the process of reproduction, amplification, and the combined amplification-reproduction functions of an amplifier. In part (a), the amplifier performs only the reproduction function. Note that the input and output signals have the same size and shape. Part (b) shows only the amplification function. In this case, the input signal is increased in amplitude. The output signal is amplified but does not necessarily resemble the input signal. In many applications amplification and reproduction must both be achieved at the same time. Part (c) shows this function of an amplifier. Depending on the application, an amplifier must be capable of developing any of these three output signals.

BIPOLAR TRANSISTOR AMPLIFIERS

The primary function of an amplifier is to prove *voltage gain, current gain,* or *power gain.* The resulting gain depends a great deal on the device being used and its circuit components. Normally, the circuit is energized by dc voltage. An ac signal is then applied to the input of the amplifier. After passing through the transistor, an ampli-

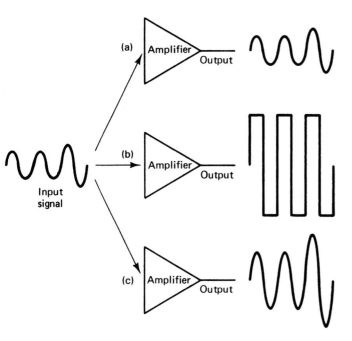

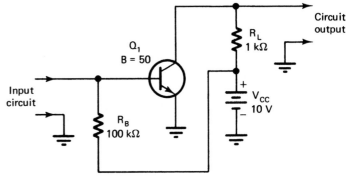

Figure 4-24. Basic bipolar amplifier.

Figure 4-23. Amplification and reproduction: (a) reproduction; (b) amplification; (c) combined amplification and reproduction.

fied version of the signal appears in the output. Operating conditions of the device and the circuit determine the level of amplification to be achieved.

In order for a bipolar transistor to respond as an amplifier, the emitter-base junction must be *forward-biased* and the collector-base junction, reverse-biased. Specific operating voltage values must then be selected that will permit amplification. If both reproduction and amplification are to be achieved, the device must operate in the center of its active region. Remember that this is between the saturation and cutoff regions of the collector family of characteristic curves. Proper selection of circuit components permits a transistor to operate in this characteristic region.

BASIC AMPLIFIERS

Suppose now that we look at the circuitry of a basic amplifier. Figure 4-24 shows a schematic diagram of this type of circuit. Notice that a schematic symbol of the NPN transistor is used in this diagram. Schematic symbols are normally used in all circuit diagrams. Keep in mind the crystal structure of the device represented by the symbol.

The basic amplifier has a number of parts that are needed to make it operational. V_{CC} is the dc energy source. The negative side of V_{CC} is connected to ground. The emitter is also connected to ground. This type of cir-

cuit configuration is called a grounded-emitter, or *common emitter, amplifier*. The emitter, one side of the input, and one side of the output are all commonly connected together. In an actual circuit these points are usually connected to ground.

Resistor R_B of the basic amplifier is connected to the positive side of V_{CC}. This connection makes the base positive with respect to the emitter. The emitter-base junction is forward-biased by R_B. Resistor R_L connects the positive side of V_{CC} to the collector. This connection reverse-biases the collector. Through the connection of R_B and R_L, the transistor is properly biased for operation. A transistor connected in this manner is considered to be in its *static, or dc operating, state*. It has the necessary dc energy applied for it to be operational. No signal is applied to the input for amplification.

Let us now consider the operation of the basic amplifier in its static state. With the given values of Figure 4-24 base current (I_B) can be calculated. In this case, I_B is limited by the value R_B and the emitter-base junction resistance. When the emitter-base junction is forward-biased, its resistance becomes very small. The value of R_B is therefore the primary limiting factor of I_B. I_B can be determined by an application of Ohm's law using the equation

$$\text{Base current} = \frac{\text{source voltage}}{\text{base resistance}} \quad \text{or}$$

$$I_B = \frac{V_{CC}}{R_B}$$

For our basic transistor amplifier, this is determined to be

$$I_B = \frac{V_{CC}}{R_B} = \frac{10 \text{ V}}{100,000 \text{ }\Omega} \quad \text{or} \quad \frac{10 \text{ V}}{100 \times 10^3 \text{ }\Omega}$$

$$= 100 \times 10^{-6} \text{ A, or } 100 \text{ }\mu\text{A}$$

The value of base current, in this case, is extremely small. Remember that only a small amount of I_B is needed to produce I_C.

The *beta* of the transistor used in Figure 4-25 has a given value. With beta and the calculated value of I_B, it is possible to determine the collector current of the circuit. Beta is the current gain of a common-emitter amplifier. Beta is described by the formula

$$\text{Beta} = \frac{\text{collector current}}{\text{base current}} \quad \text{or} \quad \beta = \frac{I_C}{I_B}$$

In our transistor amplifier, I_B has been determined and beta has a given value of 50. By transposing the beta formula, I_C can be determined by the equation

Collector current = beta × base current

or

$$I_C = \beta \times I_B$$

For the amplifier circuit, collector current is determined to be

$$
\begin{aligned}
I_C &= \beta \times I_B = 50 \times 100 \times 10^{-6} \\
&= 5 \times 10^{-3} \text{ A, or 5 mA}
\end{aligned}
$$

This means that 5 mA of I_C will flow through R_L when an I_B of 100 µA flows.

In any electric circuit, we know that current flow through a resistor will cause a corresponding voltage drop. In a basic transistor amplifier, I_C will cause a voltage drop across R_L. In the preceding step, I_C was calculated to be 5 mA. By using Ohm's law again, the voltage drop across R_L can be determined by the equation

R_L voltage = collector current × collector resistance

or

$$V_{RL} = I_C \times R_L$$

For the amplifier circuit, the voltage drop across R_L is

$$
\begin{aligned}
V_{RL} &= I_C \times R_L \\
&= 5 \times 10 \qquad\qquad \times 1 \times 10^3 \\
&= 5V
\end{aligned}
$$

This means that half or 5 V of V_{CC} will appear across R_L. With a V_{CC} value of 10 V, the other 5 V will appear across the collector-emitter of the transistor. A V_{CE} voltage of 5 V

from a V_{CC} of 10 V means that the transistor is operating near the center of its active region. Ideally, the transistor should respond as a *linear amplifier*. When a suitable signal is applied to the input, it should amplify and reproduce the signal in the output.

The operational voltage and current values of the basic transistor amplifier in its static state are summarized in Figure 4-25. Note that an I_B of 100 A causes an I_C of 5 mA. With a beta of 50, the amplifier has the capability of a rather substantial amount of output current. The collector current passing through R_L causes V_{CC} to be divided. This establishes operation in the approximate center of the collector family of characteristic curves. Ideally, this static condition should cause the amplifier to respond as a linear device when a signal is applied.

SIGNAL AMPLIFICATION

In order for the basic transistor circuit of Figure 4-26 to respond as an amplifier, it must have a *signal* applied. The signal may be either voltage or current. An applied signal causes the transistor to change from its *static* state to a *dynamic* condition. Dynamic conditions involve changing values. All ac amplifiers respond in the dynamic state. The output of this should develop an ac signal.

Figure 4-26 shows a basic transistor amplifier with an ac signal applied. A capacitor is used, in this case, to *couple* the ac signal source to the amplifier. Remember that ac passes easily through a capacitor while dc is blocked. As a result of this, the ac signal is injected into the base-emitter junction. Dc does not flow back into the signal source. The ac signal is therefore added to the dc operating voltage. The emitter-base voltage is a dc value that changes at an ac rate.

Consider now how the applied ac signal alters the emitter-base junction voltage. Figure 4-27 shows some

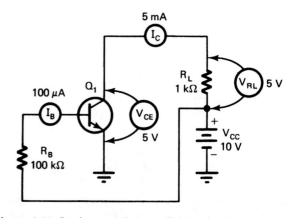

Figure 4-25. Static operating condition of a basic amplifier.

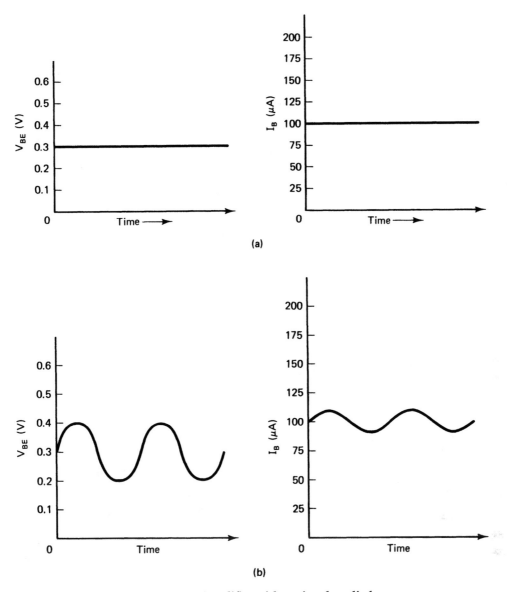

Figure 4-26. Amplifier with ac signal applied.

representative voltage and current values that appear at the emitter-base junction of the transistor. They can be measured with a voltmeter or observed with an oscilloscope. Part (a) shows the dc operating voltage and base current. This occurs when the amplifier is in its steady, or static, state. Part (b) shows the ac signal that is applied to the emitter-base junction. Note that the amplitude change of this signal is a very small value. This also shows how the resulting base current and voltage change with ac applied. The ac signal is essentially riding on the dc voltage and current.

Refer now to the schematic diagram of the basic amplifier in Figure 4-28. Note in particular the waveform inserts that appear in the diagram. These show how the current and voltage values respond when an ac signal is applied.

The ac signal applied to the *input* of our basic amplifier rises in value during the positive alternation and falls during the negative alternation. Initially, this causes an increase and decrease in the value of V_{BE}. This voltage has a dc level with an ac signal riding on it. The indicated I_B is developed as a result of this voltage. The changing value of I_B causes a corresponding change in I_C. Note that I_B and I_C both appear to be the same. There is however, a very noticeable difference in values. I_B is in microamperes, whereas I_C is in milliamperes. The resulting I_C passing through R_L *causes a corresponding voltage drop* (V_{RL}) across R_L. V_{CE} appearing across the transistor is the reverse of V_{RL}. These signals are both ac values riding upon a dc level. The output, V_O, is changed to an ac value. Capacitor C_2 blocks the dc component and passes only the ac signal.

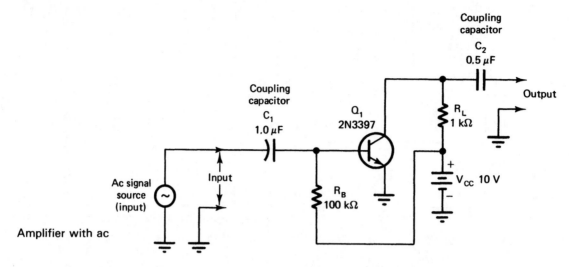

Amplifier with ac

Figure 4-27. I_B and V_{BE} conditions (a) static; (b) dynamic.

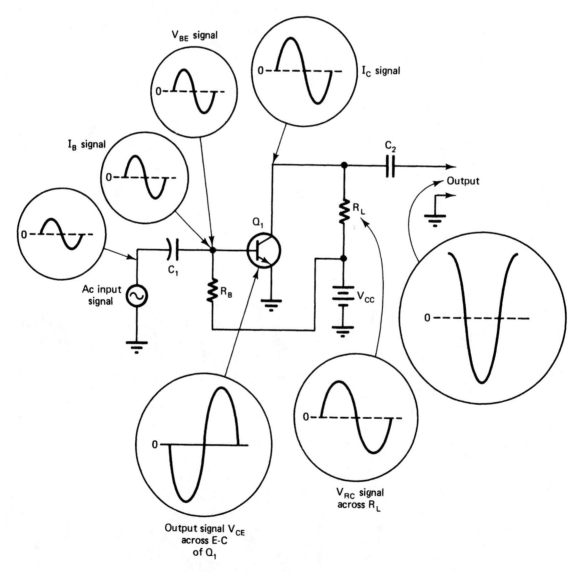

Figure 4-28. Ac amplifier operation.

It is interesting to note in Figure 4-28 that input and output signals of the amplifier are reversed. When the ac input signal rises in value, it causes the output to fall in value. The negative alternation of the input causes the output to rise in value. This condition of operation is called *phase inversion*. Phase inversion is a distinguishing characteristic of the common-emitter amplifier.

AMPLIFIER BIAS

If a transistor is to operate as a *linear amplifier*, it must be properly biased and have a suitable operating point. Its steady state of operation depends a great deal on base current, collector voltage, and collector current. Establishing a desired operating point requires proper selection of bias resistors and a load resistor to provide proper input current and collector voltage.

Stability of operation is a very important consideration. If the operating point of a transistor is permitted to shift with temperature changes, unwanted distortion may be introduced. In addition to distortion, operating point

changes may also cause damage to a transistor. Excessive collector current, for example, may cause the device to develop too much heat.

The method of *biasing* a transistor has a great deal to do with its thermal stability. Several different methods of achieving bias are in common use today. One method of biasing is considered to be *beta dependent*. Bias voltages are largely dependent on transistor beta. A problem with this method of biasing is transistor response. Biasing set up for one transistor is not necessarily the same for another transistor.

Biasing that is *independent* of beta is very important. This type of biasing responds to fixed voltage values. Beta does not alter these voltage values. As a general rule, this form of biasing is more reliable. The input and output of a transistor is very stable and the results are very predictable.

Beta-dependent Biasing

Four common methods of beta-dependent biasing are shown in Figure 4-29. The steady-state, or static, conditions of operation are a function of the transistor's beta.

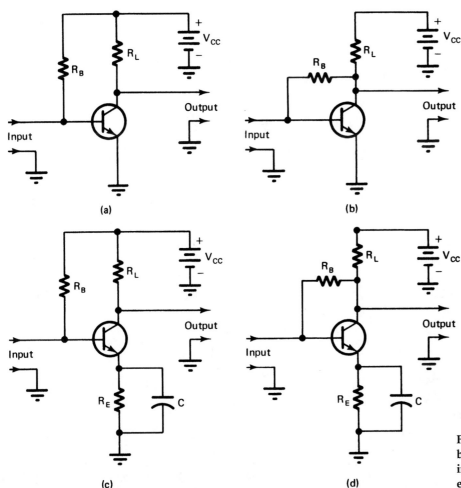

Figure 4-29. Methods of beta-dependent biasing: (a) fixed biasing; (b) self-biasing; (c) fixed-emitter biasing; (d) self-emitter biasing.

These methods are easy to achieve with a minimum of parts. They are not very widely used today.

Circuit (a) is a very simple form of biasing for a common-emitter amplifier. This method of biasing was used for our basic transistor amplifier. The base is simply connected to V_{CC} through R_B. This causes the emitter-base junction to be forward biased. The resulting collector current is beta times the value of I_B. The value of I_B is V_{CC}/R_B. As a general rule, biasing of this type is very sensitive to changes in temperature. The resulting output of the circuit is rather difficult to predict. This method of biasing is often called *fixed biasing*. It is primarily used because of its simplicity.

Circuit (b) of Figure 4-29 was developed to compensate for the temperature sensitivity of circuit (a). Bias current is used to counteract changes in temperature. In a sense, this method of biasing has negative feedback. R_B is connected to the collector rather than V_{CC}. The voltage available for base biasing is the leftover after voltage drop across the load resistor. If the temperature rises, it causes an increase in I_B and I_C. With more I_C, there is more voltage drop across R_L. Reduced V_C voltage will cause a corresponding drop in I_B. This, in turn, brings I_C back to normal. The opposite reaction will occur when transistor temperature becomes less. This method of biasing is called *self-biasing*.

Circuit (c) is an example of *emitter biasing*. Thermal stability is improved with this type of construction. I_B is again determined by the value of R_B and V_{CC}. An additional resistor is placed in series with the emitter of this circuit. Emitter current passing through R_E produces emitter voltage (V_E). This voltage opposes the base voltage developed by R_B. Proper values of R_B and R_E are selected so that I_B and I_E will flow under ordinary operating conditions. If a change in temperature should occur, V_E will increase in value. This action will oppose the base bias. As a result, collector current will drop to its normal value. The capacitor connected across R_E is called an emitter bypass capacitor. It provides an ac path for signal voltages around R_E. With C in the circuit, an average dc level is maintained at the emitter. Without C in the circuit, amplifier gain would be reduced. The value of C is dependent on the frequencies being amplified. At the lowest possible frequency being amplified, the capacitive reactance (X_C) must be 10 times as small as the resistance of R_E.

Circuit (d) of Figure 4-29 is a combination of circuits (b) and (c). It is often called *self-emitter*

bias. In the same regard, circuit (c) could be called *fixed-emitter bias*. As a general rule, emitter biasing is not very effective when used independently. Circuit (d) has good thermal stability. The output has reduced gain because of the base resistor connection.

Independent Beta Biasing

Two methods of biasing a transistor that is independent of beta are shown in Figure 4-30. These circuits are extremely important because they do not change operation with beta. As a general rule, these circuits have very reliable operating characteristics. The output is very predictable and stability is excellent.

Circuit (a) is described as the *divider method* of biasing. It is widely used today. The base voltage (V_B) is developed by a voltage-divider network made up of R_B and R_1. This network makes the circuit independent of beta changes. Voltage at the base, emitter, and collector all depend on external circuit values. By proper selection of components, the emitter-base junction is forward-biased, with the collector being reverse-biased. Normally,

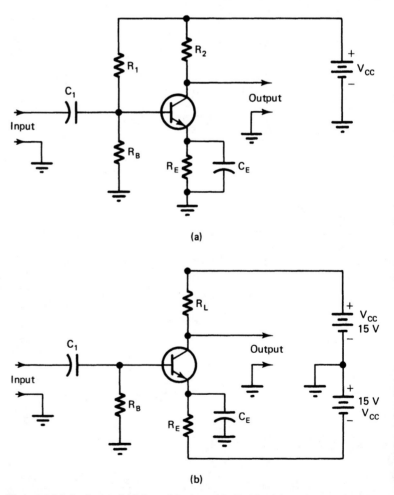

(a)

(b)

Figure 4-30. Independent beta biasing: (a) divider bias; (b) divider bias with split supply.

these bias voltages are all referenced to ground. In this case, the base is made slightly positive with respect to ground. This voltage is somewhat critical. A voltage that is too positive, for example, will drive the transistor into saturation. With proper selection of bias voltage, however, the transistor can be made to operate in any part of the active region. The temperature stability of the circuit is excellent. With proper R_E bypass capacitor selection, this method of biasing produces very high gain. The divider method of biasing is often a *universal biasing circuit*. It can be used to bias all transistor amplifier circuit configurations.

Circuit (b) is very similar in construction and operation to circuit (a). One less resistor is used. The power supply requires two voltage values with reference to ground. A *split power supply* is used as an energy source for this circuit. Note the indication of $+V_{CC}$, and $-V_{CC}$. R_B is connected to ground. In this case, the value of R_B would determine the value of I_B with only half of total supply voltage. The values of R_L and R_E are usually larger to accommodate the increased supply voltage. If R_E is properly bypassed, the gain of this circuit is very high. Thermal stability is excellent.

LOAD-LINE ANALYSIS

Earlier in this chapter, we looked at the operation of an amplifier with respect to its beta. Current and voltage were calculated and operation was related to these values. Amplifier operation can also be determined graphically. This method employs a *collector family of characteristic curves*. A *load line* is developed for the graph. With the load-line method of analysis, it is possible to predict how the circuit will respond graphically.

The *load-line method* of circuit analysis is used in designing circuits. The operation of a specific circuit can also be visualized by this method. In circuit design, a specific transistor is selected for an amplifier. Source voltage, load resistance, and input signal levels may be given values in the design of the circuit. The transistor is made to fit the limitation of the circuit.

For our application of the load line, assume that a circuit is to be analyzed. Figure 4-31 (a) shows the circuit being analyzed. A collector family of characteristic curves for the transistor is shown in Figure 4-32(b). Note that the *power dissipation rating* of the transistor is included in the diagram.

Power-Dissipation Curve

A common practice in load-line analysis is first to develop a power-dissipation curve. This gives some indication of the maximum operating limits of the transistor. *Power dissipation (Pd)* refers to maximum heat that can be given off by the base-collector junction. Usually, this value is rated at 25°C. Pd is the product of I_C and V_{CE}. In our circuit the Pd rating for the transistor is 300 mW.

To develop a Pd curve, each value of V_{CE} is used with the Pd rating to determine an I_C value. The formula is

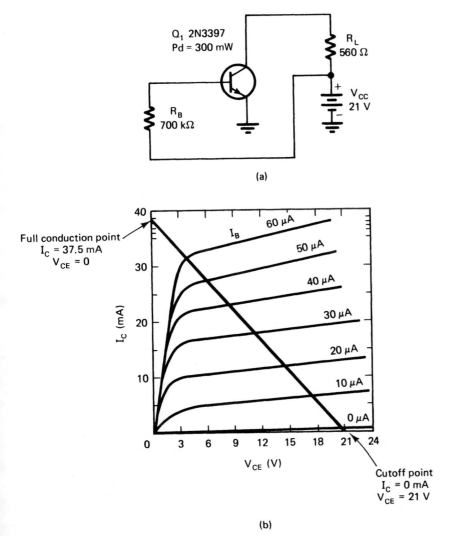

Figure 4-31. (a) Circuit and (b) characteristic curves of a circuit to be analyzed.

$$\text{Collector current} = \frac{\text{Pd}}{\text{collector-emitter voltage}}$$

or

$$I_C = \frac{\text{Pd}}{V_{CE}}$$

Using this formula, calculate the I_C value for each of the V_{CE} values of the family of curves. Using the V_{CE} value and the corresponding calculated I_C value, note the location on the curve. These points connected together are representative of the 300-mW power-dissipation curve. In practice, the load line must be located to the left of the established Pd curve. Satisfactory operation without excessive heat generation can be assured in that area of the curve.

Static Load-line Analysis

The load line of a transistor amplifier represents two extreme conditions of operation. One of these is in the cutoff region. When the transistor is cut off, there is no I_C flowing through the device. V_{CE} equals the source voltage with zero I_C. The second load-line point is in the saturation region. This point assumes full conduction of I_C. Ideally, when a transistor is fully conductive, $V_{CE} = 0$ and $I_C = V_{CC}/R_L$.

Two load-line construction points for the analysis circuit are shown on the curves of Figure 4-31. The cutoff point is located at the 0 I_C, 21 V_{CE} point. The value of V_{CC} determines this point. At cutoff, $V_{CE} = V_{CC}$. The saturation point is located at the 37.5-mA I_C and 0 V_{CE} point. The I_C value is calculated using V_{CC} and the value of R_L. The formula is

$$I_C = \frac{V_{CC}}{R_L} = \frac{21\text{ V}}{560\ \Omega}$$

$$= 0.0375\text{ A, or }37.5\text{ mA}$$

These two points are connected together with a straight line.

Figure 4-32 shows a *family of characteristic curves* with a load line for the circuit of Figure 4-31. Development of the *load line* makes it possible to determine the operating conditions of the amplifier. For linear amplification, the operating point should be located near the center of the load line. In our circuit, an operating point of 30 μA is used. In the circuit diagram the value of R_B determines I_B. It is calculated by the equation $I_B = V_{CC}/R_B$. The value is

$$I_B = \frac{V_{CC}}{R_B} = \frac{21\text{ V}}{700\text{ k}\Omega}$$

$$= 0.00003\text{ A, or}30/\text{LA}$$

The *operating point* for this value is located at the intersection of the load line and the 30-μA I_B curve. It is indicated as *point Q*. Knowing this much about an amplifier shows how it will respond in its steady, or static, state. The Q point shows how the amplifier will respond without a signal applied.

Operation of the amplifier in its static state is displayed by the family of curves in Figure 4-31. Projecting a line from the Q point to the I_C scale shows the resulting collector current. In this case I_C is 17.5 mA. The *dc beta* for the transistor at this point is determined by the formula $\beta = I_C/I_B$. This value is

$$\beta = \frac{I_C = 17.5\text{ mA}}{I_B = 30\ \mu\text{A}} = \frac{0.0175\text{ A}}{0.000030\text{ A}}$$

$$= 583.3$$

The resulting V_{CE} that will occur for the amplifier can also be determined graphically. Projecting a line directly down from the Q point shows the value of V_{CE}. In this circuit, V_{CE} is approximately 11 V. This means that 10 V (21 V – 11 V) will appear across R_L when the transistor is in its static state.

Dynamic Load-line Analysis

Dynamic load-line analysis shows how an amplifi-

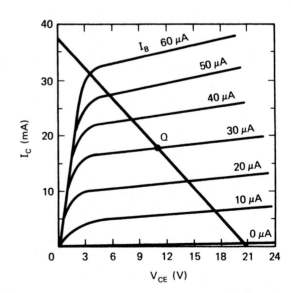

Figure 4-32. Family of characteristic curves for Figure 4-32.

er will respond to an ac signal. In this case the collector curves and circuit of Figure 4-33 are used. A load line and *Q* point for the circuit have been developed on the curves. This establishes the static operation of the amplifier. Note the values of V_{CE} and I_C in the static state.

Assume now that a 0.1-V peak-to-peak ac signal is applied to the input of the amplifier. In this case the signal will cause a 20-μA peak-to-peak change in I_B. During the positive alternation of the input, I_B will change from 30 μA to 40 μA. This is shown as point *P* on the load line. During the negative alternation, I_B will drop from 30 μA to 20 μA. This is indicated as point *N* on the load line. In effect, this means that 0.0 V p-p causes I_B to change 20 μA p-p. The I_B signal extends to the right of the load line. Its value is shown as ΔI_B.

To show how a change in I_B influences I_C lines are projected to the left of the load line. Note the projection of lines *P*, *Q*, and *N* toward the I_C values. The changing value of I_C is indicated as ΔI_C. An increase and decrease in I_B causes a corresponding increase and decrease in I_C. This shows that I_B and I_C are in phase.

The ac beta of the amplifier can be determined by ΔI_C and ΔI_B. First determine the peak-to-peak I_C and I_B values. Then divide ΔI_C by ΔI_B. Determine the ac *beta* of the amplifier circuit. Using the same procedure, deter-

mine the dc beta of the transistor at point *Q*. How do the ac and dc beta values of this circuit compare?

Projecting points *P*, *Q*, and *N* downward from the load line shows how V_{CE} changes with I_B. The value of V_{CE} is indicated as ΔV_{CE}. Note that an increase in I_B causes a decrease in the value of V_{CE}. A decrease in I_B causes V_{CE} to increase, which shows that I_B and V_{CE} are 180° out of phase. The difference in V_{CE} at any point appears across R_L.

The ac voltage gain of the amplifier can be determined from the dynamic load line. Remember that 0.1 V p-p input caused a change of 20 μA p-p in the I_B signal. The ac voltage gain can be determined by dividing ΔV_{CE} by ΔV_B. The ΔV_B value is 0.1 V p-p. Using the ΔV_{CE} value from the graph and ΔV_B the *ac voltage gain* of the amplifier circuit can be determined.

LINEAR AND NONLINEAR OPERATION

The V_{CE}, or *output*, of an amplifier should be a duplicate of the input with some gain. When a sine wave is applied, the output should develop a sine wave. When an amplifier operates in this manner it is considered to be *linear*. For linear operation to be achieved, the ampli-

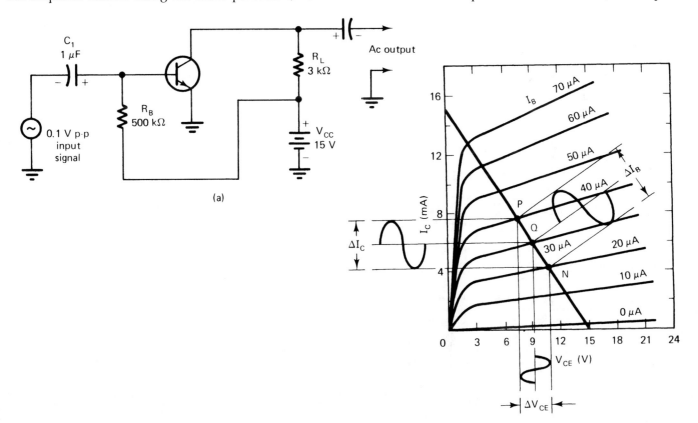

Figure 4-33. Dynamic load line: (a) circuit; (b) characteristic curves.

fier must operate in or near the center or linear area of the collector curves. As a rule, *nonlinearity* occurs near the saturation and cutoff regions. If the operating point is adjusted near these regions, it usually causes the output to be *distorted*. Normally, this is called *nonlinear distortion*.

Figure 4-34, shows how an amplifier will respond at three different operating points. Part (a) shows *linear* operation. The input and output are duplicates in this case. Part (b) shows operation in the *saturation* region. Note that the top of the input wave distorts the negative alternation of the output voltage. Part (c) shows operation near the *cutoff* region. The lower part of the input wave causes distortion of the positive alternation. As a general rule, nonlinear distortion can be tolerated in some applications. In other applications nonlinear distortion is very obvious.

An input signal that is too large can also produce distortion. Figure 4-35 shows how an amplifier will respond to a *large signal*. Note that the input signal swings into the saturation region during the positive alternation and into the cutoff region during the negative alternation. This condition causes distortion of both alternations of the output. Normally, this condition is described as *overdriving*. In a radio receiver or stereo amplifier, overdriving causes poor quality in speech or music. This usually occurs when the volume control is turned too high. The au-

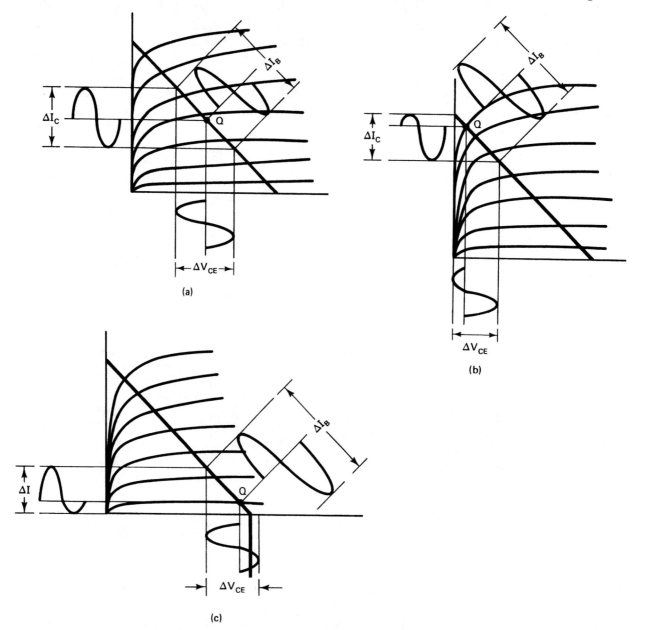

Figure 4-34. Amplifier operation: (a) linear operation; (b) operation near saturation; (c) operation near cutoff.

dio or sound signal has its peaks *clipped*. The volume level may be higher, but the quality of reproduction is usually very poor when this occurs. It is interesting to note that the operating point is near the center of the active region. Overdriving can occur even with a properly selected operating point.

CLASSES OF AMPLIFICATION

Transistor amplifiers are frequently classified according to their bias operating point. This means of classification describes the shape of the output wave. Three general groups of amplifiers are class A, class B, and class C. Figure 4-36 shows a graphical display of these three amplifier classes.

Class A amplifiers generally have linear operation. The bias operating point is set near the center of the active region. With a sine wave applied to the input, the output is a complete sine wave. Figure 4-36(a) shows the input-output waveforms of a class A amplifier. This type

of amplifier is used when a true reproduction of the input signal is required.

Figure 4-36(b) shows the input-output waveforms of a *class B amplifier*. The operating point of this amplifier is adjusted near the cutoff point. With a sine wave applied to the input, only one alternation of the signal is reproduced. When using class B amplifiers it is possible to get a large change in output for one alternation. Two class B amplifiers working together can each amplify one alternation of a sine wave. When the wave is restored, a complete sine wave is developed. Class B amplifiers are commonly used in push-pull audio output circuits. These amplifiers are usually concerned with power amplification. A class AB amplifier has its operating point between class A and class B. It reproduces all of one input alternation and only part of the other alternation at its output.

A class C amplifier is shown in Figure 4-36(c). In this amplifier the bias operating point is below cutoff. With a sine wave applied to the input, the output is less than half of one alternation. Radio-frequency circuits are the primary application of class C amplifiers today. The op-

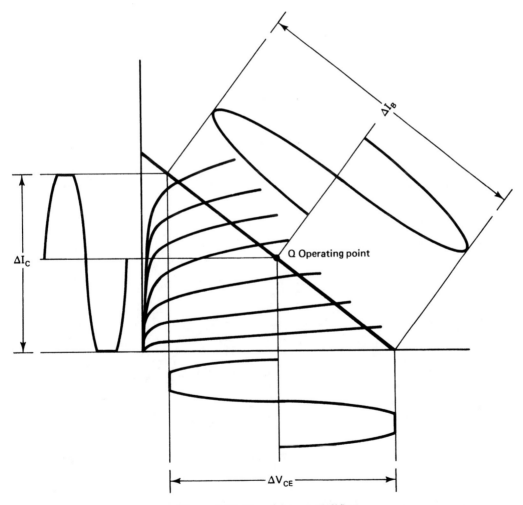

Figure 4-35. Overdriven amplifier.

erational efficiency of this amplifier is quite high. It consumes energy only for a small portion of the applied sine wave.

TRANSISTOR CIRCUIT CONFIGURATIONS

The elements of a transistor can be connected together in one of three different circuit configurations. These are usually described as common-emitter, common-base, and common-collector circuits. Of the three transistor leads, one is connected to the input and one to the output. The third lead is commonly connected to both the input and output. The common lead is generally used as a reference point for the circuit. It is usually connected to the circuit *ground,* or *common* point. This has brought about the

terms grounded emitter, grounded base, and grounded collector. The terms common and ground mean the same thing. In some circuit configurations, the emitter, base, or collector may be connected directly to ground. When this occurs, the lead is at both dc and ac ground potential. When the lead goes to ground through a battery or resistor that is bypassed by a capacitor, it is at an *ac ground.*

Common-emitter Amplifiers

The *common-emitter amplifier* is a very important transistor circuit. A very high percentage of all amplifiers in use today are of the common-emitter type. The *input* signal of this amplifier is applied to the base and the *output* is taken from the collector. The emitter is the common, or grounded, element.

Figure 4-37 shows a circuit diagram of a common-

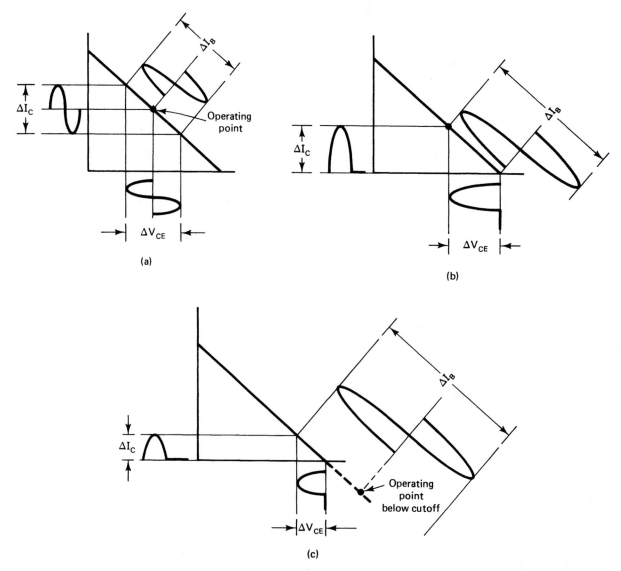

Figure 4-36. Classes of amplification: (a) class A; (b) class B; (c) class C.

emitter amplifier. This circuit is very similar to the basic amplifier used in the first part of the unit. In effect, this circuit is used to acquaint you with amplifiers in general. It is presented here to make a comparison with the other circuit configurations.

The signal being amplified by the common-emitter amplifier is applied to the emitter-base junction. This signal is superimposed on the dc bias of the emitter-base. The base current then varies at an ac rate. This action causes a corresponding change in collector current. The output voltage developed across the collector-emitter is *inverted* 180°. The current gain of the circuit is determined by beta. Typical beta values are in the range of 50. Voltage gain of the circuit ranges from 250 to 500. Power gain is in the range of 20,000. Input impedance is moderately high, typical values being 100 Ω. Output impedance is moderate, typical values being 2000. In general, common-emitter amplifiers are used in small-signal applications or as voltage amplifiers.

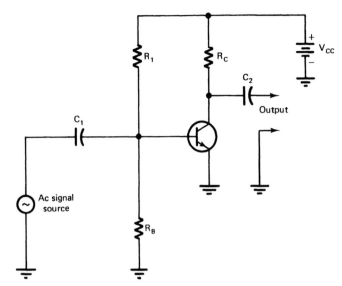

Figure 4-37. Common-emitter amplifier.

Common-base Amplifiers

A *common-base amplifier* is shown in Figure 4-38. In this type of amplifier, the emitter-base junction is forward-biased and the collector-base junction is reverse-biased. In this circuit configuration, the emitter is the input. An applied input signal changes the circuit value of I_E. The output signal is developed across R_L by changes in collector current. For each value change in I_E, there is a corresponding change in I_C.

In a common-base amplifier the current gain is called alpha. *Alpha* is determined by the formula I_C/I_E. In a common-base amplifier the gain is always less than

a value of 1. Remember that $I_E = I_B + I_C$. Thus, I_C will always be less than I_E by the value of I_B. Typical values of alpha are 0.98 to 0.99.

In Figure 4-38, V_{EE} forward-biases the emitter-base, whereas V_{CC} reverse-biases the collector-base. Resistor R_E is an emitter-current-limiting resistor, and R_L is the load resistor. Note that the transistor is a PNP type.

When an ac signal is applied to the *input*, it is added to the dc operating value of V_E. The positive alternation therefore adds to the forward-bias voltage of V_E. This condition causes an increase in I_E. A corresponding increase in I_C also takes place. With more I_C through R_L, there is an increase in its voltage drop. The collector therefore be-

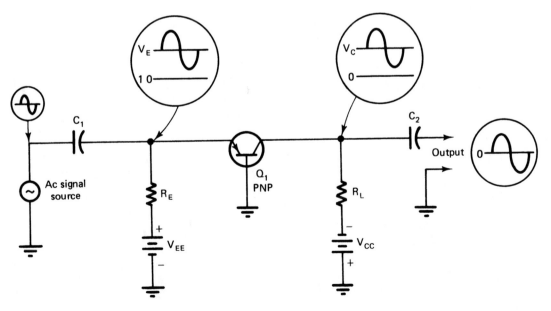

Figure 4-38. Common-base amplifier.

comes less negative or swings positive. In effect, the positive alternation of the input produces a positive alternation in the output. The input and output of this amplifier are *in phase*. See the waveform inserts in the diagram. The negative alternation causes the same reaction, the only difference being a reverse in polarity.

A common-base amplifier has a number of rather unique characteristics that should be reviewed. Its current gain is called alpha. The current gain is always less than 1. Typical values are 0.98 to 0.99. Voltage gain is usually very high. Typical values range from 100 to 2500 depending on the value of R_L. Power gain is the same as the voltage gain. The input impedance of the circuit is quite low. Values of 10 to 200 Ω are very common. The *output impedance* of the amplifier is somewhat moderate. Values range from 10 to 40 kΩ. This amplifier does not invert the applied signal.

Common-base amplifiers are used primarily to *match* a low-impedance input device to a circuit. This type of circuit configuration is also used in radio-frequency amplifier applications. As a general rule, the common-base amplifier is not used very commonly today.

Common-collector Amplifiers

A *common-collector amplifier* is shown in Figure 4-39. In this circuit the base serves as the signal input point. The input of this amplifier is primarily the same as the common-emitter circuit. The collector is connected to ground through V_{CC}. Note that the input, output, and collector are all commonly connected. The unique part of this circuit is the output. It is developed across the load resistor (R_L) in the emitter. There is no resistor in the collector circuit.

When an ac signal is applied to the input, it adds to the base current. The positive alternation increases the value of I_B above its static operating point. An increase in I_B causes a corresponding increase in I_E and I_C. With an increase in I_E, there is more voltage drop across R_L. The top side of R_L, the emitter, becomes more positive. A positive input alternation therefore causes a positive output alternation across R_L. Essentially, this means that the input and output are in phase. The negative alternation reduces I_B, I_E, and I_C. With less I_E through R_L, the output swings negative. The output is again in phase for this alternation.

The common-emitter amplifier is capable of current gain. A small change in I_B causes a large change in I_E. This current gain does not have a descriptive term such as alpha or beta. In general, the current gain is assumed to be greater than 1. Values approximate those of the common-emitter amplifier.

Other factors of importance are voltage gain, power gain, input impedance, and output impedance. Voltage

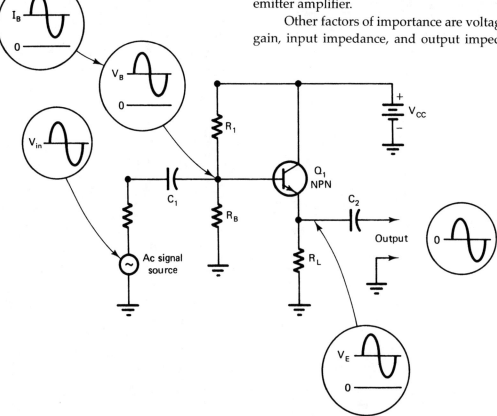

Figure 4-39. Common-collector amplifier.

gain is less than 1. Power gain is moderately high. Typical values are in the range of 50. Input impedance is in the range of 50 kΩ. The output impedance is quite low because of the location of R_L. Typical values are 50 Ω.

Common-emitter amplifiers are used primarily as impedance-matching devices, matching a high-impedance device to a low-impedance load. Practical applications include preamplifiers operated from a high-impedance microphone or phonograph pickup. Common-collector amplifiers are also used as driver transistors for the last stage of amplification. In this application power transistors generally require large amounts of input current to deliver maximum power to the load device. Since the emitter output follows the base, this type of amplifier is often called an *emitter* follower.

FIELD-EFFECT TRANSISTOR AMPLIFIERS

The primary function of an amplifier is to reproduce the applied signal and provide some level of amplification. Unipolar transistors can be used to achieve this function. JFET and MOSFET are examples of unipolar

amplifying devices. Amplification depends on the device being used and its circuit components. The composite circuit is normally energized by DC voltage. As ac signal is then applied to the input of the amplifier. After passing through the device, an amplified version of the signal appears in the output. Operating conditions of the device and the circuit combine to determine the level of amplification to be achieved.

For an FET to respond as an amplifier, it must be energized with dc voltage. Specific voltage values and polarities must be applied to each type of device. Forward- and reverse-biasing are not as meaningful in FET circuits. Remember that current conduction is through a channel. The value of voltage and its polarity determines the amount of control being achieved.

An illustration of the supply voltages and polarities needed for different types of FETs is shown in Figure 4-40. In JFETs the gate-source junction must be reverse-biased to achieve control. The gate voltage of an *enhancement MOSFET* is opposite that of the source. *Depletion MOSFETs* have some flexibility in the polarity of the gate voltage. Some of these devices have a zero-value V_G and swing positive or negative during operation.

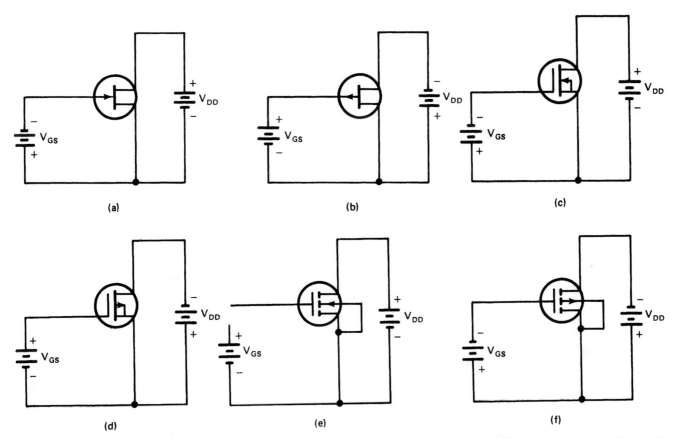

Figure 4-40. FET operational voltages: (a) N-channel JFET; (b) P-channel JFET; (c) N-channel D-MOSFET; (d) P-channel D-MOSFET; (e) N-channel E-MOSFET; (f) P-channel E-MOSFET.

BASIC FET AMPLIFIER OPERATION

To look at the circuitry of a basic *FET amplifier,* an N-channel JFET is used. A P-channel device could also have been selected. The primary difference in operation is the polarity of the source voltage and the type of current carrier in the channel.

Figure 4-41 shows an N-channel JFET amplifier connected in the common-source configuration. The source of this device is common to both the input signal and the output. This circuit is very similar to the common-emitter amplifier. V_{DD} serves as the dc energy source for the source and drain. V_{GS} reverse-biases the gate with respect to the source. The value of V_{GS} establishes the static operating point of the circuit. R_g is normally extremely high, and no gate current flows through R_g. The input signal is applied to the gate through capacitor C. Selecting a large value of R_g keeps the input impedance of the amplifier high. If R_g is 1 MΩ, the signal source sees an impedance of 1 MΩ.

To consider the operation of the JFET amplifier of Figure 4-41 in its static state, the family of drain curves of Figure 4-42 will be used. A load line for the circuit must be established. Remember that the two extreme conditions of operation are needed for the load line. In a JFET, these extremes are *full condition* and *cutoff* At the cutoff point, no current passes through the channel. The V_{DD} voltage all appears across V_{DS}. Full conduction occurs when a maximum value of I_D flows through R_L. This value is determined by the formula

$$\text{Drain current} = \frac{\text{source voltage}}{\text{load resistance}}$$

$$= I_D = \frac{V_{DD}}{R_L}$$

For the amplifier circuit, the maximum value of I_D is

$$I_D = \frac{25V}{10 \text{ K}\Omega} = \frac{25V}{10 \times 10^3}$$

$$= 0.0025 \text{ A, or} 2.5 \text{ mA}$$

With this conduction through R_L, V_{DS} is zero. The two locating points of the load line are $V_{DS} = 25$ V with 0 mA or I_D and $V_{DS} = 0$ V at 2.5 mA I_D. Check these two location points on the load line. A straight line connects the points.

With the load line developed, it is possible to see how the JFET will respond in its static state. Static operation occurs without a signal applied. For linear operation, the amplifier should respond near the center of the active region. In the basic circuit, V_{GS} is − 1.5 V. Point Q on the load line denotes the operating point. With an appropriate input signal, the amplifier should respond in the *linear* region.

To see how the JFET responds in its static state, lines are projected from the Q point. Projecting a line from the Q point to the I_D scale shows the drain-source voltage. V_{DS} is approximately 10 V. This means that 15 V appears across R_L at this point of operation. The dc voltage gain can be determined with this value. Voltage amplification (A_V) equals the drain-source voltage (V_{DS}) divided by the

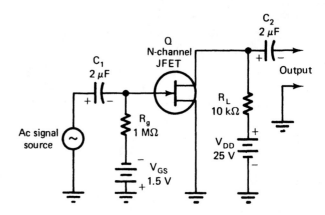

Figure 4-41. Common-source N-channel JFET amplifier.

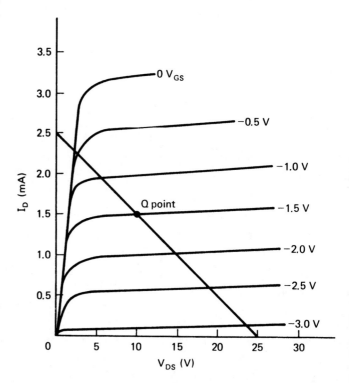

Figure 42. Family of drain curves.

gate-source voltage. For our circuit the gain is

$$A_V = \frac{V_{DS}}{V_{GS}} = \frac{10\ V}{1.5\ V}$$

$$= 6.667$$

DYNAMIC JFET AMPLIFIER ANALYSIS

Dynamic JFET amplifier analysis shows how the device will respond when an ac *signal* is applied. For this explanation we will use the *drain curves* and circuit of Figure 4-43. A load line and Q point for the circuit has been developed on the curves. This shows how the JFET responds in its static state. Check the load-line location points and Q point for accuracy.

The circuit diagram shows a 1-V p-p ac *signal* applied to the input of the JFET amplifier. With this signal the V_{GS} operating point will change in value from –2.0 V to –3.0 V. For the positive alternation, V_{GS} will swing from –2.5 V to –2.0 V. This change is shown as point P on the load line. During the negative alternation, V will drop from –2.5 V to –3 V. This value is shown as point N on the load line. In effect, this means that a 1-V p-p input signal will cause V_{GS} to change from –2.0 V to –3.0 V. This is considered to be the ΔV_{GS} value.

To show how ΔV_{GS} alters I_D, lines are projected to the left of points P, Q, and N. Note the projection of these points toward the I_D values in Figure 4-43. The changing value of I_D is shown as ΔI_D. An increase or decrease in

V_{GS} causes a corresponding change in I_D. This shows that V_{GS} and I_D are in phase.

Projecting lines P, Q, and N downward from the load line shows how V_{DS} changes with V_{GS}. The value of is shown as ΔV_{DS}. Note that an increase in V_{GS} causes a decrease in V_{DS}. This tells us that V_{GS} and V_{DS} are 180°

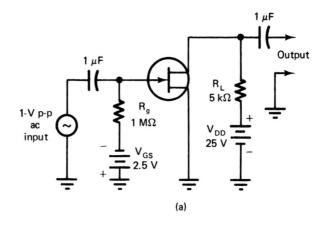

(a)

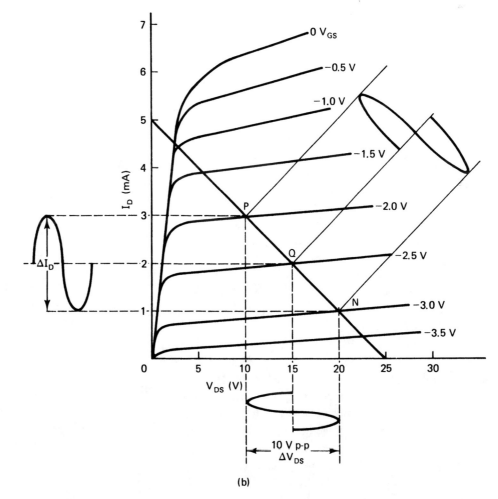

(b)

Figure 4-43. JFET amplifier: (a) N-channel amplifier; (b) drain family of curves.

out of phase. The value difference in V_{DS} and V_{DD} appears across the load resistor as V_{RL}.

The voltage gain of a JFET amplifier can be readily observed by load-line data. A ΔV_{GS} of 1 V p-p causes a ΔV_{DS} of approximately 18 V p-p. This shows an obvious gain of 18. Using the formula,

$$A_V = \frac{\Delta V_{DS}}{\Delta V_{GS}} = \frac{18 \text{ V p-p}}{1 \text{ V p-p}}$$

$$= 18$$

FET CIRCUIT CONFIGURATIONS

FET amplifiers, like their bipolar counterparts, can be connected in three different circuit configurations. These are described as common-source, common-gate, and common-drain circuits. One lead is connected to the input with a second lead connected to the output. The third lead is commonly connected to both input and output. This lead is used as a circuit reference point. It is often connected to the circuit ground, which results in use of the terms grounded source, grounded gate, and grounded drain. The terms common and ground refer to the same type of connection.

Common-source Amplifiers

The *common-source amplifier* is the most widely used FET circuit configuration. This circuit is similar in many respects to the common-emitter bipolar amplifier. The input signal is applied to the gate-source and the output signal is taken from the drain-source. The source lead is common to both input and output.

A practical common-source amplifier is shown in Figure 4-44. This circuit is essentially the same as the one used in our basic JFET amplifier. It is shown here so that a comparison can be made with the other circuit configurations. The device can be a JFET, a D-MOSFET, or an E-MOSFET. Circuit characteristics are primarily the same for all three devices.

The signal being processed by the common-source amplifier is applied to the gate-source. *Self-biasing* of the circuit is achieved by the source resistor R_2. This voltage establishes the static operating point. The incoming signal voltage is superimposed on the gate voltage. This causes the gate voltage to vary at an ac rate. This action causes a corresponding change in drain current. The output voltage developed across the source and drain is inverted 180°. Voltage gain A_V is V_{DS}/V_{GS}. Typical A_V values are 5 to 10. The input impedance is extremely high for nearly any signal source. One to several megohms is

common. The output impedance (Z_O) is moderately high. Typical values are in the range of 2 to 10 kΩ. Z_O is dependent primarily on the value of R_L.

A common-source amplifier has a very good ratio between input and output impedance. Circuits of this type are extremely valuable as impedance-matching devices. Common-source amplifiers are used almost exclusively as voltage amplifiers. They respond well in radio-frequency signal applications.

Common-gate Amplifiers

The *common-gate FET amplifier* is similar in many respects to the common-base bipolar transistor circuit. The input signal is applied to the source-gate and the output appears across the drain-gate. The common-gate amplifier is capable of *voltage gain* but has a current gain capability that is less than 1. JFETs, E-MOSFETs, and D-MOSFETs may all be used as common-gate amplifiers.

A practical common-gate amplifier is shown in Figure 4-45. This circuit employs an N-channel JFET. The operating voltages are the same as those of the common-source circuit. Self-biasing is achieved by the source resistor R_1. This voltage is used to establish the static operating point. An input signal is applied to R_1 through capacitor C_1. A variation in signal voltage causes a corresponding change in source voltage. Making the source more positive has the same effect on I_D as making the gate more negative. The positive alternation of the input signal will make the source more positive. This, in turn, reduces drain current. With less I_D there is a smaller voltage drop across the load resistor R_2. The drain, or output voltage, therefore swings positive. The negative alternation of the input reduces the source voltage by an equal amount. This is the same as making the gate less negative. As a result, I_D is increased. The voltage drop across R_2 similarly increases. This, in turn, causes the drain or output to be less positive or negative going. The input signal is therefore in phase with the output signal.

The common-gate amplifier has a number of rather unusual characteristics. Its voltage gain is somewhat less than that of the common-source amplifier. Representative values are 2 to 5. The common-gate circuit has very low input impedance (Z_{in}). The output impedance (Z_{out}) is rather moderate. Typical Z_{in} values are 200 to 1500 Ω with Z_{out} being 5 to 15 kΩ. This type of circuit configuration is often used to amplify radio-frequency signals. Amplification levels are very stable for RF, without feedback between the input and output.

Common-drain Amplifiers

A *common-drain amplifier* has the input signal ap-

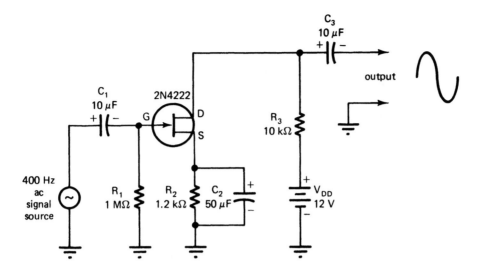

Figure 4-44. Common-source JFET amplifier.

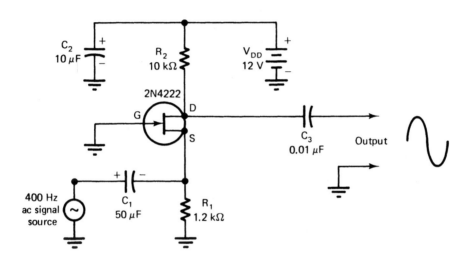

Figure 4-45. Common-gate JFET amplifier.

plied to the gate and the output signal removed from the source. The drain is commonly connected to one side of the input and output. Common-drain amplifiers are also called *source followers*. This circuit has similar characteristics to those of the common-collector bipolar amplifier.

Figure 4-46 shows a practical common-drain amplifier using an N-channel JFET. The *input* of this amplifier is primarily the same as that of the common-source amplifier. The input impedance is therefore very high. Z_{in} is determined largely by the value of R_1. The operating point of the amplifier is determined by the value of R_2. Essentially, this circuit has the same operating point as our other circuit configurations. Resistor R_3 has been switched from the drain to the source in this circuit. Resistors R_2 and R_3 are combined to form the load resistance. The output impedance is based on this value.

When an ac signal is applied to the input of the amplifier it changes the gate voltage. The dc operating point is established by the value of source resistor R_2. The positive alternation of an input signal makes the gate less negative. This causes an N-channel device to be more conductive. With more current through R_3 and R_2, the source swings positive. The negative alternation of the input then makes the gate more negative. This action causes the channel to be less conductive. A smaller current therefore causes the source to swing negative. In effect, this means that the input and output voltage values of the amplifier are in step with each other. These signals are in phase in a common-drain amplifier.

Common-drain amplifiers are primarily used today as impedance-matching devices, matching a high-impedance device to a low-impedance load. This type of circuit

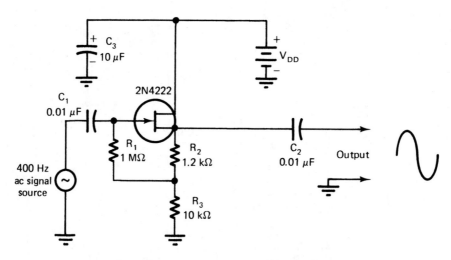

Figure 4-46. Common-drain JFET amplifier.

is capable of handling a high-input signal level without causing distortion. The input impedance places a minimum load on the signal source. Common-drain amplifiers are used to match high-impedance devices such as microphones to the input of an audio amplifier.

AMPLIFYING SYSTEM FUNCTIONS

Regardless of its application, an amplifying system has a number of primary functions that must be performed. An understanding of these functions is an extremely important part of operational theory. A block di-

agram of an amplifying system is shown in Figure 4-47. The triangular-shaped items of the diagram show where amplification is performed. A stage of amplification is represented by each triangle. Three amplifiers are included in this particular system. A stage of amplification consists of an active device and all its associated components. *Small-signal amplifiers* are used in the first three stages of this system. The amplifier on the right side of the diagram is an output stage. A large signal is needed to control the output amplifier. An output stage is generally called a *large-signal amplifier*, or a *power amplifier*. In effect, this amplifier is used to control a large amount of current and voltage. Remember that power is the prod-

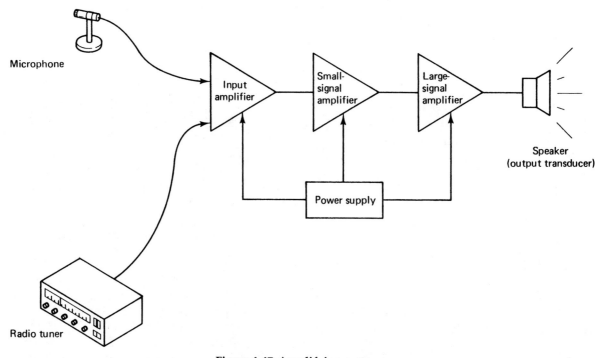

Figure 4-47. Amplifying system.

uct of current and voltage.

In an amplifying system, a signal must be developed and applied to the input. The source of this signal varies a great deal with different systems. In a stereophonic (stereo) phonograph, the signal is generated by a phonograph cartridge. Variations in the groove of a record are changed into electrical energy. This signal is then applied to the input of the system. The amplitude level of the signal is increased to a suitable value by the amplifying devices. A variety of different input signal sources may be applied to the input of an amplifying system. In other systems the signal may also be developed by the input. *Transducers* are responsible for this function. An *input transducer* changes energy of one form into energy of a different type. Microphones are typical input transducers. Input signals may also be received through the air. Antenna coils and networks serve as input transducers for these types of systems. An antenna changes electromagnetic waves into RF voltage signals. The signal is then processed through the remainder of the system.

Signals processed by an amplifying system are ultimately applied to an *output transducer*. This type of transducer changes electrical energy into another form of energy. In a sound system, the *speaker* is an output transducer because it changes electrical energy into sound energy. Work is performed by the speaker when it achieves this function. Lamps, motors, relays, transformers, and inductors are transducers. An output transducer is also considered to be the load of a system.

For an amplifying system to be operational, it must be supplied with electrical energy. A dc power supply performs this function. A relatively pure form of dc must be supplied to each amplifying device. In most amplifying systems, ac is the primary energy source. Ac is changed into dc, filtered, and—in some systems—regulated before being applied to the amplifiers. The reproduction quality of the amplifier is very dependent on the dc supply voltage.

Amplifier Gain

The *gain* of an amplifier system can be expressed in a variety of ways. Voltage, current, power, and (in some systems) decibels are expressed as gain. Nearly all input amplifier stages are *voltage amplifiers*. These amplifiers are designed to increase the voltage level of the signal. Several voltage amplifiers may be used in the front end of an amplifier system. The voltage value of the input signal usually determines the level of amplification being achieved.

A three-stage voltage amplifier is shown in Figure 4-

48. These three amplifiers are connected in *cascade*, which refers to a series of amplifiers where the output of one stage is connected to the input of the next amplifier. The voltage gain of each stage can be observed with an oscilloscope. The waveform shows representative signal levels. Note the voltage-level change in the signal and the amplification factor of each stage.

The first stage has a voltage gain of 5 V. With a 0.25-V p-p input, the output is 1.25 V p-p. The second stage also has a voltage gain of 5. With a 1.25-V p-p input, the output is 6.25 V p-p. The output stage has a gain factor of 4. With a 6.25-V p-p input, the output is 25 V p-p.

The total gain of the amplifier is 5 × 5 × 4, or 100. With a 0.25-V p-p input the output is 25 V p-p. Output is the *product* of the individual amplifier gains. *Voltage gain* (A_V) is an expression of output voltage (V_{out}) divided by input voltage (V_{in}). For the amplifier system, A_V is expressed by the formula

$$A_V = \frac{V_{out}}{V_{in}} = \frac{25 \text{ V p-p}}{0.25 \text{Vp-p}} = 100\text{V}$$

Note that the units of voltage cancel each other in the problem. Voltage gain is therefore expressed as a pure number, such as 100.

Power gain is generally used to describe the operation of the last stage of amplification. Power amplification (A_P) is equal to power output (P_{out}) divided by power input (P_{in}). If the last stage of amplification in Figure 4-49 were a power amplifier, its gain would be expressed in watts rather than volts. In this case, the gain would be

$$A_P = \frac{P_{out}}{P_{in}} = \frac{25 \text{ W}}{6.25\text{W}} = 4$$

Note that the power units of this problem also cancel each other. Power amplification is also expressed as a pure number, such as 4.

Amplifier Coupling

In many systems one stage of amplification will not provide the desired level of signal output. Two or more *stages* of amplification are therefore connected together to increase the overall gain. These stages must be properly coupled for the system to be operational. *Coupling* refers to a method of connecting amplifiers together. The signal being processed is transferred between stages by the coupling component.

Several methods of *interstage amplifier coupling* are in use today. Three very common methods include capacitive coupling, direct coupling, and transformer coupling.

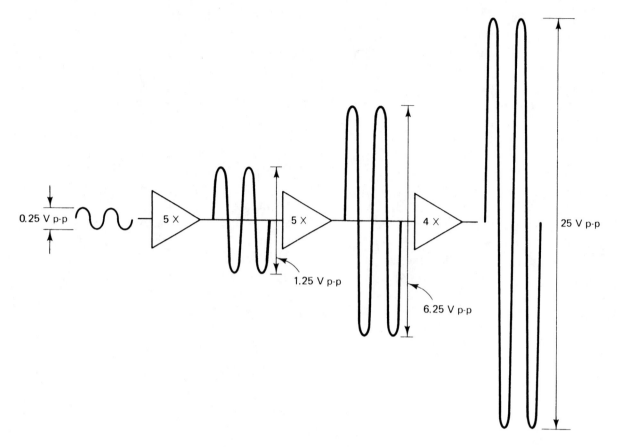

Figure 4-48. Three-stage voltage amplifier.

Each type of coupling has specific applications. In most amplifying devices, the output voltage is greater than the input voltage. If the output of one stage is connected directly to the input of the next stage, this voltage difference could cause a problem. Signal distortion and component damage might take place. Proper coupling procedures reduce this type of problem.

Capacitive Coupling

Capacitive coupling is particularly useful when amplifier systems are designed to pass ac signals. A capacitor will pass an ac signal and block dc voltages. The capacitor selected must have low capacitive reactance (X_C) at its lowest operating frequency. This is done to ensure amplification over a wide range of frequency. In general, the capacitive reactance of a good capacitor is extremely high at low frequency. At zero frequency or dc, the X_C is nearly infinite. Capacitive coupling has some difficulty in passing low-frequency ac signals. Large capacitance values are selected when good low-frequency response is desired.

A two-stage amplifier employing capacitor coupling is shown in Figure 4-49. Transistor Q_1 is the input amplifier. Its collector voltage is 8.5 V. At the base of the second

amplifier stage (Q_2), the voltage is approximately 1.7 V. The voltage difference across C_1 is 8.5 V – 1.7 V, or 6.8 V. The capacitor isolates these two operating voltages. It must have a dc working voltage value that will withstand this difference in potential.

The value of a coupling capacitor is a very important circuit consideration. In low-frequency amplifying systems, large values of electrolytic capacitors are normally used. These capacitors respond well to low-frequency ac. In high-frequency amplifier applications, small capacitor values are very common. Selection of a specific coupling capacitor for an amplifier is dependent on the frequency being processed.

Direct Coupling

In *direct coupling* the output of one amplifier is connected directly to the input of the next stage. In a circuit of this type a connecting wire or conductor couples the two stages together. Circuit design must take into account device voltage values. This type of coupling is an outgrowth of the divider method of biasing. A transistor acts as one resistor in a divider network. The output voltage of one amplifier is the same as the input voltage of the second amplifier. When the circuit is designed to take this into

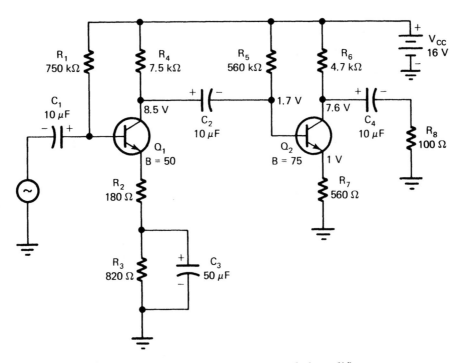

Figure 4-49. Two-stage capacitor-coupled amplifier.

account, the two can be connected together without isolation.

Figure 4-50 shows a two-stage direct-coupled transistor amplifier. Notice that the signal passes directly from the collector into the base. The base current (I_B) of transistor Q_2 is developed without a base resistor. Any I_B needed for Q_2 also passes through the load resistor R_3. The collector voltage V_C of Q_1 remains fairly constant when the two transistors are connected. Note also that the emitter bias voltage of Q_2 is quite large (7.2 V). The base-emitter voltage (V_{BE}) of Q_2 is the voltage difference between V_B and V_E. This is 7.8 V_B − 7.2 V_E, or 0.6 V_{BE}. This value of V_{BE} forward biases the base-emitter junction of Q_2. In direct-coupled amplifiers each stage has a different operating point based on

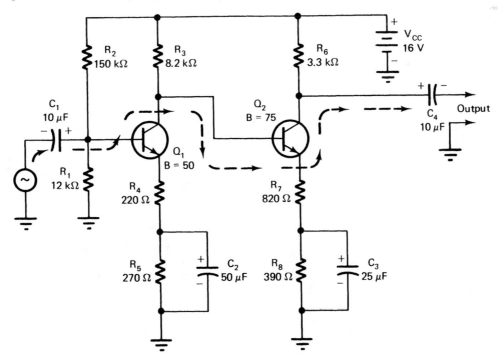

Figure 4-50. Two-stage direct-coupled amplifier.

a common voltage source. Several direct-coupled stages supplied by the same source are rather difficult to achieve. As a general rule, only two stages are coupled together by this method in an amplifying system.

Direct-coupled amplifiers are very sensitive to changes in temperature. The beta of a transistor will, for example, change rather significantly with temperature. An increase in temperature causes an increase in beta and leakage current. This tends to shift the operating point of a transistor. All stages that follow amplify according to the operating point shift. Changes in the operating point can cause *nonlinear distortion.*

Two transistors directly coupled form what is often called a *Darlington amplifier.* The transistors are usually called a *Darlington pair.* Two transistors connected in this manner are also manufactured in a single case. This type of unit has three leads. It is generally called a Darlington transistor. The gain produced by this device is the product of the two transistor beta values. If each transistor has a beta of 100, the total current gain is 100 × 100, or 10,000. Figure 4-51 shows a Darlington transistor amplifier circuit.

Darlington amplifiers are used in a system where high current gain and high input impedance are needed. Only a small input signal is needed to control the gain of a Darlington amplifier. In effect, this means that the amplifier does not load down the input signal source. The output impedance of this amplifier is quite low and is developed across the emitter resistor R_3. A Darlington amplifier has an emitter-follower output. Several different transistor combinations are used in Darlington amplifiers today.

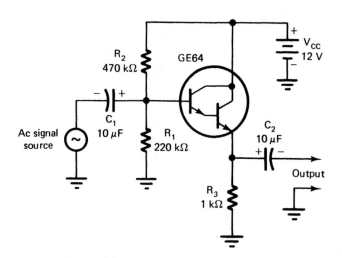

Figure 4-51. Darlington transistor amplifier.

Transformer Coupling

In *transformer coupling,* the output of one amplifier is connected to the input of the next amplifier by mutual inductance. Depending on the frequency being amplified, a *coupling transformer* may use a metal core or an air core. The output of one stage is connected to the primary winding and the input of the next stage is connected to secondary winding. The number of primary and secondary turns determines the *impedance ratio* of the respective windings. The input and output impedance of an amplifier stage can easily be matched with a transformer. Ac signals pass easily through the transformer windings. Dc voltages are isolated by the two windings.

Figure 4-52 shows a two-stage transformer-coupled

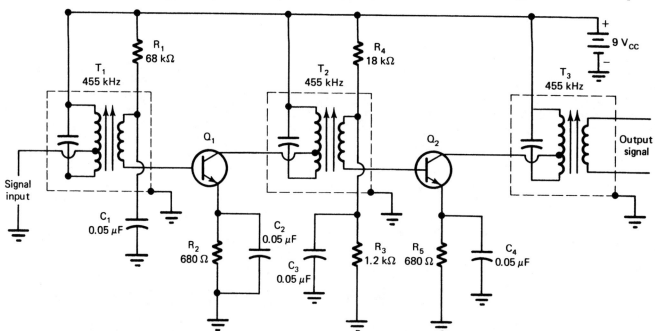

Figure 4-52. Two-stage transformer-coupled RF amplifier.

RF amplifier. This particular circuit is used in an *amplitude-modulated* (AM) radio receiver. Circuits of this type operate very well when they are built on a printed circuit board. The operational frequency of this amplifier is 455 kHz.

The operation of a transformer-coupled circuit is very similar to that of a capacitive-coupled amplifier. Biasing for each transistor element is achieved by resistance and transformer impedance. The primary impedance of T_2 serves as the load resistor of Q_1. The secondary winding of the transformer and R_4 serves as the input impedance for Q_2. The low output impedance of Q_1 is matched to the high input impedance of Q_2 by the transformer. The primary tap connection is used to assure proper load impedance of Q_1. The impedance of each winding is dependent on the primary and secondary turns ratio. This particular transformer-coupled amplifier is tuned to pass a specific frequency. Selection of this frequency is achieved by coil inductance (L) and capacitance (C). Tuned transformer-coupled amplifiers are widely used in radio receivers and TV circuits. The dashed line surrounding each transformer indicates that the transformer is housed in a metal can. This is done purposely to isolate the transformers from one another.

Transformers are also used to couple an amplifier to a load device. Figure 4-53 shows a transistor circuit with a coupling transformer. In this application the coupling device is called an *output transformer*. The primary transformer winding serves as a collector load for the transistor. The secondary winding couples transistor output to the load device. The transformer serves as an impedance-matching device, since its primary impedance matches the collector load. In a common-emitter amplifier, typical load resistance values are in the range of 1000 Ω. The output device in this application is a speaker. The speaker impedance is 10 Ω. To get *maximum transfer of power* from the transistor to the speaker, the impedances must match. The primary winding matches the transistor output impedance. The secondary winding matches the speaker impedance. Maximum power is transferred from the transistor to the speaker through the transformer impedance.

The number of turns of wire on a particular coil of a transformer determines its impedance. Assume that the output transformer of Figure 4-56 has 1000 turns on the primary and 100 turns on the secondary. Its *turns ratio* is

$$\text{Turns ratio} = \frac{\text{primary turns } (N_P)}{\text{secondary turns } (N_S)}$$

$$= \frac{1000}{100} = 10$$

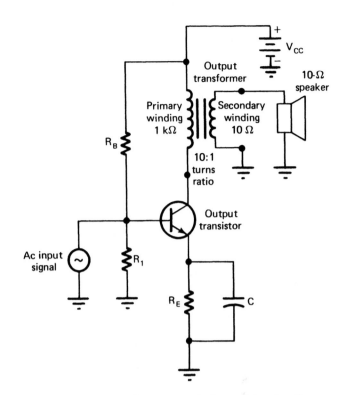

Figure 4-53. Transformer-coupled transistor circuit.

The *impedance ratio* of a transformer is the square of its turns ratio. For the output transformer, the impedance ratio is 10^2, or 100 Ω. This means the impedance of the collector is 100 times as great as the impedance of the load device. If the load (speaker) of our circuit has an impedance of 10 Ω, the input impedance is therefore 100 × 10, or 1000 Ω.

Transformer coupling has a number of problems. In sound system applications, the impedance increases with frequency. As a result of this, the high-frequency response of a sound signal is rather poor. The physical size of a transformer must also be quite large when high-power signals are being amplified. Because of this, low- and medium-power amplifying circuits use only transformer coupling. In addition to this, good-quality transformers are rather expensive.

POWER AMPLIFIERS

The last stage of an amplifying system is nearly always a *power amplifier*. This particular type of amplifier is generally a low-impedance device. It controls a great deal of current and voltage. Power is the product of current and voltage. Normally, a power amplifier dissipates a great deal more heat than does a comparable signal amplifier. In transistor systems, when 1 W or more of power is handled by a device, it is considered to be a power amplifier. Power amplifiers are generally designed to operate over the entire active region of the device. With careful selection of the device and circuit components, it is possible to achieve power gain with a minimum of distortion.

Single-ended Power Amplifiers

An amplifier system that has only one active device that develops output power is called a *single-ended amplifier*. This type of amplifier is very similar in many respects to a small-signal amplifier. It is normally operated as a class A amplifier. An ac signal applied to the input is fully reproduced in the output. The operating point of the amplifier is near the center of its active region. As a general rule, this type of amplifier operates with a minimum of distortion. The operational efficiency of this amplifier is, however, quite low. Efficiency levels are in the range of 30%. *Efficiency* refers to a ratio of developed output power to dc supply power. The equation for efficiency is

$$\text{Efficiency (\%)} = \frac{\text{ac power output}}{\text{applied dc operating power}}$$

$$= \frac{P_O}{P_{in}} \times 100$$

For a class A amplifier it takes 9 W of dc power to develop 3 W of output signal when an amplifier is 33% efficient. Because of their low efficiency rating, signal-ended amplifiers are generally used only to develop low-power output signals.

Figure 4-54 shows a *single-ended audio-output amplifier*. The term *audio* refers to the range of human hearing. *Audio frequency (AF)* refers to frequencies that are from 15 Hz to 15 kHz. This particular amplifier will respond to

signals somewhere within this range.

A family of characteristic curves for the transistor used in the single-ended power amplifier is shown in Figure 4-55. Note that the power dissipation curve of this amplifier is 5 W. The developed load lines for the amplifier are well below the indicated P_D curve. This means that the transistor can operate effectively without becoming overheated.

Investigation of the family of collector curves shows that two load lines are plotted. The static or dc load line is based on the resistance of the transformer primary. For a primary impedance of 50 Ω, the resistance would be approximately 10 Ω. The dc load line is therefore based on 12.5 V_{CC} and 1.25 A of I_C. This means that the dc load line will extend almost vertically from the 12.5 V_{CC} point on the curve. The ac load line is determined with the primary winding impedance (50 Ω) and twice the value of V_{CC} (25 Ω). The ac output signal is therefore twice the value of V. The operational points are 25 V_{CC} and 500 mA of I_C. Note the location of these operating points on the ac load line.

When an ac signal is applied to the input, it can cause a swing 12.5 V above and below V_{CC}. In this case, the transformer accounts for the voltage difference. Since a transformer is an inductive device, when its electromagnetic field collapses, a voltage is generated. This voltage is added to the source voltage. In an ac load line, V_{CE} will therefore swing to twice the value of the source voltage. This usually occurs only in transformer-coupled amplifier circuits. Note that the operating point (Q point) is near the center of the ac load line.

With the load line of the amplifier established, the base resistance value must be determined. For this circuit the emitter voltage is zero. The emitter-base junction of the transistor, being silicon, has 0.7 V across it when it is in conduction. The base resistor therefore has 12.5 V − 0.7,

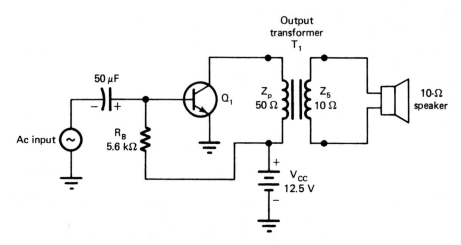

Figure 4-54. Single-ended audio amplifier.

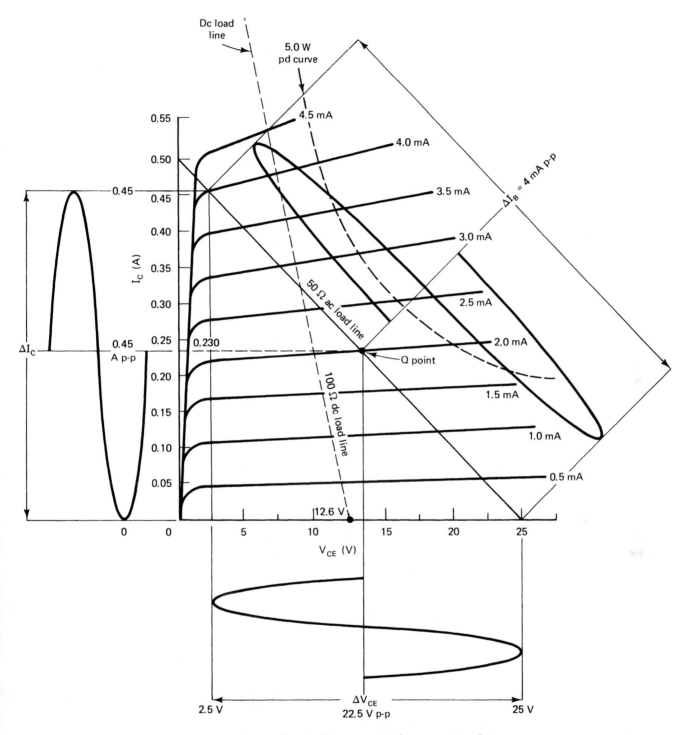

Figure 4-55. Family of collector curves for power transistor.

or 11.8 V, across it at the Q point. By Ohm's law, the value of R_B is 11.8 V ÷ 2 mA, or 5900 Ω. A standard resistor value of 5.6 kΩ would be selected for R_B. Notice the location of this resistor in the circuit.

With the available data, it is possible to look at the operational efficiency of the single-ended amplifier. The *dc power* supplied to the circuit for operation is

Applied dc operating power =
dc source voltage × collector current

or

$$P_{in} = V_{CC} \times I_C$$
$$= 12.6 \text{ V} \times 0.230 \text{ A}$$
$$= 2.898 \text{ W}$$

Push-pull Amplifiers

Power amplifiers can also be connected in a push-pull circuit configuration. Two amplifying devices are needed to achieve this type of output. The operational *efficiency* of this circuit is much higher than that of a single-ended amplifier. Each transistor handles only one alternation of the sine-wave input signal. *Class B amplifier* operation is used to accomplish this. A class B amplifier has sine-wave input and half-sine-wave output. The bias operating point is at cutoff. This means that no power is consumed unless a signal is applied. The operational efficiency of a push-pull amplifier is approximately 80%.

A simplified *transformer-coupled push-pull transistor amplifier* is shown in Figure 4-56. The circuit is somewhat like two single-ended amplifiers placed back to back. The input and output transformers both have a common connection. The V_{CC} supply voltage is connected to this common point. For operation, the base voltage must rise slightly above zero to produce conduction. The turn-on voltage is usually about 0.6 V. Base voltage is developed by the incoming signal between the center tap and outside ends of the transformer. The transistor does not conduct until the base voltage rises above 0.6 V. The secondary winding resistance of the input transformer (T_1) is several thousand ohms. The divided transformer winding means that each transistor is alternately fed an input signal, so the signals are the same value but are 180° out of phase.

A class B push-pull amplifier produces an output signal only when an input signal is applied. For the first alternation of the input, Q_1 will swing positive and Q_2 negative. Only Q will go into conduction with this input. Note the resulting I_C waveform. At the same time, Q_2 is at cutoff. The negative alternation drives it further into cut-

off. The resulting I_C from Q_1 flows into the top part of primary winding of T_2. This causes a corresponding change in the electromagnetic field. Voltage is induced in the secondary winding through this action. Note the polarity of the secondary voltage for this alternation.

For the next alternation of the input signal, the process is reversed. The base of Q_1 swings negative and becomes nonconductive. Q_2 swings positive at the same point in time, so Q_2 goes into conduction. Note the indicated I_C of Q_2 for this alternation. The I_C of Q_2 flows into the lower side of T_2. It, in turn, causes the electromagnetic field to change. Voltage is again induced into the secondary winding. Note the polarity of the secondary voltage for this alternation. It is reversed because the primary current is in an opposite direction to that of the first alternation.

Figure 4-57 shows a combined collector family of characteristic curves for Q_1 and Q_2. These transistors are operated as class B amplifiers. With each transistor operating at cutoff (point Q), only one alternation of output will occur. The entire active region of each transistor can be used in this case to amplify the applied alternation. The combined alternations make a noticeable increase in power output.

COMPLEMENTARY SYMMETRY AMPLIFIERS

The availability of *complementary transistors* made the design of transformerless power amplifiers possible. Complementary transistors are PNP and NPN types with similar characteristics. Power amplifiers designed with these transistors have an operating efficiency and linearity that is equal to that of a conventional push-pull cir-

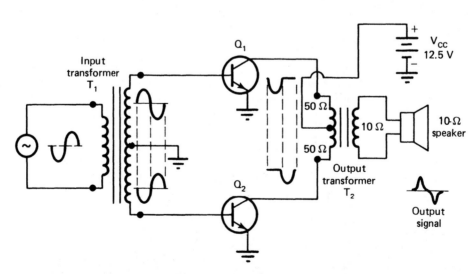

Figure 4-56. Transformer-coupled class B push-pull amplifier.

Figure 4-57. Combined collector curves for push-pull operation.

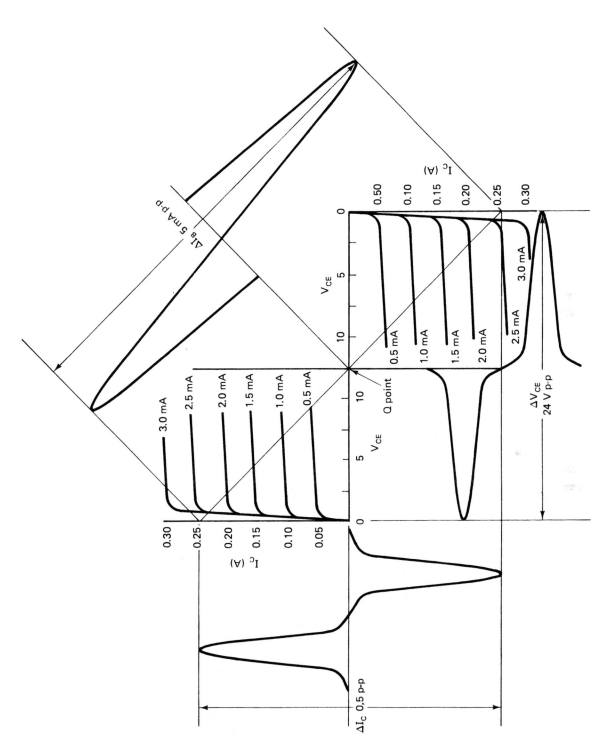

Figure 4-57. Combined collector curves for push-pull operation.

cuit. Transformerless amplifiers have the high- and low-frequency response of the output extended. Component cost and weight are also reduced with this type of circuitry.

An elementary complementary symmetry amplifier is shown in Figure 4-58. This circuit responds as a simple voltage divider. Transistors Q_1 and Q_2 serve as variable resistors with the load resistor (R_L) *connected to their common center point. When the positive alternation of the input signal occurs, it is applied to both Q and Q_2.* This alternation causes Q_1 to be conductive and Q_2 to be cut off. Conduction of Q_1 causes it to be low resistant, which connects the load resistor to the + V_{CC} supply. Capacitor C_2 charges in the direction indicated by the solid-line arrows. Note the direction of the resulting current through the load resistor. The polarity of voltage developed across R_L is the same as the input alternation. This means that the input and output signals are *in phase*. This is typical of a common-collector or emitter-follower amplifier.

INTEGRATED-CIRCUIT AMPLIFYING SYSTEMS

A number of manufacturers market amplifying systems in integrated-circuit packages. A wide range of different ICs are available. Some of these are designed to perform one specific system function. This includes such things as *preamplifiers, linear signal amplifiers,* and *power amplifiers*. Other ICs may perform as a complete system. Electrical power for operation and only a limited number of external components are needed to complete the system. Low-power audio amplifiers and stereo amplifiers are examples of these devices. Two IC applications are presented here.

Integrated-circuit power amplifiers are used in sound-amplifying systems. An example of such a circuit is shown in Figure 4-59. An internal circuit diagram of the IC is shown in part (a). A pin-out diagram of the unit is shown in part (b). This particular chip has a differential amplifier input. Darlington pairs are used to increase each input resistance. Inverting and noninverting inputs are both available. As a general rule, when a signal is applied to one input, the other is grounded. This IC will respond only to a difference in input signal levels. It has a fixed gain of 20 and variable gain capabilities up to 200. Without any external components connected to pins 1 and 8, the device has a gain of 20. A 10-mF capacitor connected between these two pins causes a gain of 200. The output of the unit is a Darlington-pair power amplifier. The supply voltage is automatically set to $V_{CC}/2$ at the output transistor common point.

An application of the IC power amplifier is shown in Figure 4-60. Pins 1 and 8 have a capacitor-resistor combination. This causes the gain to be less than the maximum value of 200. A decoupling capacitor is also connected between pin 7 and ground. R_1 serves as a volume control for the circuit. The amplitude level of the input signal is controlled by this potentiometer. The power output of the IC is 500 mW into a 1641 load resistor. Applications include radio amplifiers and TV sound systems.

A dual 6-W audio amplifier is shown in Figure 4-61. This particular IC offers high-quality performance for stereo phonographs, tape players, recorders, and AM-FM stereo receivers. The internal circuitry of this IC has 52

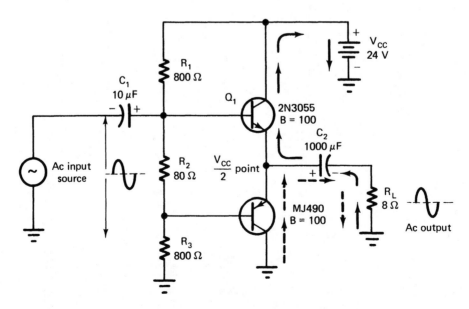

Figure 4-58. Complementary-symmetry amplifier.

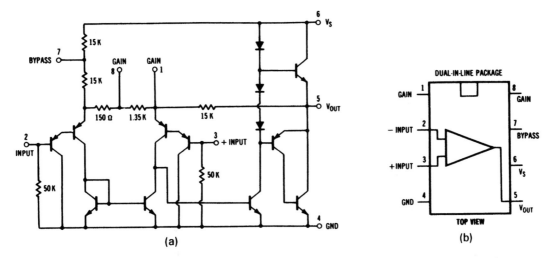

Figure 4- 59. Low-voltage audio power amplifier IC: (a) internal circuit diagram; (b) pin diagram.

transistors, 48 resistors, a zener diode, and two silicon diodes. Two triangle amplifier symbols are used to represent the internal components of this device. This particular amplifier is designed to operate with a minimum of external components. It contains internal bias regulation circuitry for each amplifier section. Overload protection consists of internal current-limiting and thermal shutdown circuits.

The *LM 379* has high gain capabilities. The supply voltage can be from 10 to 35 V. Approximately 0.5 A of current is needed for each amplifier. This particular IC must be mounted to a heat sink. A hole is provided in the housing for a bolted connection to the heat sink. A number of companies manufacture amplifying system ICs to-

day. As a general rule, a person should review the technical data developed by these manufacturers before attempting to use a device.

OPERATIONAL AMPLIFIERS

An *operational amplifier* or *op-amp* is a high-performance, directly coupled amplifier circuit containing several transistor devices. The entire assembly is built on a small silicon substrate and packaged as an integrated circuit (IC). ICs of this type are capable of high-gain signal amplification from dc to several million hertz. An op-amp is a modular, multistage amplifying device.

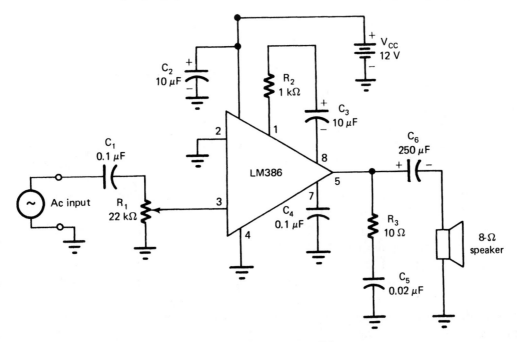

Figure 4-60. IC power amplifier.

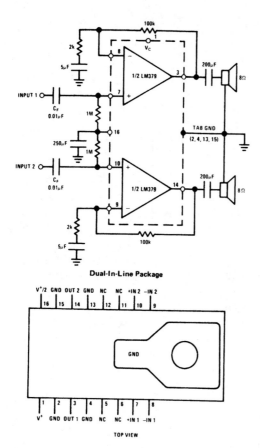

Figure 4- 61. LM 379 dual 6-W audio amplifier.

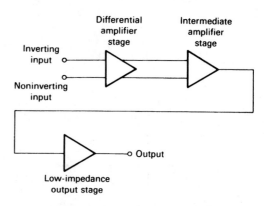

Figure 4-62. Op-amp diagram.

ally called *stages* of amplification. A stage of amplification contains one or more active devices and all the associated components needed to achieve amplification.

The first stage, or input, of an op-amp is usually a *differential amplifier*. This amplifier has two inputs, which are labeled V_1 and V_2. It provides high gain of the signal difference supplied to the two inputs and low gain for common signals applied to both inputs simultaneously. The input impedance is high to any applied signal. The output of the amplifier is generally two signals of equal amplitude and 180° out of phase. This could be described as a *push-pull* input and output.

One or more intermediate stages of amplification follow the differential amplifier. Figure 4-62 shows an op-amp with only one intermediate stage. This amplifier is designed to shift the operating point to a zero level at the output and has high current and voltage gain capabilities. Increased gain is needed to drive the output stage without loading down the input. The intermediate stage generally has two inputs and a single-ended output.

The output stage of an op-amp has a rather low output impedance and is responsible for developing the current needed to drive an external load. Its input impedance must be great enough that it does not load down the output of the intermediate amplifier. The output stage can be emitter-follower amplifier or two transistors connected in a complementary-symmetry configuration. Voltage gain is rather low in this stage with a sizable amount of current gain.

An op-amp has at least five terminals or connections in its construction. Two of these are for the power-supply voltage, two are for differential input, and one is for the output. There may be other terminals in the makeup of this device, depending on its internal construction or intended function. Each terminal is general-

Operational amplifiers have several important properties. They have *open-loop gain* capabilities in the range of 200,000 with an input impedance of approximately 2 MΩ. The output impedance is rather low, with values in the range of 50 Ω or less. Their bandwidth, or ability to amplify different frequencies, is rather good. The gain does, however, have a tendency to drop, or *roll off*, as the frequency increases.

The internal construction of an op-amp is quite complex and usually contains a large number of discrete components. A person working, with an op-amp does not ordinarily need to be concerned with its internal construction. It is helpful, however, to have some general understanding of what the internal circuitry accomplishes. This permits the user to see how the device performs and indicates some of its limitations as a functioning unit.

The internal circuitry of an op-amp can be divided into three functional units. Figure 4-62 shows a simplified diagram of the internal functions of an op-amp. Notice that each function is enclosed in a triangle. Electronic schematics use the triangle to denote the amplification function. This diagram shows that the op-amp has three basic amplification functions. These functions are generly attached to a schematic symbol at some convenient location. Numbers located near each terminal of the symbol indicate pin designations.

The schematic symbol of an op-amp is generally displayed as a triangular-shaped wedge. The triangle symbol in this case denotes the amplification function. Figure 4-63 shows a typical symbol with its terminals labeled. The point, or apex, of the triangle identifies the output. The two leads labeled minus (–) and plus (+) identify the differential input terminals. The minus sign indicates inverting input and the plus sign denotes noninverting input. A signal applied to the minus input will be inverted 180° at the output. Standard op-amp symbols usually have the inverting input located in the upper-left corner. A signal applied to the plus input will not be inverted and remains in phase with the input. The plus input is located in the lower-left corner of the symbol. In all cases, the two inputs are clearly identified as plus and minus inside the triangle symbol.

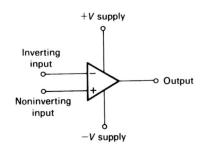

Figure 4-63. Op-amp schematic symbol.

Connections or terminals on the sides of the triangle symbol are used to identify a variety of functions. The most significant of these are the two power supply terminals. Normally, the *positive voltage terminal* (+V) is positioned on the top side and the *negative voltage* (–V) is on the bottom side. In practice, most op-amps are supplied by a *split*, or *divided, power supply*. This supply has +V, ground, and –V terminals. It is important that the correct voltage polarity be supplied to the appropriate terminals, or the device may be permanently damaged. A good rule to follow for most op-amps is not to connect the ground lead of the power supply to –V. An exception to this rule is the *current-differencing amplifier* (CDA) op-amp. These op-amps are made to be compatible with digital logic ICs and are supplied by a straight 5-V voltage source.

In the actual schematic symbol of an op-amp, the terminals are generally numbered and the element names are omitted. There is a pin-out key to identify the name of each terminal. The manufacturer of the device supplies information sheets that identify the pin-out and operating data. Figure 4-64 shows part of the data sheet for a uA741 linear op-amp. Note the package styles, symbol location, and pin-out key for the device. This sheet also shows absolute maximum ratings, electrical characteristics, and typical performance curves.

Figure 4-65 shows the schematic symbol of a uA741 connected as an open-loop amplifier. The power-supply voltage terminals are labeled +V, –V, and ground. Typical supply voltages are ±15, ±12, and ±6 V. The absolute maximum voltage that can be applied to this particular op-amp is 36 V between +V and –V, or ±18 V. The output of an op-amp circuit is generally connected through an external load resistor to ground. In this case, R_L serves as an external load for the uA741. All the output voltage developed by the op-amp appears across R_L. This type of connection is described as a single-ended output. The operational load current for this device is a value between 10 and 15 mA. The maximum output, or saturation voltage, that can be developed is approximately 90% of the supply voltage. For a +15 V supply, the saturation voltage is + 13.5 V. An ac output would be + 13.5 and – 13.5, or 27 V p-p. The current and voltage limits of the output restrict the load resistance to a value of approximately 1 kΩ. Most op-amps have a built-in load current-limiting feature. The uA741 is limited to a maximum short-circuit current of 25 mA. This feature prevents the device from being destroyed in the event of a direct short in the load.

The inputs of an op-amp are labeled minus (–) for inverting and plus (+) for the noninverting function, which also denotes the fact that the inputs are differentially related. Essentially, this means that the polarity of the developed output is based on the voltage difference between the two inputs. In Figure 4-65(a), the output voltage is positive with respect to ground. This is the result of the inverting input being made negative with respect to the noninverting input. Reversing the input voltage, as in Figure 4-65(b), causes the output to be negative with respect to ground. For this connection, the noninverting input is negative when the inverting input is positive. This condition of operation holds true for both input possibilities. In some applications one input may be grounded; the difference voltage is then made with respect to ground. The inputs of Figure 4-65 are connected in a *floating* configuration that does not employ a ground. For either type of input, the resulting output polarity is always based on the voltage difference. This characteristic of the input makes the op-amp an extremely versatile amplifying device.

schematic and connection diagrams

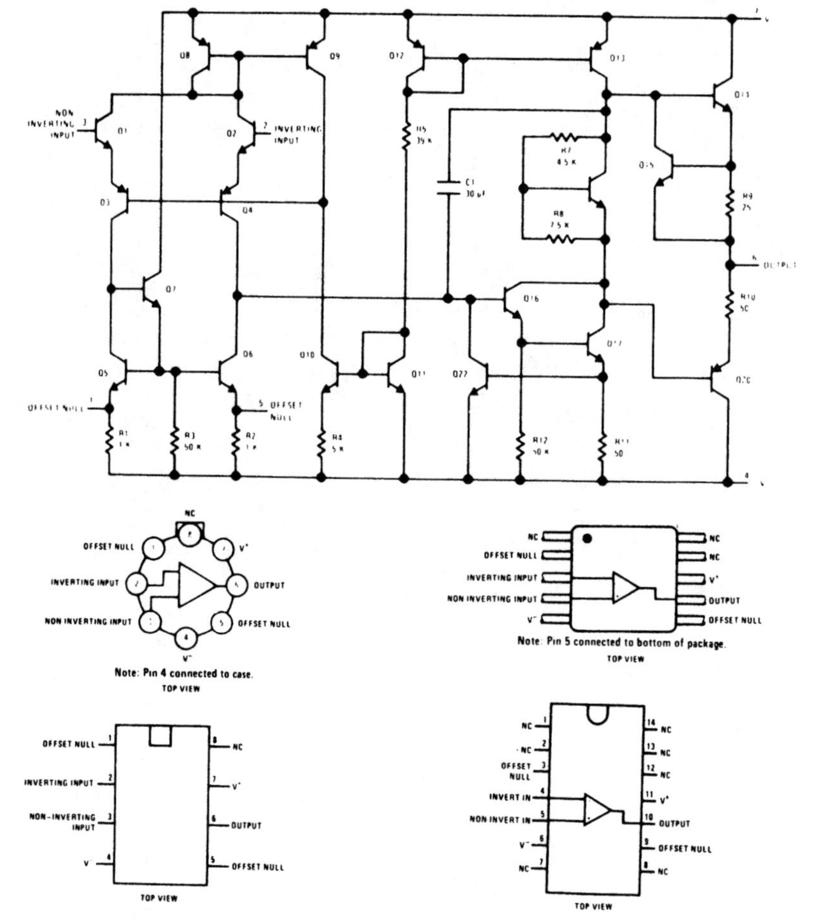

The LM741 and LM741C are general purpose operational amplifiers which feature improved performance over industry standards like the LM709. They are direct, plug-in replacements for the 709C, LM201, MC1439 and 748 in most applications.

The offset voltage and offset current are guaranteed over the entire common mode range. The amplifiers also offer many features which make their application nearly foolproof: overload protection on the input and output, no latch-up when the common mode range is exceeded, as well as freedom from oscillations.

The LM741C is identical to the LM741 except that the LM741C has its performance guaranteed over a 0°C to 70°C temperature range, instead of -55°C to 125°C.

Figure 4-64. Schematic diagram of an operational amplifier. (Courtesy of National Semiconductor Corp.)

OSCILLATOR CIRCUITS

There are several possible ways of describing an oscillator: according to generated frequency, operational stability, power output, signal waveforms, and components. In this unit it is convenient to divide oscillators according to their methods of operation, including feedback oscillator and relaxation oscillators. Each group has a number of distinguishing features.

In *feedback oscillators* a portion of the output power is returned to the input circuit. This type of oscillator usu-

ally employs a tuned *LC* circuit. The operating frequency of the oscillator is established by this circuit. Most sine-wave oscillators are of this type. The frequency range is from a few hertz to millions of hertz. Applications of the feedback oscillator are widely used in radio and television receiver tuning circuits and in transmitters. Feedback oscillators respond very well to RF generator applications.

Relaxation oscillators respond to an electronic device that goes into conduction for a certain time and then turns off for a period of time. This condition of operation repeats itself on a continuous basis. This oscillator usu-

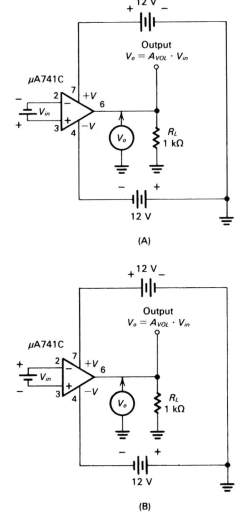

Figure 4-65.Inverting dc amplifiers.

ally responds to the charge and discharge of an *RC* or an *RL* network. Oscillators of this type usually generate square- or triangular-shaped waves. The active device of the oscillator is "triggered" into conduction by a change in voltage. Applications include the vertical and horizontal sweep generators of a television receiver and computer clock circuits. Relaxation oscillators respond extremely well to low-frequency applications.

Feedback Oscillators

Feedback is a process by which a portion of the output signal of a circuit is returned to the input. This function is an important characteristic of any oscillator circuit. Feedback may be accomplished by inductance, capacitance, or resistance coupling. A variety of different circuit techniques have been developed today. As a general rule, each technique has certain characteristics that distinguishes it from others. This accounts for the large number

of different circuit variations.

In many public gatherings where sound systems are used and a microphone is placed too close to a speaker, feedback can occur. Sound from the speaker is picked up by the microphone and returned to the amplifier. It then comes out of the speaker and is returned again to the microphone. The process continues until a hum or high-pitched howl is produced. Figure 4-66 shows an example of sound-system feedback. This condition is considered to be mechanical feedback. In sound-system operation, feedback is an undesirable feature. In an oscillator circuit, operational feedback is a necessity.

Oscillator Fundamentals

An *oscillator* is an electronic circuit that generates a continuously repetitive output signal. The output signal may be alternating current or some form of varying dc. The unique feature of an oscillator is its generation function. It does not receive an external input signal. It develops the resulting output signal from the dc operating power provided by the power supply. An oscillator in a sense is an electronic power converter. It changes dc power into ac or some useful form of varying dc. A block diagram of a feedback oscillator is shown in Figure 4-67.

An oscillator has an amplifying device, a feedback network, frequency-determining components, and a dc power source. The *amplifier* is used primarily to increase the output signal to a usable level. Any device that has signal gain capabilities can be used for this function. The *feedback network* can be inductive, capacitive, or resistive. It is responsible for returning a portion of the amplifier's output back to the input. The feedback signal must be of the correct phase and value in order to cause oscillations to occur. In-phase, or *regenerative*, feedback is essential in an oscillator. An inductor-capacitor (*LC*) network determines the frequency of the oscillator. The charge and discharge action of this network establishes the oscillating voltage. This signal is then applied to the input of the amplifier. In a sense, the *LC* network is energized by the feedback signal. This energy is needed to overcome the internal resistance of the *LC* network. With suitable feedback, a continuous ac signal can be generated. The output of a good oscillator must be uniform and should not vary in frequency of amplitude.

LC Circuit Operation

The operating frequency of a feedback oscillator is usually determined by an inductance-capacitance network. An *LC* network is sometimes called a *tank circuit*. The tank, in this case, has a storage capability. It stores an ac voltage that occurs at its *resonant frequency*. This volt-

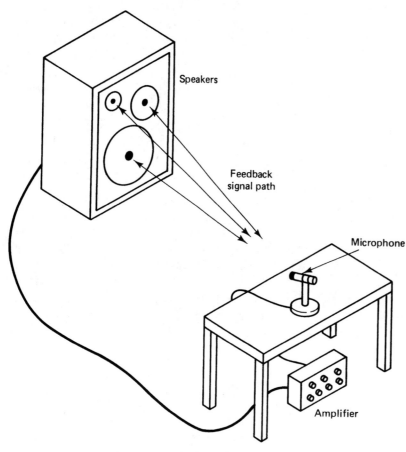

Figure 4-66. Sound system feedback.

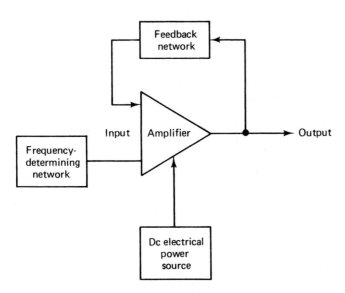

Figure 4-67. Fundamental oscillator parts.

age can be stored in the tank for a short period of time. In an oscillator, the tank circuit is responsible for the frequency of the ac voltage that is being generated.

In order to see how ac is produced from dc, let us take a look at an *LC* tank circuit. Figure 4-68(a) shows a simple tank circuit connected to a battery. In Figure 4-68(b), the switch is closed momentarily. This action causes the capacitor to charge to the value of the battery voltage. Note the direction of the charging current. After a short time the switch is opened. Figure 4-68(c) shows the accumulated charge voltage on the capacitor.

To develop a sine-wave voltage, assume that the capacitor of Figure 4-69(a) has been charged to a desired voltage value by momentarily connecting it to a battery. Figure 4-69(b) shows the capacitor discharging through the inductor. Discharge current flowing through *L* causes an electromagnetic field to expand around the inductor. Figure 4-69(c) shows the capacitor charge depleted.' The field around the inductor then collapses. This causes a continuation of current flow for a short time. Figure 4-69(d) shows the capacitor being charged by the induced voltage of the collapsing field. The capacitor once again begins to discharge through *L*. In Figure 4-69(e), note that the direction of the discharge current is reversed. The electromagnetic field again expands around *L*. Its polarity is reversed. Figure 4-69(f) shows the capacitor with its charge depleted. The field around *L* collapses at this time. This causes a continuation of current flow. Figure 4-69(g) shows *C* being charged by the induced voltage of the collapsing field. This causes *C* to be charged to the same polarity as the beginning step. The process is then repeated. The charge and discharge of *C* through *L* causes an ac voltage to occur in the circuit.

The frequency of the ac voltage generated by a tank circuit is based on the value of *L* and *C*. This is called the *resonant frequency* of a tank circuit. The formula for resonant frequency is

$$\text{Frequency } (f) = \frac{1}{2\,\pi\sqrt{LC}}$$

where *f* is in hertz, *L* is the inductance in henries, and *C* is the capacitance in farads. This formula is extremely important because it applies to all *LC* sine-wave oscillators. Resonance occurs when the inductive reactance (X_L) equals the capacitive reactance (X_C). A tank circuit will oscillate at this frequency for a rather long period of time without a great deal of circuit opposition.

At resonant frequency an *LC* tank always has some circuit resistance. This resistance tends to oppose the ac

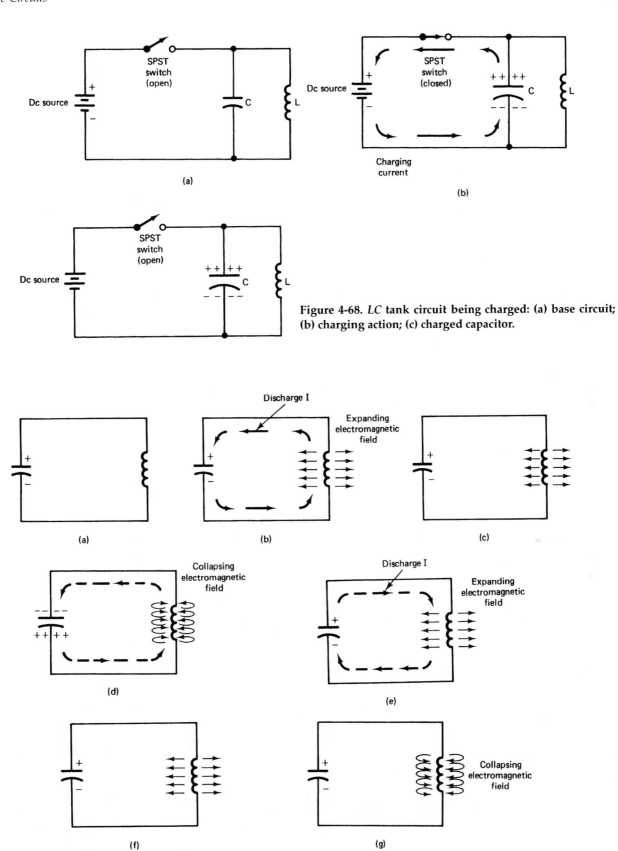

Figure 4-68. *LC* tank circuit being charged: (a) base circuit; (b) charging action; (c) charged capacitor.

Figure 4-69. Charge-discharge action of an *LC* circuit.

that circulates through the circuit. The ac voltage therefore decreases in amplitude after a few cycles of operation. Figure 4-70(a) shows the resulting wave of a tank circuit. Note the gradual decrease in signal amplitude. This is called a *damped sine wave*. A tank circuit produces a wave of this type.

A tank circuit, as we have seen, generates an ac signal when it is energized with dc. The generated wave gradually diminishes in amplitude. After a few cycles of operation, the lost energy is transformed into heat. When the lost energy is great enough, the signal stops oscillating.

The oscillations of a tank circuit could be made continuous if energy were periodically added to the circuit. The added energy would restore that lost by the circuit through heat. The added energy must be in step with the circulating tank energy. If the two are additive, the circulating signal will have a continuous amplitude. Figure 4-70(b) shows the *continuous wave* (CW) of an energized tank circuit.

Our original tank circuit was energized by momentarily connecting a dc source to the capacitor. This action caused the capacitor to become charged by manual switching action. To keep a tank circuit oscillating by manual switch operation is physically impossible. Switching

of this type is normally achieved by an electronic device, such as transistors. The output of the transistor is applied to the tank circuit in proper step with the circulating energy. When this is achieved the oscillator generates a CW signal of a fixed frequency.

The inductance (coil) of a tank circuit varies a great deal depending on the frequency being generated. Most applications of *LC* oscillators are in the RF range._Coil inductance is usually a variable. Inductance is changed by moving the position of a *ferrite core* inside the coil. Adjustment of the inductance will permit some change in the frequency of the tank circuit.

Armstrong Oscillator

The Armstrong oscillator of Figure 4-71 is used to show the basic operation of an *LC* oscillator. The characteristic curve of the transistor and its ac load line are also used to explain its operation. Note that conventional bias voltages are applied to the transistor. The emitter-base junction is forward-biased and the collector is reverse-biased. Emitter biasing is achieved by R_3, and R_1 and R_2 serve as a divider network for the base voltage.

The output of the Armstrong oscillator of Figure 4-71 can be changed by altering the value of R_3. Gain is highest when the variable arm of R_3 is at the top of its

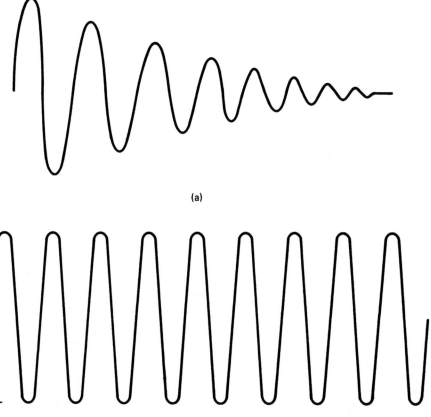

Figure 4-70. Wave types: (a) damped oscillatory wave; (b) continuous wave.

(a)

(b)

range. This adjustment may cause a great deal of signal distortion. In some cases, the output voltage may appear as a square wave instead of the sine wave. When the position of the variable arm is moved downward, the output wave becomes a sine wave. Adjusting R_3 to the bottom of its range may cause oscillations to stop. The gain of Q_1 may not be capable of developing enough feedback for regeneration at this point. Normally 20 to 30% of output must be returned to the base to assure continuous oscillation.

Hartley Oscillator

The Hartley oscillator of Figure 4-72 is used in AM and FM radio receivers. The resonant frequency of the circuit is determined by the values of T_1 and C_1. Capacitor C_2 is used to couple ac to the base of Q_1. Biasing for Q is provided by R_2 and R_1. Capacitor C4 couples ac variations in collector voltage to the lower side of T_1. The RF *choke coil* (L_1) prevents ac from going into the power supply. L_1 also serves as the load for the circuit. Q_1 is an NPN transistor connected in a common-emitter circuit configuration.

A distinguishing feature of the Hartley oscillator is its *tapped coil*. A number of circuit variations are possible. The coil may be placed in series with the collector. Collector current flows through the coil in normal operation. A variation of this type is called a *series-fed Hartley oscillator*. The circuit of Figure 4-72 is a *shunt-fed Hartley*

oscillator. I_C does not flow through T_1. The coil is connected in parallel or shunt with the dc voltage source. Only ac flows through the lower part of T_1. Shunt-fed Hartley oscillators tend to produce more stable output.

Colpitts Oscillator

A *Colpitts oscillator* is very similar to the shunt-fed Hartley oscillator. The primary difference is in the tank circuit structure. A Colpitts oscillator uses two capacitors instead of a divided coil. Feedback is developed by an *electrostatic field* across the capacitor divider network. Frequency is determined by two capacitors in series and the inductor.

Figure 4-73 shows a schematic of the Colpitts oscillator. Bias voltage for the base is provided by resistors R_1 and R_2. The emitter is biased by R_4. The collector is reverse biased by connection to the positive side of through R_3. This resistor also serves as the collector load. The transistor is connected in a common-emitter circuit configuration. Feedback energy is added to the tank circuit momentarily during each alternation. As a general rule the Colpitts circuit is a very reliable oscillator.

The amount of feedback developed by the Colpitts oscillator is based on the *capacitance ratio* of C_1 and C_2. The value of C_1 in our circuit is much smaller than C_2. The capacitive reactance (X_C) of C_1 will therefore be much greater for C_1 than for C_2. The voltage across C_1 will be much greater than that across C_2. By making the value

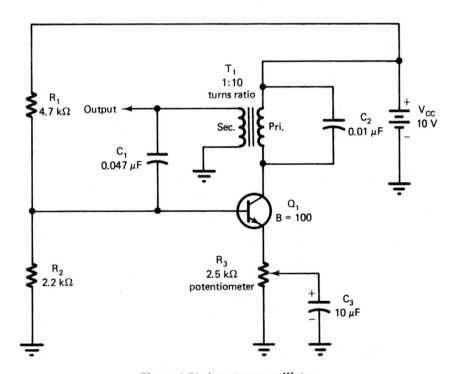

Figure 4-71. Armstrong oscillator.

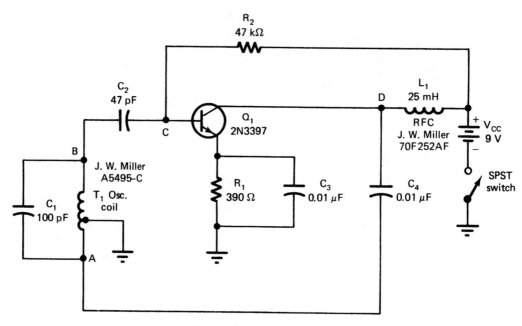

Figure 4-72. Hartley oscillator.

of C_2 smaller, the feedback voltage can be increased. As a general rule, large amounts of feedback may cause the generated wave to be distorted. Small values of feedback will not permit the circuit to oscillate. In practice, 10 to 50% of the collector voltage is returned to the tank circuit as feedback energy.

Crystal Oscillator

When extremely high-frequency stability is desired, *crystal oscillators* are used. The crystal of an oscillator is a finely ground wafer of quartz or Rochelle salt. It can change electrical energy into vibrations of mechanical energy and can also change mechanical vibrations into electrical energy. These two conditions of operation are called the *piezoelectrical effect.*

The crystal of an oscillator is placed between two metal plates. Contact is made to each surface of the crystal by these plates. The entire assembly is then mounted in a holding case. Connection to each plate is made through terminal posts. When a crystal is placed in a circuit, it is plugged into a socket.

A crystal by itself behaves like a series resonant circuit. It has inductance (L), capacitance (C), and resistance (R). Figure 4-74(a) shows an equivalent circuit. The L function is determined by the mass of the crystal. Capacitance deals with its ability

to change mechanically. Resistance corresponds to the electrical equivalent of mechanical friction. The series resonant equivalent circuit changes a great deal when the crystal is placed in a holder. Capacitance due to the metal support plates is added in parallel with the crystal equivalent circuit. Figure 4-74(b) shows the equivalent circuit of a crystal placed in a holder. In a sense, a crystal can have either a series resonant or a parallel resonant characteristic.

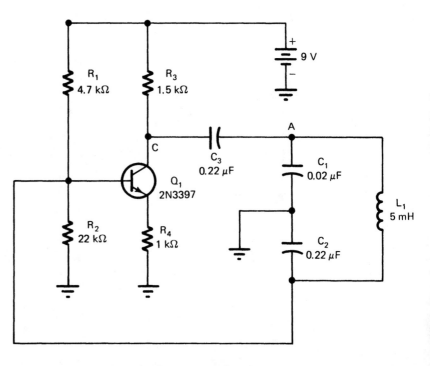

Figure 4-73. Colpitts oscillator.

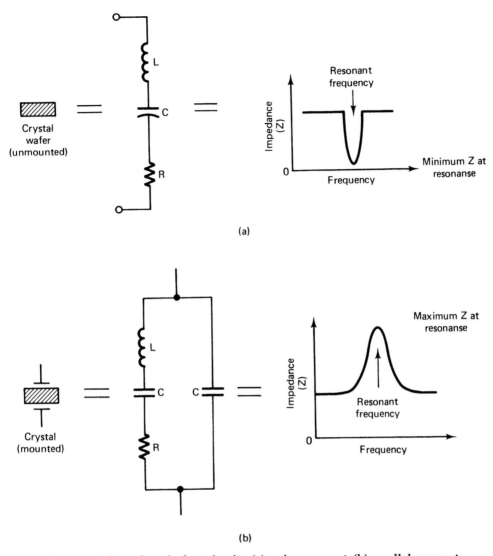

Figure 4-74. Crystal equivalent circuits: (a) series resonant; (b) parallel resonant.

In an oscillator, the placement of a crystal in the circuit has a great deal to do with how it will respond. If placed in the tank circuit, it responds as a parallel resonant device. In some applications the crystal serves as a tank circuit. When a crystal is placed in the feedback path it responds as a series resonant device. It essentially responds as a sharp filter. It only permits feedback of the desired frequency. Hartley and Colpitts can be modified to accommodate a crystal of this type. The operating stability of this type of oscillator is much better than the circuit without a crystal. Figure 4-75 shows crystal-controlled versions of the Hartley and Colpitts. Operation is essentially the same.

The Pierce oscillator of Figure 4-76 employs a crystal as its tank circuit. In this circuit the crystal responds as a parallel resonant circuit. In a sense, the Pierce oscillator is a modification of the basic Colpitts oscillator.

The crystal is used in place of the tank circuit inductor. A specific crystal is selected for the desired frequency to be generated. The parallel resonant frequency of the crystal is slightly higher than its equivalent series resonant frequency.

Operation of the Pierce oscillator is based on feedback from the collector to the base through C_1 and C_2. These two capacitors provide a combined 180° phase shift. The output of the common-emitter amplifier is therefore inverted to achieve in-phase, or regenerative, feedback. The value ratio of C_1 and C_2 determines the level of feedback voltage. From 10 to 50% of the output must be fed back to energize the crystal. When it is properly energized, the resonant frequency response of the crystal is extremely sharp. It will vibrate only over a narrow range of frequency. The output at this frequency is very stable. The output of a Pierce oscillator is usually

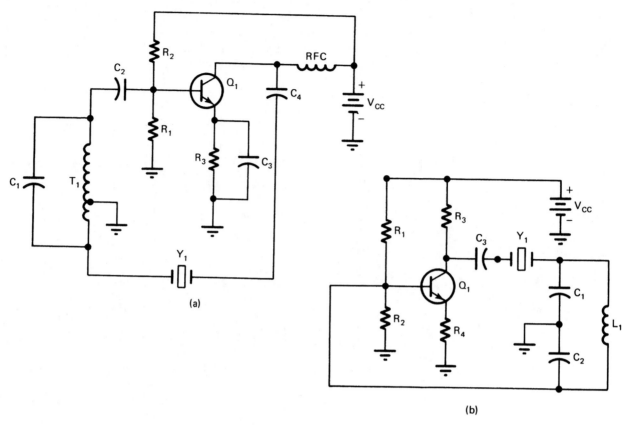

Figure 4-75. Crystal-controlled oscillators: (a) Hartley; (b) Colpitts.

quite small. A crystal can be damaged by excessive mechanical strains and by heat caused by excessive power.

RELAXATION OSCILLATORS

Relaxation oscillators are primarily responsible for the generation of *nonsinusodial* waveforms. *Sawtooth, rectangular,* and a variety of irregular-shaped waves are included in this classification. These oscillators generally depend on the charge and discharge of a capacitor-resistor network for their operation. Voltage changes from the network are used to alter the conduction of an electronic device. Transistors, unijunction transistors (UJTs), and ICs can be used to perform the control function of this oscillator.

RC Circuits

When a source of dc is connected in series with a resistor and a capacitor, we have an *RC* circuit (see Figure 4-77(a)). The SPDT switch is used to charge and discharge the capacitor. In the charge position, the voltage source is

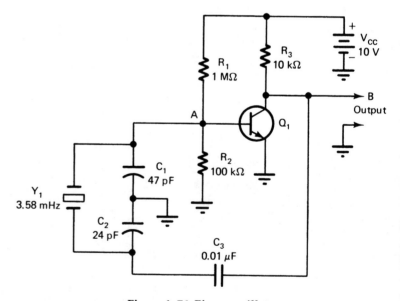

Figure 4- 76. Pierce oscillator.

applied to the *RC* circuit, and *C* charges through *R*. In the discharge position, the source is removed from the circuit and *C* discharges through *R*.

Individual *time constant curves* are shown in Figure 4-1(b) for each component of the *RC* circuit. V shows the

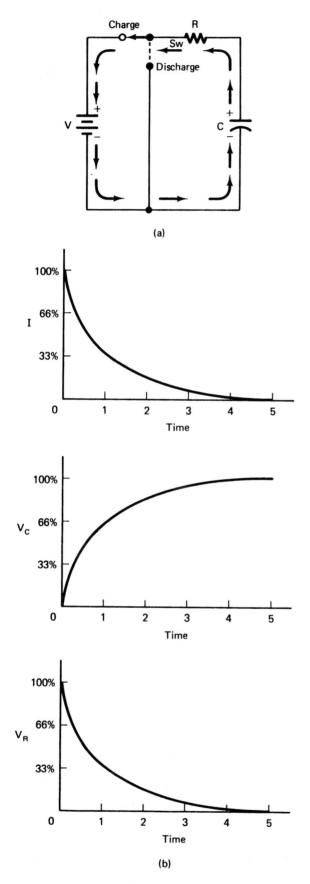

(a)

(b)

Figure 4-77. *RC* charging action: (a) circuit; (b) circuit values.

capacitor voltage with respect to time. V_R is the resistor voltage. Circuit current flow is displayed by the I_C curve. Note how the value of these curves change with respect to time.

The circuit switch is initially placed in the charge position. Note the direction of charge current indicated by the circuit arrows. In one time constant the V curve rises to 63% of the source voltage. Beyond this point there is a smaller change in the value of V. After five time constants the capacitor is considered to be fully charged. Note also how the circuit current and the resistor voltage change with respect to time. Initially, I_C and V_R rise to maximum values. After one time constant, only 37% of I_C flows. V_E, which is current-dependent, follows the change in I_C. After five time constants, there is zero current flow. This shows that C has been fully charged. With C charged there is no circuit current flow, and V_R is zero. The circuit remains in this state as long as the switch remains in the charge position.

Assume now that the capacitor of the circuit of Figure 4-78(a) has been fully charged. The switch is now placed in the discharge position. In this situation, the dc source is removed from the circuit. The resistor is now connected across the capacitor, and C discharges through R. Current flows from the lower plate of C through R to the upper plate of C. Note the direction of the discharge current path indicated by circuit arrows.

The initial discharge current is of a maximum value. See the individual time constant curves for V, I_C, and V_R in Figure 4-78(b). The discharge current shows I_C to be a maximum value initially. After one time constant, it drops to 37% of the maximum value. After five time constants I drops to zero. V_C and V_R follow I in the same manner. After five time constants, C is discharged. There is no circuit I, and V_C and V_R are both zero. The circuit remains in this state as long as the switch remains in the discharge position.

UJT Oscillator

The charge and discharge of a capacitor through a resistor can be used to generate a *sawtooth* waveform. The charge-discharge switch of Figs. 4-77 and 4-78 can be replaced with an active device. Transistors or ICs can be used to accomplish the switching action. Conduction and nonconduction of the active device regulate the charge and discharge of the *RC* network. Circuits connected in this manner are classified as *relaxation oscillators*. When the circuit device is conductive, it is active. When it is not conductive, it is relaxed. The active device of a relaxation oscillator switches states between conduction and nonconduction. This regulates the charge and discharge rate

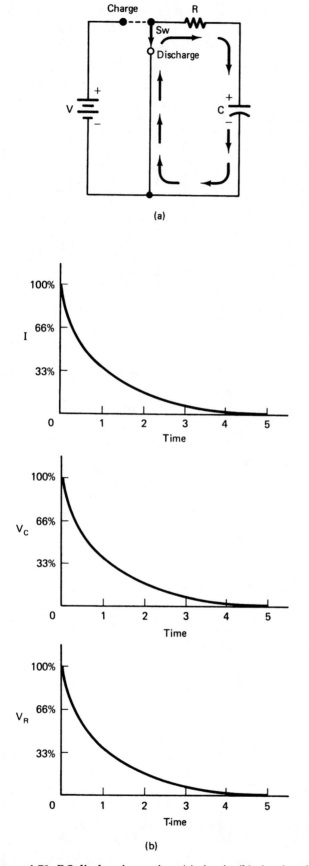

(a)

(b)

Figure 4-78. *RC discharging action: (a) circuit; (b) circuit values.*

of the capacitor. A sawtooth waveform appears across the capacitor of this type of oscillator.

A UJT is used in the relaxation oscillator of Figure 4-79. The *RC* network is composed of R_1 and C_1. The junction of the network is connected to the emitter (*E*) of the UJT. The UJT will not go into conduction until a certain voltage value is reached. When conduction occurs, the emitter-base 1 junction becomes low resistant. This provides a low-resistant discharge path for *C*. Current flows through R_3 only when the UJT is conducting. R_3 refers to the resistance of the speaker in this circuit.

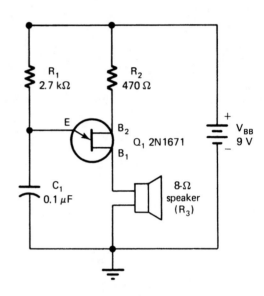

Figure 4-79. UJT oscillator.

Assume now that power is applied to the UJT oscillator circuit. The UJT is in a nonconductive state initially. The emitter-B_1 junction is reverse-biased. C_1 begins to accumulate charge voltage after a short period of time. Time equals $R \times C$ in this case. The capacitor eventually charges to a voltage value that will cause the $E-B_1$ junction to become conductive. When this point is reached, the $E-B_1$ junction becomes low resistant. C_1 discharges immediately through the low-resistant $E-B_1$ junction. This action removes the forward bias voltage from the emitter. The UJT immediately becomes nonconductive. C_1 begins to charge again through R_1. The process is then repeated on a continuous basis.

UJT oscillators are found in applications that require a signal with a slow rise time and a rapid fall time. The $E-B_1$ junction of the UJT has this type of output. Between B_1 and circuit ground, the UJT produces a *spiked pulse*. This type of output is frequently used in timing circuits and counting applications. The time that it takes for the waveform to repeat itself is called the *pulse repetition rate* (*PRR*).

This term is very similar to the hertz designation for frequency. As a general rule, UJT oscillators are very stable and are quite accurate when used with time constants of one or less.

Astable Multivibrator

Multivibrators are a very important classification of relaxation oscillator. This type of circuit employs an *RC* network in its physical makeup. A rectangular-shaped wave is developed by the output. Astable multivibrators are commonly used in television receivers to control the electron-beam deflection of the picture tube. Computers use this type of generator to develop timing pulses.

Multivibrators are considered to be either *triggered* devices or *free running*. A triggered multivibrator requires an input signal or timing pulse to be made operational. The output of this multivibrator is controlled or *synchronized* by an input signal. The electron-beam deflection oscillators of a television receiver are triggered into operation. When the receiver is tuned to an operating channel, its oscillators are synchronized by the incoming signal. When it is tuned to an unused channel the oscillators are free running. Free-running oscillators are self-starting. They operate continuously as long as electrical power is supplied. The shape and frequency of the waveform is determined by component section. An astable multivibrator is a free-running oscillator.

A multivibrator is composed of two amplifiers that are *cross-coupled*. The output of amplifier 1 goes to the input of amplifier 2. The output of amplifier 2 is coupled back to the input of amplifier 1. Since each amplifier inverts the polarity of the input signal, the combined effect is a positive feedback signal. With positive feedback an oscillator is regenerative and produces continuous output.

Figure 4-80, shows a representative multivibrator using bipolar transistors. These amplifiers are connected in a common-emitter circuit configuration. R_2 and R_3 provide forward bias voltage for the base of each transistor. Capacitor C_1 couples the collector of transistors Q_1 to the base of Q_2. Capacitor C_2 couples the collector of Q_2 to the base of Q_1. Because of the cross-coupling, one transistor will be conductive and one will be cut off. After a short period of time the two transistors will change states. The conducting transistor will be cut off and the off transistor will become conductive. The circuit changes back and forth between these two states. The output of the circuit will be a rectangular-shaped wave. An output signal can be obtained from the collector of either transistor. As a rule the output is labeled Q or $\bar{Q}$. This denotes that outputs are of an opposite polarity.

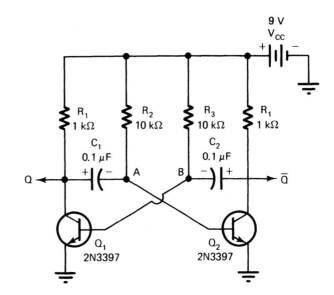

Figure 4-80. Astable multivibrator.

The oscillation frequency of a multivibrator is determined by the time constant of R_2–C_1 and R_3–C_2. The values of R_2 and R_3 are usually selected to cause each transistor to reach saturation. C_1 and C_2 are then chosen to develop the desired operating frequency. If C_1 equals C_2 and R_2 equals RB_3, the output is symmetrical. This means that each transistor will be on and off for an equal amount of time. The output frequency f of a *symmetrical multivibrator* is determined by the formula

$$f = \frac{1}{1.4RC}$$

If the resistor and capacitor values are unequal, the output will not be symmetrical. One transistor could be on for a long period with the alternate transistor being on for only a short period. The output of a nonsymmetrical multivibration is described as a rectangular wave.

Monostable Multivibrator

A *monostable multivibrator* has one stable state of operation. It is often called a *one-shot* multivibrator. One trigger pulse causes the oscillator to change its operational state. After a short period of time, however, the oscillator returns to its original starting state. The *RC* time constant of this circuit determines the time period of the state change. A monostable multivibrator always returns to its original state. No operational change occurs until a trigger pulse is applied. A monostable multivibrator is considered to be a *triggered* oscillator.

Figure 4-81 shows a schematic of a monostable multivibrator. This circuit has two operational states. Its stable state is based on conduction of Q_2 with Q_1 cut off. The stable state occurs when Q_1 is conductive and Q_2 is cut off. The circuit relaxes in its stable state when no trigger pulse is supplied. The unstable state is initiated by a trigger pulse. When a trigger pulse arrives at the input, the circuit changes from its stable state to the unstable state. After the time of $0.7 \times R_2C_1$, the circuit goes back to its stable state. No current occurs until another trigger is applied to the input.

Bistable Multivibrator

A *bistable multivibrator* has two stable states of operation. A trigger pulse applied to the input causes the circuit to assume one stable state. A second pulse causes it to switch to the alternate stable state. This type of multivibrator changes states only when a trigger pulse is applied. It is often called a flip-flop. It flips to one state when triggered and flops back to the other state when triggered. The circuit becomes stable in either state. It does not change states, or *toggle*, until commanded to do so by a trigger pulse. Figure 4-82 shows a schematic of a bistable multivibrator using bipolar transistors.

IC Waveform Generators

The *NE/SE 555* IC is a multifunction device that is widely used today. It can be modified to respond as an *astable multivibrator*. This particular circuit can be achieved with a minimum of components and a power

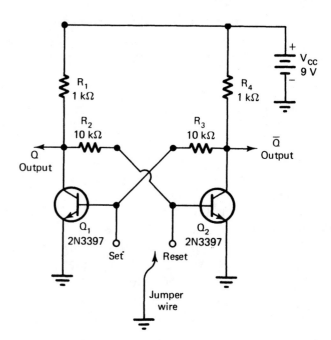

Figure 4-82. Bistable multivibrator.

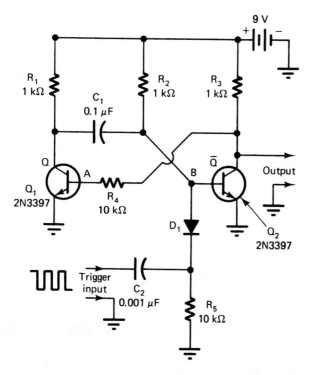

Figure 4-81. Monostable multivibrator.

source. Circuit design is easy to accomplish, and operation is very reliable through a number of manufacturers. As a rule, the number 555 usually appears in the manufacturer's part identification number. Some of the common part numbers for this chip are SN72555, MC14555, SE555, LM555, XR555, and CA555.

The internal circuitry of a 555 IC is generally viewed in functional blocks. In this regard, the chip has two comparators, a bistable flip-flop, a resistive divider, a discharge transistor, and an output stage. Figure 4-83 shows the functional blocks of a 555 IC.

The *voltage divider* of the IC consists of three 5-kΩ resistors. The network is connected internally across the $+V_{CC}$ and ground-supply source. Voltage developed by the lower resistor is one-third of V_{CC}. The middle divider point is two-thirds of the value of V_{CC}. This connection is terminated at pin 5. Pin 5 is designated as the control voltage.

The two *comparators* of the 555 respond as an amplifying switch circuit. A reference voltage is applied to one input of each comparator. A voltage value applied to the other input initiates a change in output when it differs with the reference value. Comparator is referenced at two thirds of V_{CC} at its negative input, which is where pin 5 is connected to the middle divider resistor. The other input is terminated at pin 6. This pin is called the *threshold* terminal. When the voltage at pin 6 rises above two-thirds of V_{CC}, the output of the comparator swings positive. This is then applied to the *reset* input of the flip-flop.

Comparator 2 is referenced to one third of V_{CC}. The

positive input of comparator 2 is connected to the lower divider network resistor. External pin connection 2 is applied to the negative input of comparator 2. This is called the *trigger* input. If the voltage of the trigger drops below one third of V_{CC} the comparator output will swing positive. This is applied to the *set* input of the flip-flop.

The flip-flop of the 555 IC is a bistable multivibrator. It has reset and set inputs and one output. When the reset input is positive, the output goes positive. A positive voltage to the set input causes the output to go negative. The output of the flip-flop is dependent on the status of the two comparator inputs.

The output of the flip-flop is applied to both the output stage and the discharge transistor. The output stage is terminated at pin 3. The discharge transistor is connected to terminal 7. The output stage is a power amplifier and a signal inverter. A load device connected to terminal 3 will see either $+V_{CC}$ or ground, depending on the state of the

input signal. The output terminal switches between these two values. Load current values of up to 200 mA can be controlled by the output terminal. A load device connected to $+V_{CC}$ is energized when pin 3 goes to ground. When the output goes to $+V_{CC}$ the output is off. A load device connected to ground turns on when the output goes to $+V_{CC}$. It is off when the output goes to ground. The output switches back and forth between these two states.

Transistor Q_1 is called a discharge transistor. The output of the flip-flop is applied to the base of Q. When the flip-flop is reset (positive), it forward biases Q. Pin 7 connects to ground through Q_1. This causes pin 7 to be grounded. When the flip-flop is set (negative), it reverse biases Q_1. This causes pin 7 to be infinite, or open, with respect to ground. Pin 7 therefore has two states, shorted to ground or open.

We will now see how the internal circuitry of the 555 responds as a multivibrator.

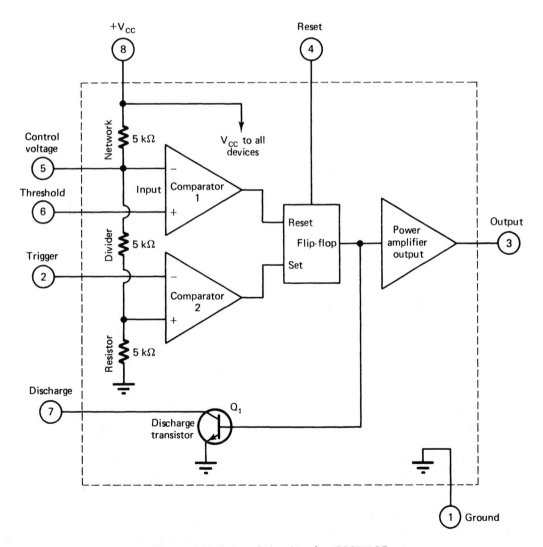

Figure 4-83. Internal circuits of an LM555 IC.

IC Astable Multivibrator

When used as an astable multivibrator, the 555 is an *RC oscillator*. The shape of the waveform and its frequency are determined primarily by an *RC* network. The astable multivibrator circuit is self-starting and operates continuously for long periods of time. Figure 4-84(a) shows the LM 555 connected as an astable multivibrator. A common

put to the input. An iron core transformer is used in this circuit because the generated frequency is 60 Hz.

A transistor vertical blocking oscillator is shown in Figure 4-85. A sawtooth-forming capacitor (C_1) charges when the source voltage is initially applied. It discharges through the transistor when it becomes conductive. The output across the capacitor is a sawtooth waveform.

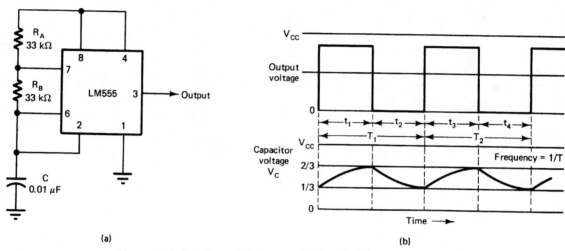

(a)   (b)

Figure 4-84. Astable multivibrator: (a) circuit; (b) waveforms.

application of this circuit is the time-base generator for clock circuits and computers.

Connection of the 555 IC as an astable multivibrator requires two resistors, a capacitor, and a power source. The output of the circuit is connected to pin 3. Pin 8 is $+V_{CC}$ and pin 1 is ground. The supply voltage can be from 5 to 15 V dc. Resistor *RA* is connected between $+V_{CC}$ and the discharge terminal (pin 7). Resistor R_B is connected between pin 7 and the threshold terminal (pin 6). A capacitor is connected between the threshold and ground. The trigger (pin 2) and threshold (pin 6) are connected together.

Blocking Oscillators

Blocking oscillators are an example of the relaxation principle. The active device, which is ordinarily a transistor, is cut off during most of the operational cycle. It turns on for a short operational period of time to discharge an *RC* network. In appearance this circuit closely resembles the Armstrong oscillator. Feedback from the output to the input is needed to achieve the blocking function. The output of the oscillator is used to change the shape of a waveform. This circuit was commonly used as the vertical oscillator of a television receiver. The vertical *blocking oscillator (VBO)* transformer provides regenerative feedback from the out-

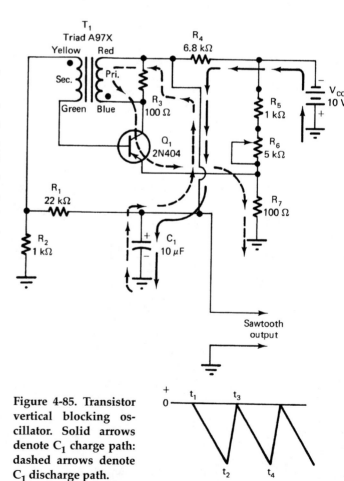

Figure 4-85. Transistor vertical blocking oscillator. Solid arrows denote C_1 charge path: dashed arrows denote C_1 discharge path.

Chapter 5

Digital Electronics

INTRODUCTION

A circuit that employs a numerical signal in its operation is classified as a digital circuit. Computers, pocket calculators, digital instruments, and numerical control (NC) equipment are common applications of digital circuits. Practically unlimited quantities of digital information can be processed in short periods of time electronically. With operational speed of prime importance in electronics today, digital circuits are used more frequently.

In this chapter, digital circuit applications are discussed. There are many types of digital circuits that have applications in electronics, including logic circuits, flip-flop circuits, counting circuits, and many others. The first sections of this unit discuss the number systems that are basic to digital circuit understanding. The remainder of the chapter introduces some of the types of digital circuits and explains Boolean algebra as it is applied to logic circuits.

OBJECTIVES

Upon completion of this unit, you will be able to:

1. Explain the difference between digital and analog systems.
2. Identify the functional operation of different logic gates according to their input and output.
3. Verify the operation of logic gates through a truth table.
4. Count with binary, octal, and hexadecimal numbers.
5. Convert decimal numbers to equivalent binary, octal, and hexadecimal numbers.
6. Convert binary, octal, and hexadecimal numbers to equivalent decimal values.
7. Define Boolean algebra and show how it is used in digital electronics.
8. Describe the operation of a digital counting circuit.

DIGITAL NUMBER SYSTEMS

The most common number system used today is the *decimal* system, in which 10 digits are used for counting. The number of digits in the system is called its *base* (or radix). The decimal system, therefore, has a base of 10.

Numbering systems have a *place value*, which refers to the placement of a digit with respect to others in the counting process. The largest digit that can be used in a specific place or location is determined by the base of the system. In the decimal system the first position to the left of the decimal point is called the *units place*. Any digit from 0 to 9 can be used in this place. When number values greater than 9 are used, they must be expressed with two or more places. The next position to the left of the units place in a decimal system is the tens place. The number 99 is the largest digital value that can be expressed by two places in the decimal system. Each place added to the left extends the number system by a power of 10.

Any number can be expressed as a sum of weighted place values. The decimal number 2583, for example, is expressed as $(2 \times 1000) + (5 \times 100) + (8 \times 10) + (3 \times 1)$. The decimal number 5362 is expressed in Figure 5-1.

The decimal number system is commonly used in our daily lives. Electronically, however, it is rather difficult to use. Each digit of a base 10 system would require a specific value associated with it, so it would not be practical.

BINARY NUMBER SYSTEM

Electronic digital systems are ordinarily of the *binary* type, which has 2 as its base. Only the numbers 0 or 1 are used in the binary system. Electronically, the value of 0 can be associated with a low-voltage value or no voltage. The number 1 can then be associated with a voltage value larger than 0. Binary systems that use these voltage values are said to have *positive logic*. *Negative logic*, by comparison, has a voltage assigned to 0 and no voltage value assigned to 1. Positive logic is used in this chapter.

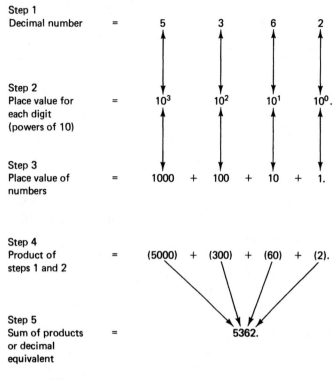

Figure 5-1. Expressing a decimal number.

The two operational states of a binary system, 1 and 0, are natural circuit conditions. When a circuit is turned off or has no voltage applied, it is in the off, or 0, state. An electrical circuit that has voltage applied is in the on, or 1, state. By using transistors or ICs, it is electronically pos-

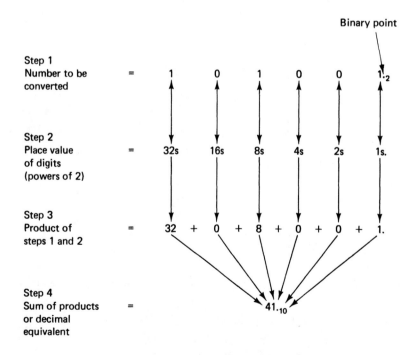

Figure 5-2. Binary-to-decimal conversion.

sible to change states in less than a microsecond. Electronic devices make it possible to manipulate millions of 0s and is in a second and thus to process information quickly.

The basic principles of numbering used in decimal numbers apply in general to binary numbers. The base of the binary system is 2, meaning that only the digits 0 and 1 are used to express place value. The first place to the left of the binary point, or starting point, represents the units, or is, location. Places to the left of the binary point are the powers of 2. Some of the place values in base 2 are $2^0 = 1$, $21 = 2, 22 = 4, 2 = 8, 2 = 16, 2 = 32$, and $26 = 64$.

When bases other than 10 are used, the numbers should have a subscript to identify the base used. The number 100_2 is an example.

The number 100_2 (read "one, zero, zero, base 2") is equivalent to 4 in base 10, or 4_{10}. Starting with the first digit to the left of the binary point, this number has value $(0 \times 2^0) + (0 \times 2^1) + (1 \times 2^2)$ or $0 + 0_0 + 4_{10} = 4_{10}$. The conversion of a binary number to an equivalent decimal number is shown in Figure 5-2. In this method of conversion, write down the binary number first. Starting at the binary point, indicate the decimal equivalent for each binary place location where a 1 is indicated. For each 0 in the binary number leave a blank space or indicate a 0. Add the place values and then record the decimal equivalent.

The conversion of a decimal number to a binary equivalent is achieved by repetitive steps of division by the number 2. When the quotient is even with no remainder, a 0 is recorded. When the quotient has a remainder, a 1 is recorded. The steps to convert a decimal number to binary number are shown in Figure 5-3. This conversion is achieved by writing down the decimal number (45). Divide this number of the base of the system (2). Record the quotient and remainder as indicated. Move the quotient of step 1 to step 2 and repeat the process. The division process continues until the quotient is 0. The binary equivalent consists of the remainder values in the order last to first.

BINARY-CODED DECIMAL (BCD) NUMBER SYSTEM

When large numbers are indicated by binary numbers, they are difficult to use. For this reason, the *binary-coded decimal* (BCD) method of counting was devised. In this system four binary digits are used to represent each decimal digit. To illustrate this procedure, the

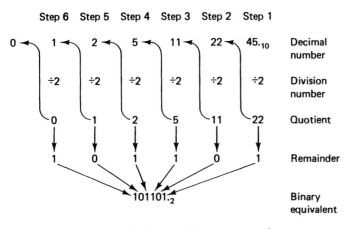

Figure 5-3. Decimal-to-binary conversion.

number 105, is converted to a BCD number. In binary numbers, $105_{10} = 1000101_2$.

To apply the BCD conversion process, the base 10 number is first divided into digits according to place values (see Figure 5-4). The number 105_{10} gives the digits 1-0-5. Converting each digit to binary gives 0001-0000-0101_{BCD}. Decimal numbers up to 999_{10} may be displayed by this process with only 12 binary numbers. The hyphen between each group of digits is important when displaying BCD numbers.

The largest digit to be displayed by any group of BCD numbers is 9. Six digits of a number-coding group are not used at all in this system. Because of this, the octal (base 8) and the hexadecimal (base 16) systems were devised. Digital circuits process numbers in binary form but usually display them in BCD, octal, or hexadecimal form.

OCTAL NUMBER SYSTEM

The octal (base 8) number system is used to process large numbers by digital circuits. The *octal system* of num-

bers uses the same basic principles as the decimal and binary systems.

The octal number system has a base of 8. The largest number used in a base 8 system is 7. The place values starting at the left of the octal point are the powers of eight: $8^0 = 1$, $8^1 = 8$, $8^2 = 64$, $8^3 = 512$, $8^4 = 4096$, and so on.

The process of converting an octal number to a decimal number is the same as that used in the binary-to-decimal conversion process. In this method, however, the powers of 8 are used instead of the powers of 2. The procedure for changing 382_8 to an equivalent decimal number is outlined in Figure 5-5.

Converting an octal number to an equivalent binary number is similar to the BCD conversion process. The octal number is first divided into digits according to place value. Each octal digit is then converted into an equivalent binary number using only three digits. The steps of this procedure are shown in Figure 5-6.

Converting a decimal number to an octal number is a process of repetitive division by the number 8. After the

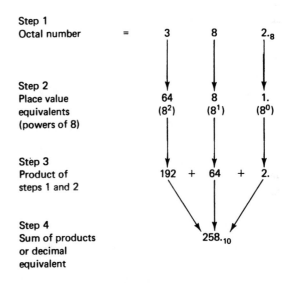

Figure 5-5. Octal-to-decimal conversion.

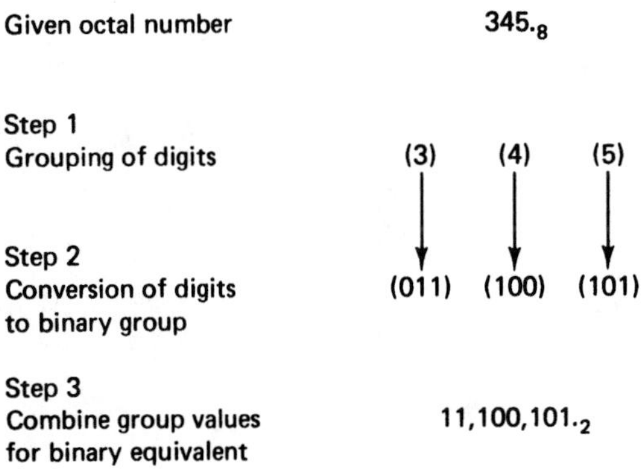

Given decimal number			105.$_{10}$
Step 1 Grouping of digits	(1)	(0)	(5)
Step 2 Conversation of each digit to binary group	(0001)	(0000)	(0101)
Step 3 Combine group values	0001/0000/0101.$_{BCD}$		

Figure 5-4. Decimal-to-BCD conversion.

Given octal number		345.$_8$	
Step 1 Grouping of digits	(3)	(4)	(5)
Step 2 Conversion of digits to binary group	(011)	(100)	(101)
Step 3 Combine group values for binary equivalent	11,100,101.$_2$		

Figure 5-6. Octal-to-binary conversion.

quotient has been determined, the remainder is brought down as the place value. When the quotient is even with no remainder, a 0 is transferred to the place position. The procedure for converting 4098_{10} to base 8 is outlined in Figure 5-7.

Converting a binary number to an octal number is an important conversion process of digital circuits. Binary numbers are first processed at a very high speed. An output circuit then accepts this signal and converts it to an octal signal displayed on a readout device.

Assume that the number 110100100_2 is to be changed to an equivalent octal number. The digits must first be divided into groups of three, starting at the octal point. Each binary group is then converted into an equivalent octal number. These numbers are then combined, while remaining in their same respective places, to represent the equivalent octal number. The conversion steps are outlined in Figure 5-8.

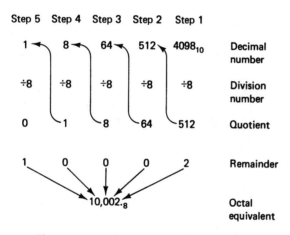

Figure 5-7. Decimal-to-octal conversion.

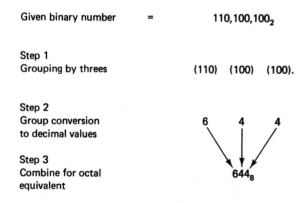

Figure 5-8. Binary-to-octal conversion.

HEXADECIMAL NUMBER SYSTEM

The *hexadecimal* number system is used in digital systems to process large number values. The base of this system is 16, which means that the largest number used in a place is 15. Digits used by this system are the numbers 0-9 and the letters *A-F* The letters *A-F* are used to denote the digits 10-15, respectively. The place values to the left of the hexadecimal point are the powers of 16: $16^0 = 1$, $16^1 = 16$, $16^2 = 256$, $16^3 = 4096$, $16^4 = 65,536$, and so on.

The process of changing a hexadecimal number to a decimal number is similar to that outlined for other conversions. Initially, a hexadecimal number is recorded in proper digital order, as shown in Figure 5-9. The place values, or powers of the base, are then positioned under the respective digits in step 2. In step 3, the value of each digit is recorded. The values in steps 2 and 3 are then multiplied together and added in step 4. The sum (step 5) gives the decimal equivalent value of a hexadecimal number.

The process of changing a hexadecimal number to a binary equivalent is a simple grouping operation. Figure 5-10 shows the steps for making this conversion. Initially, the hexadecimal number is separated into digits in step 1. Each digit is then converted to a binary number using four digits per group. Step 3 shows the binary group combined to form the equivalent binary number.

The conversion of a decimal number to a hexadecimal number is achieved by repetitive division, as with other number systems. In this procedure the division is by 16 and remainders can be as large as 15. Figure 5-11 shows the steps for this conversion.

Converting a binary number to a hexadecimal equivalent is the reverse of the hexadecimal to binary process. Figure 5-12 shows the steps of this procedure. Initially, the binary number is divided in groups of four digits, starting at the hexadecimal point. Each number group is then converted to a hexadecimal value and combined to form the hexadecimal equivalent number.

BINARY LOGIC CIRCUITS

In digital circuit-design applications binary signals are far superior to those of the octal, decimal, or hexadecimal systems. Binary signals can be processed very easily through electronic circuitry, since they can be represented by two stable states of operation. These states can be easily defined as on or off, 1 or 0, up or down, voltage or no voltage, right or left, or any other two-condition states. There must be no in-between state.

Step 1:	Hexadecimal number	=	1	2	C	D.$_{16}$
Step 2:	Place value equivalents (powers of 16)	=	4096s	256s	16s	1s.
Step 3:	Place value digits	=	1	2	12.	13.
Step 4:	Product of steps 2 and 3	=	4096 +	512 +	192 +	13.
Step 5:	Sum of products or decimal equivalent	=	4813_{10}			

Figure 5-9. Hexadecimal-to-decimal conversion.

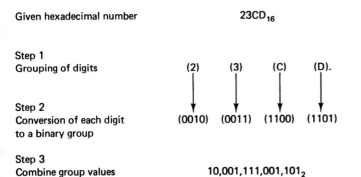

Given hexadecimal number	$23CD_{16}$			
Step 1 Grouping of digits	(2)	(3)	(C)	(D).
	↓	↓	↓	↓
Step 2 Conversion of each digit to a binary group	(0010)	(0011)	(1100)	(1101)
Step 3 Combine group values	$10{,}001{,}111{,}001{,}101_2$			

Figure 5-10. Hexadecimal-to-binary conversion.

Step 4	Step 3	Step 2	Step 1	
1	18	298	4780_{10}	Decimal number
÷16	÷16	÷16	÷16	Divisor
0	1	18	298	Quotient
1	2	10	12	Remainder
1	2	A	C_{16}	Hexadecimal equivalent

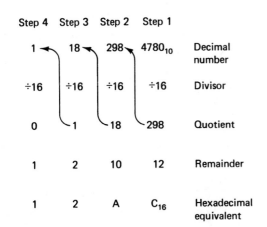

Figure 5-11. Decimal-to-hexadecimal conversion.

The symbols used to define the operational state of a binary system are very important. In positive binary logic, the state of voltage, on, true, or a letter designation (such as A) is used to denote the operational state 1. No voltage, off, false, and the letter A are commonly used to denote the 0 condition. A circuit can be set to either state

Given binary number	$1{,}001{,}101{,}101{,}010_2$			
Step 1 Grouping of fours	(0001)	(0011)	(0110)	(1010)
Step 2 Group conversion to hexadecimal values	1 ↓ 1	3 ↓ 3	6 ↓ 6	10 ↓ A
Step 3 Combine for hexadecimal equivalent	$136A_{16}$			

Figure 5-12. Binary-to-hexadecimal conversion.

and will remain in that state until it is caused to change conditions.

Any electronic device that can be set in one of two operational states or conditions by an outside signal is said to be *bistable*. Relays, lamps, switches, transistors, diodes, and ICs may be used for this purpose. A bistable device has the capability of storing one binary digit or bit of information. By using many of these devices, it is possible to build an electronic circuit that will make decisions based upon the applied input signals. The output of this circuit is a decision based upon the operational conditions of the input. Since the application of bistable devices in digital circuits makes logical decisions, they are commonly called *binary logic circuits*.

Three basic circuits of this type are used to make simple logic decisions. These are the *AND circuit*, *OR circuit*, and the *NOT circuit*. The logic decision made by each circuit is different from the others.

Electronic circuits designed to perform logic functions are called *gates*. This term refers to the capability of a circuit to pass or block specific digital signals. An IF-THEN type of sentence is often used to describe the basic operation of a logic state. For example, if the inputs applied to an AND gate are all 1, then the output will be 1. If a 1 is applied to any input of an OR gate, then the output will be 1. If an input is applied to a NOT gate, then the output will be the opposite or inverse.

AND GATES

An AND gate has two or more inputs and one output. If all inputs are in the 1 state simultaneously, then there will be a 1 at the output. Figure 5-13 shows a switch-and-lamp analogy of the AND gate, its symbol, and a truth table. In Figure 5-13(a), when a switch is turned on, it represents a 1 condition; off represents a 0. The lamp displays the condition of being a 1 when it is on and 0 when turned off. The switches are labeled *A* and *B* and the output lamp is *C*.

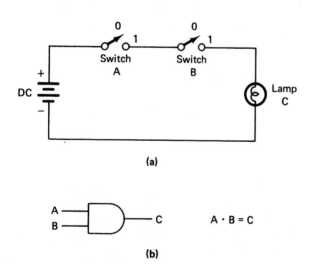

(a)

(b)

Input switch A	Input switch B	Output lamp C
0	0	0
0	1	0
1	0	0
1	1	1

(c)

Figure 5-13. The AND gate.

The operation of a gate is simplified by describing the input-output relationships in a table. The table in Figure 5-13(c) shows the alternatives at the inputs and the corresponding outputs that occur as a result. This description of a gate is called a *truth table*. It shows the predictable operating conditions of a logic circuit.

Each input to an AND gate has two operational states of 1 and 0. A two-input AND gate has 2^2, or 4, possible combinations that influence the output. A three-input gate has 2^3, or 8, combinations, whereas a four-input has 2^4, or 16, combinations. These combinations are normally placed in the truth table in a binary progression. For a two-input gate such a progression is 00, 01, 10, and 11, which shows the binary count of 0, 1, 2, and 3 in order.

The AND gate of Figure 5-13(a) produces only a 1 output when switches A and B are both 1. Mathematically, this action is described as $A \cdot B = C$. This expression shows the multiplication operation. The symbol of an AND gate is shown in Figure 5-13(b).

OR GATES

An OR gate has two or more inputs and one output. Like the AND gate, each input to the OR gate has two possible states: 1 or 0. The output of this gate produces a 1 when either or both inputs are 1. Figure 5-14 shows a lamp-and-switch analogy of the OR gate, its symbol, and a truth table.

An OR gate produces an output of 1 when both switches are 1 or when either switch A or B is 1. Mathematically, this action is described as $A + B = C$. This expression shows OR addition. This gate is used to make logic decisions of whether or not a 1 appears at either input. The truth table of Figure 10-14(c) shows that if any input is a 1, the output is 1.

NOT GATES

A NOT gate has one input and one output. The output of a NOT gate is opposite to that of the input state. Figure 5-15 shows a switch and lamp NOT gate analogy, its symbol, and truth table. When the switch of Figure 5-15(a) is on (in the 1 state), it shorts out the lamp. Placing the switch in the off condition (0) causes the lamp to be on, or in the 1 state. NOT gates are also called *inverters*.

COMBINATION LOGIC GATES

When a NOT gate is combined with an AN gate or an OR gate, it is called a combination logic gate. A NOT-AND

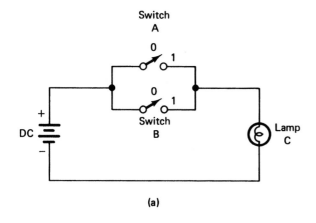

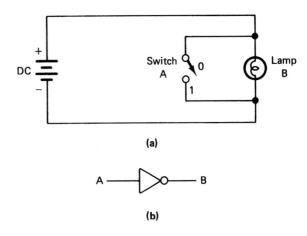

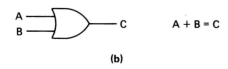

$$A + B = C$$

(b)

(a)

A —▷o— B

(b)

Input switch A	Input switch B	Output lamp C
0	0	0
0	1	1
1	0	1
1	1	1

(c)

Figure 5-14. The OR gate.

Input switch A	Output lamp B
0	1
1	0

(c)

Figure 5-15. The NOT gate.

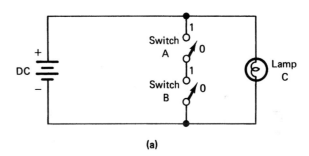

(a)

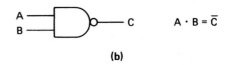 $A \cdot B = \bar{C}$

(b)

gate is called a NAND *gate,* which is an inverted AND gate. Figure 5-16 shows a simple switch-and-lamp circuit analogy of the NAND gate, its symbol, and truth table.

The NAND gate is an inversion of the AND gate. When switches A and B are both on (in the 1 state), lamp C is off (0). When either or both switches are off, lamp C is in the on, or 1, state. Mathematically, the operation of a NAND gate is $A \cdot B = \bar{C}$. The bar over C denotes the inversion, or negative function, of the gate.

A combination NOT-OR, or NOR, gate produces a negation of the OR function. Figure 5-17 shows a switch-and-lamp circuit analogy of this gate, its symbol, and truth table. Mathematically, the operation of a NOR gate is $A + B = \bar{C}$. A 1 appears at the output only when A is 0 and B is 0.

The logic gates discussed here illustrate basic gate operation. The switch-and-lamp input-output analogies are simple ways of showing gate operation. In actual digital electronic applications, solid-state components are ordinarily used to accomplish gate functions.

Input switch A	Input switch B	Output lamp C
0	0	1
0	1	1
1	0	1
1	1	0

(c)

Figure 5-16. The NAND gate.

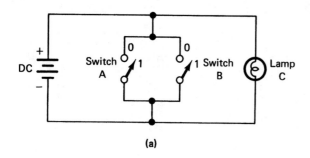

(a)

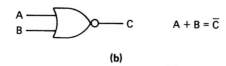

(b)

Input switch A	Input switch B	Output lamp C
0	0	1
0	1	0
1	0	0
1	1	0

(c)

Figure 5-17 The NOR gate.

BOOLEAN ALGEBRA

Boolean algebra is a special form of algebra that was designed to show the relationships of logic operations. This form of algebra is ideally suited for analysis and design of binary logic systems. Through the use of Boolean algebra, it is possible to write mathematical expressions that describe specific logic functions. Boolean expressions are more meaningful than complex word statements or elaborate truth tables. The laws that apply to Boolean algebra are used to simplify complex expressions. Through this type of operation, it may be possible to reduce the number of logic gates needed to achieve a specific function before the circuits are designed.

In Boolean algebra the variables of an equation are assigned letters of the alphabet. Each variable then exists in states of 1 or 0 according to its condition. The 1, or true state, is normally represented by a single letter such as A, B, or C. The opposite state or condition is then described as 0, or false, and is represented by $\bar{A}$ or A'. This is described as NOT A, A negated, or A complemented.

Boolean algebra is somewhat different from conven-

tional algebra with respect to mathematical operations. The Boolean operations are expressed as follows:

Multiplication: A AND B, AB, $A \cdot B$

OR addition: A OR B, $A + B$

Negation, or complementing: NOT A, $\bar{A}$, A'

Assume that a digital logic circuit has three input variables, A, B, and C. The output circuit should operate when only C is on by itself or when A, B, and C are all on at the same time. Expressing this statement in a Boolean expression describes the desired output. Eight (2^3) different combinations of A, B, and C exist in this expression because there are three inputs. Only two of those combinations should cause a signal that will actuate the output. When a variable is not on (0), it is expressed as a negated letter. The original statement is expressed as follows: With A, B, and C on or with A off, B off, and C on, an output (X) will occur:

$$ABC + \overline{ABC} = X$$

A truth table illustrates if this expression is achieved or not. Figure 5-18 shows a truth table for this equation. First, ABC is determined by multiplying the three inputs together. A 1 appears only when the A, B, and C inputs are all 1. Next the negated inputs A and B are determined. Then the products of inputs C, A, and B are listed. The next column shows the addition of ABC and $\overline{ABC}$. The output of this equation shows that output 1 is produced only when $\overline{ABC}$ is 1 or when ABC is 1.

A logic circuit to accomplish this Boolean expression is shown in Figure 5-19. Initially the equation is analyzed to determine its primary operational function. Step 1 shows the original equation. The primary function is

$$ABC + \overline{ABC} = X$$

Inputs A B C	ABC	$\bar{A}$	$\bar{B}$	$\overline{ABC}$	ABC + $\overline{ABC}$ Output
0 0 0	0	1	1	0	0
0 0 1	0	1	1	1	1
0 1 0	0	1	0	0	0
0 1 1	0	1	0	0	0
1 0 0	0	0	1	0	0
1 0 1	0	0	1	0	0
1 1 0	0	0	0	0	0
1 1 1	1	0	0	0	1

Figure 5-18. Truth table for Boolean equation: $ABC + \overline{ABC} = X$.

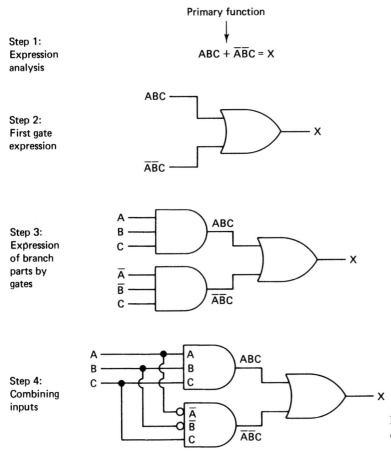

Step 1:
Expression
analysis

Primary function
↓

$ABC + \overline{A}\overline{B}C = X$

Step 2:
First gate
expression

Step 3:
Expression
of branch
parts by
gates

Step 4:
Combining
inputs

Figure 5-19. Logic circuit to accomplish Boolean equation: $ABC + \overline{A}\overline{B}C = X$.

addition, since it influences all parts of the equation in some way. Step 2 shows the primary function changed to a logic gate diagram. Step 3 shows the branch parts of the equation expressed by logic diagrams, with AND gates used to combine terms. Step 4 completes the process by connecting all inputs together. The circles at inputs $\overline{A}\overline{B}$ of the lower AND gate are used to achieve the negative function of these branch parts.

The general rules for changing a Boolean equation into a logic circuit diagram are very similar to those outlined. Initially the original equation must be analyzed for its primary mathematical function. This is then changed into a gate diagram that is inputted by branch parts of the equation. Each branch operation is then analyzed and expressed in gate form. The process continues until all branches are completely expressed in diagram form. Common inputs are then connected together.

TIMING AND STORAGE ELEMENTS

Digital electronics involves a number of items that are not classified as gates. Circuits or devices of this type

have a unique role to play in the operation of a system. Included in this are such things as timing devices, storage elements, counters, decoders, memory, and registers. Truth tables, symbols, operational characteristics, and applications of these items will be presented here. Today, these circuits or devices are built primarily on an IC chip. The internal construction of the chip cannot be effectively altered. Operation is controlled by the application of an external signal to the input. As a rule, very little can be done to control operation other than altering the input signal.

FLIP-FLOPS

In Chapter 4, bistable multivibrators were discussed. This type of device was used to generate a square wave. It could also be triggered to change states when an input signal is applied. A bistable multivibrator, in a strict sense, is a flip-flop. When it is turned on, it assumes a particular operational state. It does not change states until the input is altered. A flip-flop has two outputs. These are generally labeled Q and $\overline{Q}$. They are always of an opposite polarity. Two inputs are usually needed to alter the state of a flip-flop. A variety of names are used for the inputs. These vary a great deal between different flip-flops.

Figure 5-20 shows the truth table and logic symbol of an R-S flip-flop. R stands for the reset input and S represents the set input. Notice that the truth table is somewhat more complex than that of a gate. It shows, for example, the applied input, previous output, and resulting output. Two of the input conditions cause an unpredictable output status.

To understand the operation of an R-S flip-flop, we must first look at the previous outputs. This is the status of the output before a change is applied to the input. The first four items of the previous outputs are $Q1$ and $\bar{Q}0$. The second four states have $Q0$ and $\bar{Q}1$.

In the second line of the truth table note the previous outputs of Q and $\bar{Q}$. When the S is 0 and R is 1, the resulting output does not change. Refer now to the third row. Again note the previous output status of Q and $\bar{Q}$. When S is 1 and R is 0, the resulting outputs change status. The resulting output has three predictable states and one unpredictable state for each of the four previous input combinations.

A variety of different flip-flops are used in digital electronic systems today. In general, each flip-flop type has some unique characteristic to distinguish it from the others. An R-S-T flip-flop, for example, is a triggered R-S flip-flop. It will not change states when the R and S inputs assume a value until a trigger pulse is applied. This would permit a large number of flip-flops to change states all at the same time. Figure 5-21 shows the logic symbol and truth table of the R-S-T flip-flop.

Another very important flip-flop has J-T-K inputs. A J-K flip-flop of this type does not have an unpredictable output state. The J and K inputs have set and clear input capabilities. These inputs must be present for a short time before the clock or trigger input pulse arrives at T In addition to this, J-K flip-flops may employ preset and preclear functions. This is used to establish sequential timing operations. Figure 5-22 shows the logic symbol and truth table of a J-K flip-flop.

A flip-flop has a memory, it can be used in switching operations, and it can count pulses. A series of interconnected flip-flops is generally called a *register*. Each register can store one binary digit or bit of data. Several flip-flops connected form a counter. Counting is a fundamental digital electronic function.

For an electronic circuit to count, a number of things must be achieved. Basically, the circuit must be supplied with some form of data or information that is suitable for

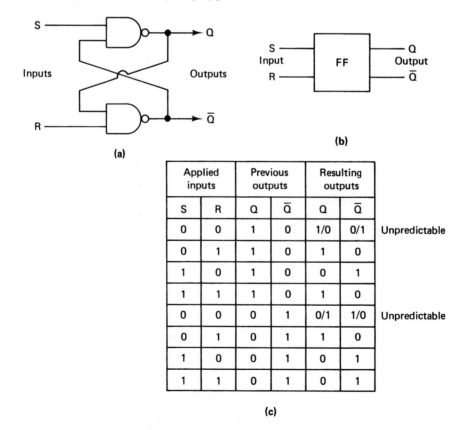

(a)

(b)

Applied inputs		Previous outputs		Resulting outputs		
S	R	Q	$\bar{Q}$	Q	$\bar{Q}$	
0	0	1	0	1/0	0/1	Unpredictable
0	1	1	0	1	0	
1	0	1	0	0	1	
1	1	1	0	1	0	
0	0	0	1	0/1	1/0	Unpredictable
0	1	0	1	1	0	
1	0	0	1	0	1	
1	1	0	1	0	1	

(c)

Figure 5-20. The R-S flip-flop.

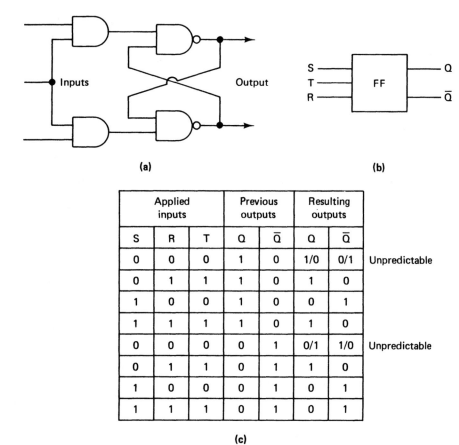

Figure 5-21. The *R-S-T* flip-flop.

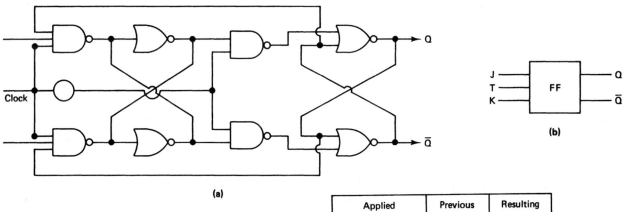

Figure 5-22. The *J-K* flip-flop.

processing. Typically, electrical pulses that turn on and off are applied to the input of a counter. These pulses must initiate a state change in the circuit when they are received. The circuit must also be able to recognize where it is in counting sequence at any particular time. This requires some form of memory. The counter must also be able to respond to the next number in the sequence. In digital electronic systems flip-flops are primarily used to achieve counting. This type of device is capable of changing states when a pulse is applied, has memory, and will generate an output pulse.

There are several types of counters used in digital circuitry today. Probably the most common of these is the binary counter. This particular counter is designed to process two-state or binary information. *J-K* flip-flops are commonly used in binary counters.

Refer now to the single *J-K* flip-flop of Figure 5-22. In its *toggle* state this flip-flop is capable of achieving counting. First, assume that the flip-flop is in its reset state. This would cause Q to be 0 and $\bar{Q}$ to be 1. Normally, we are concerned only with Q output in counting operations. The flip-flop is now connected for operation in the toggle mode. *J* and *K* must both be made high or in the 1 state. When a pulse is applied to the *T*, or clock, input, Q changes to 1. This means that with one pulse applied, a 1 is generated in the output. The flip-flop has, therefore, counted one time. When the next pulse arrives, Q resets, or changes to 0. Essentially, this means that two input pulses produce only one output pulse. This is a divide-by-two function. For binary numbers, counting is achieved by a number of divide-by-two flip-flops.

To count more than one pulse, additional flip-flops must be employed. For each flip-flop added to the counter, its capacity is increased by- the power of 2. With one flip-flop the maximum count was 2^0, or 1. For two flip-flops it would count two places, such as 2^0 and 2^1. This would reach a count of 3 or a binary number of 11. The count would be 00, 01, 10, and 11. The counter would then clear and return to 00. In effect, this counts four state changes. Three flip-flops would count three places, or 2^0, 2^1, and 2^2. This would permit a total count of eight state changes. The binary values are 000, 001, 010, 011, 100, 101, 110, and 111. The maximum count is seven, or 111. Four flip-flops would count four places, or 2^0, 2^1, 2^2, and 2^3. The total count would make 16 state

changes. The maximum count would be 15, or the binary number 1111. Each additional flip-flop would cause this to increase one binary place.

Figure 5-23 shows a four-bit binary counter. Four *J-K* flip-flops are included in this circuit. Sixteen counts can be made with this circuit. The counts are listed under the diagram. A light-emitting diode is used as an output display for each flip-flop. When a particular LED is on, it indicates the 1 state. An off LED indicates the 0 state.

Operation of the counter is based on the number of clock pulses applied to its input. The flip-flops will only trigger or change states on the negative-going part of the clock pulse. Note this on the input waveform. The output of flip-flop one (FF1) will change back and forth between 1 and 0 for each pulse. One pulse applied to the input of FF1 will cause its Q output to be 1. The 2^0 LED will turn on, indicating a 1 count. The display will be 0001. The second pulse will turn off FF1. The negative-going output of Q will turn on FF2. The Q output of FF2 will then become a 1. This will light LED-2, or the 2^1 display. The entire display will now be 0010, or the binary 2 count. The next pulse will trigger FF1. LED-1 will light again because the Q output of FF1 is 1 again. The display will show 0011,

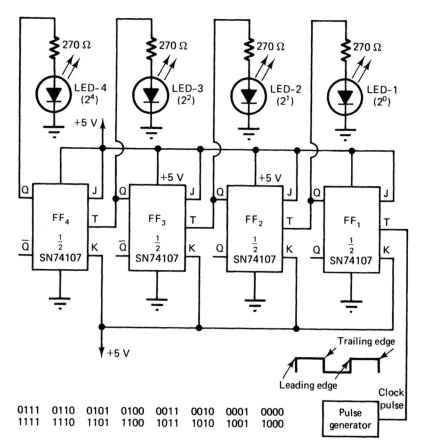

| 0111 | 0110 | 0101 | 0100 | 0011 | 0010 | 0001 | 0000 |
| 1111 | 1110 | 1101 | 1100 | 1011 | 1010 | 1001 | 1000 |

Figure 5-23. Four-bit binary counter.

or the binary 3 count. The fourth pulse will reset FF1 and FF2 will set FF3. The Q output of FF3 will go positive and turn on LED-3. The display will show 0100, or the binary 4 count. The operation is then repeated, the counts being 0101, 0110, and 0111. The next pulse will clear FF1, FF2, and FF3. The state change of FF3 will set FF4. The count will now be 1000, or the binary 8 count. The operation is again repeated. The count display will be 1001, 1010, 1011, 1100, 1101, 1110, and 1111. This reaches the maximum count that can be achieved with four flip-flops. The next pulse will clear all four flip-flops. The LED display will then be 0000. The entire process will repeat itself starting with the first pulse.

A four-bit IC binary counter is shown in Figure 5-24. This particular circuit achieves the same thing as the counter of Figure 5-23. Four J-K flip-flops are built into this chip. The Q output of each flip-flop appears at terminals A, B, C, and D. These serve as the 2^0, 2^1, 2^2, and 2^3 or 1, 2, 4, and 8 outputs. LEDs connected to each output will respond as the display. This chip essentially has one input and four outputs.

DECADE COUNTERS

Decade counters are widely used in digital electronic systems. This type of counter has ten counting states in its output. The output is in binary numbers. In practice, this is more commonly called a BCD counter. Its output is 000, 0001, 0010, 0011, 0100, 0101, 0110, 0111, 1000, and 1001. It is used to change binary numbers into decimal form. One BCD counter is needed for each place of a decimal number. One counter of this type can only count to 9. Two would be required to count to 99. Three are needed for 999. The place value increases with each additional counter.

BCD counters are primarily binary counters that clear all flip-flops after the 9 count, or 1001. The first seven counts appear in the normal binary sequence. Figure 5-25 shows how the J-K flip-flops of the BCD counter are connected. Notice that FFA, FFB, and FFC are connected the same as those of Figure 5-23. The resulting binary count is shown under the diagram.

At count 7, or binary 0111, a 1 is applied to the AND gate from the Q outputs of FFC and FFB. Two 1s applied to an AND gate causes the J of FFD to go 1. The next pulse clears FFA, FFB, and FFC and sets FFD. The counter now reads 8, or a binary 1000. Note that the $\bar{Q}$ output of FFD is returned to FFB. With FFD set to a 1, $\bar{Q}$ is 0. This prevents FFB from further triggering until FFD is cleared.

Arrival of the next clock pulse causes FFA to be set. The count is now 9, or binary 1001. The next pulse clears FFA and FFD at the same time. The count is now 0, or binary 0000. With FFB and FFC already clear, the counter is reset to zero and is ready for the next count.

Figure 5-26 shows a BCD counter constructed from an IC chip. The ABCD output will be in binary code for numbers 0 through 9. The input is applied to pin 14. As a general rule, the output of a BCD counter is usually applied to a decoding device. The decoder changes binary data into an output that drives a display device.

DIGITAL SYSTEM DISPLAY

Many people who need to read the output of a digital system may not be familiar with the BCD method of number display. The system must therefore change its output to a more practical method of display. A variety of display devices have been developed to perform this function. One that is primarily used for numbers is called a seven-segment display. Calculators, watches, and test instruments employ this type of display. It is widely used in digital electronic systems today.

A seven-segment number display divides the decimal numbers 0 through 9 into lines. Only four

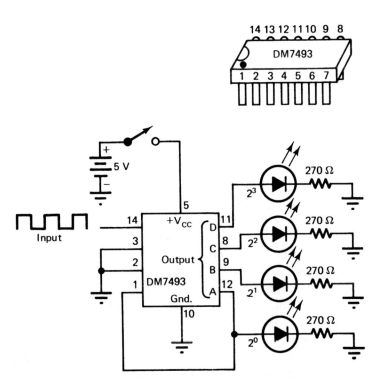

Figure 5-24. Four-bit IC binary counter.

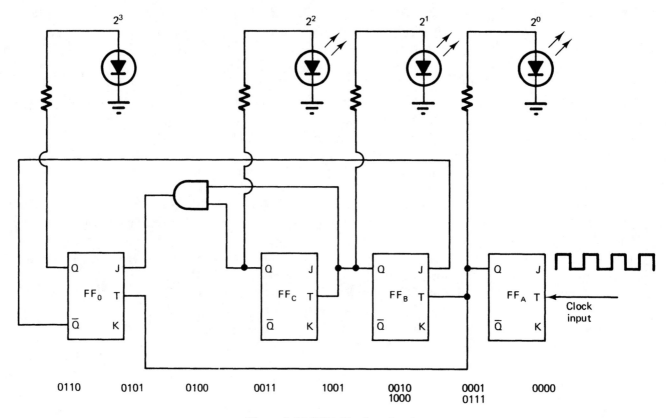

Figure 5-25. BCD flip-flop circuit.

vertical lines and three horizontal lines are needed to display a specific number. Each line or segment is designed to be illuminated separately. Figure 5-27(a) shows a typical seven-segment display. Each segment may contain several LEDs in a common bar. The size of the display generally determines the number of LEDs in a specific segment. All the LEDs in a segment are energized at the same time.

Individual lines or bars of the display are labeled a, b, c, d, e, f, and g. A seven-bit binary number is applied to the display. Each segment is fed by a specific part of the binary number. In a common-anode type of display, when the input goes to ground, or 0, it illuminates the segments. In a common-cathode display, the input must be 1 or +5 V to illuminate a segment. The truth table of Figure 5-27(b) is for a common-anode display. For a common-cathode display the polarity of each segment would be inverted.

A common-anode seven-segment display test circuit is shown in Figure 5-28. Pins 14, 9, and 3 are the anode connections. These are connected to the positive side of the power source. Pins 1, 2, 6, 7, 8, 10, 11, and 13 are the segment connections. Note that each segment is connected to a resistor. This limits the current to the LEDs in each segment to a reasonable value. Connecting one resistor to

the ground will energize a specific segment. Connecting them all to ground will produce an 8. display. Seven segments and a decimal point are included in this display.

DECODING

For a display to be operational, it must receive an appropriate input signal. This signal is normally in binary form. The input of a seven-segment display requires seven bits of data. The number 2, for example, is 0010010. See the data for the decimal number 2 in Figure 5-27. This does not correspond directly to the output of a counter. A counter usually contains binary data in a specific sequence. A BCD output for the number 2 would be 0010. These data therefore need to be changed into the seven-segment code to cause a 2 to appear on the display. The process of changing data of one form into that of another form is called decoding.

A logic diagram of a BCD-to-seven-segment decoder-driver is shown in Figure 5-29. The BCD input is applied to terminals A, B, C, and D. The seven-segment output appears at a, b, c, d, e, f, and g. The circuit also has a lamp test input, blanking input, and ripple blanking input. A low or "0" applied to the lamp test will cause all

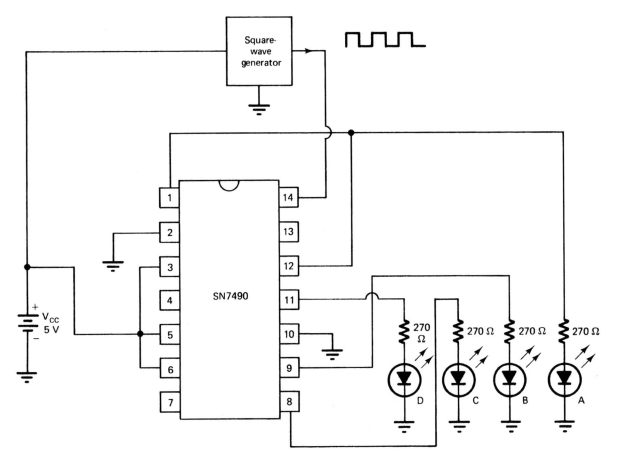

Figure 5-26. BCD counter.

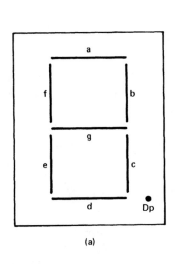

(a)

Decimal number	Display	a	b	c	d	e	f	g
0		0	0	0	0	0	0	1
1		1	0	0	1	1	1	1
2		0	0	1	0	0	1	0
3		0	0	0	0	1	1	0
4		1	0	0	1	1	0	0
5		0	1	0	0	1	0	0
6		1	1	0	0	0	0	0
7		0	0	0	1	1	1	1
8		0	0	0	0	0	0	0
9		0	0	0	1	1	0	0

Figure 5-27. Seven-segment display data: (a) display; (b) truth table.

(b)

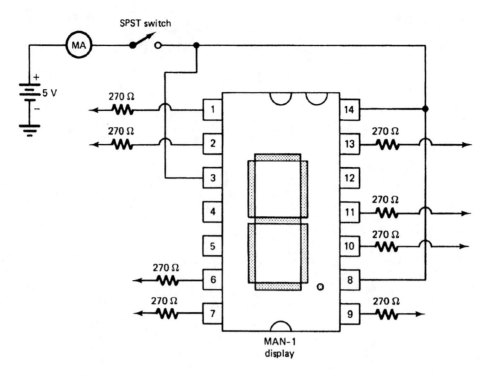

Figure 5-28. Common-anode seven-segment display test circuit.

segments of the display to be illuminated. A low or "0" applied to the blanking input turns off all segments. This control is also used when several displays are grouped together for a multidigit number. It blanks out all leading zeros automatically. If this were not used, a multidigit display would show a number of unnecessary zeros. For example, the number 52 on an eight-digit display would be 00000052. With automatic leading zero blanking, the number displayed, would be 52.

For driving a common-anode display the outputs of the decoder must go low or to ground to light a segment. Common-cathode displays must go high or to + VCC to produce segment illumination. The selected decoder/driver chip would depend on the type of display employed. Note the truth table of Figure 5-29. The 0 indicates the low or grounded output. The 1 output condition represents an open circuit. The decoder shown here is used to drive a common-anode type of display. The number zero, for example, illuminates all segments except g. A representative pin connection, diagram of a decoder-driver is also shown in the diagram.

A counting timer circuit with a digital display is shown in Figure 5-30. This circuit demonstrates the operation of a decoder-driver and the seven-segment display. The 555 square-wave generator can be adjusted to produce one pulse per second or minute. The display will count 0 to 9 as an indication of the pulse rate. RI is adjusted for time in seconds or minutes. Operational time is

based on the formula

$$Time = 0.693(RA + 2RB)C$$

For 1-s counting, C should be a value of 2 mF. One-minute counting is achieved with C in the range 100 mF.

The 0 to 9 counter can be extended to a 0 to 99 counter by adding an additional BCD counter, decoder, and display. Pin 11 of the low-order or units display is connected to the input (pin 14) of the tens display. The common or ground of each display unit must also be connected together. Each additional counter would be connected in the, same manner. The count would be 0, 1, 2, 3, 4, 5, 6, 7, 8, 9, 10, 11 and continue up to 99. The next count would clear the counter to 00.

DIGITAL COUNTING SYSTEMS

Digital counting systems are very common today. This application of the counting process is used to determine a number of pulses occurring in a given unit of time. Each pulse simply triggers the input when it occurs. The input source generates a two-state data signal. In its simplest form, a light beam can be broken as an indication of a signal change. Figure 5-31 shows a production line counter circuit that employs this principle in its operation.

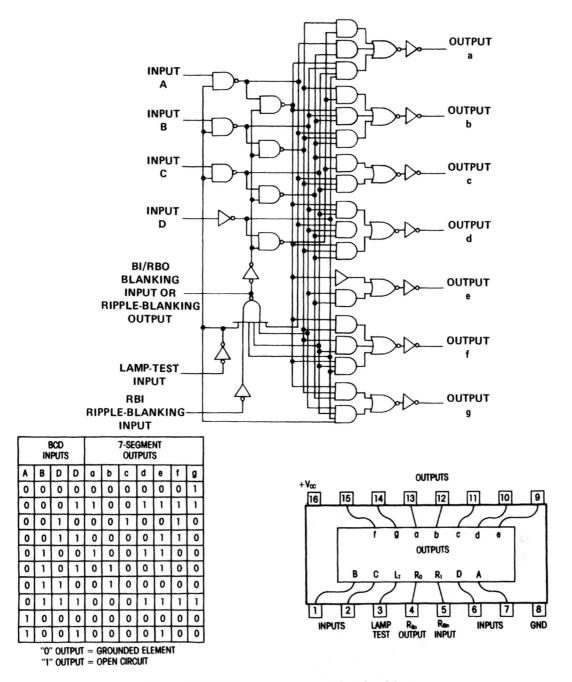

Figure 5-29. BCD-to-seven segment decoder-driver.

The fundamental parts of a counting system are the basis of nearly all digital display devices. The primary difference in the system is the input device. The display or readout device is essentially the same for all digital instruments. A BCD counter, decoder, and a display are used for each digit. In operation, a two-state or (on and off) signal is simply applied to the input of the counter assembly. Each on pulse, or 1, causes the counter to advance one count. The output of the counter is then applied to a decoder-driver. The display is driven by the decoder. One input pulse change will cause a 1 to appear on the display. The is, or unit counter, can progress only from 0 to 9 counts. A count of 10 or more would activate both the 1s and 10s display units. Counts in excess of 99 would require three or more displays.

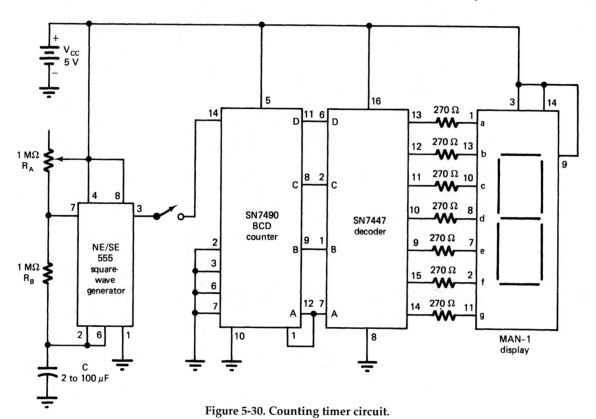

Figure 5-30. Counting timer circuit.

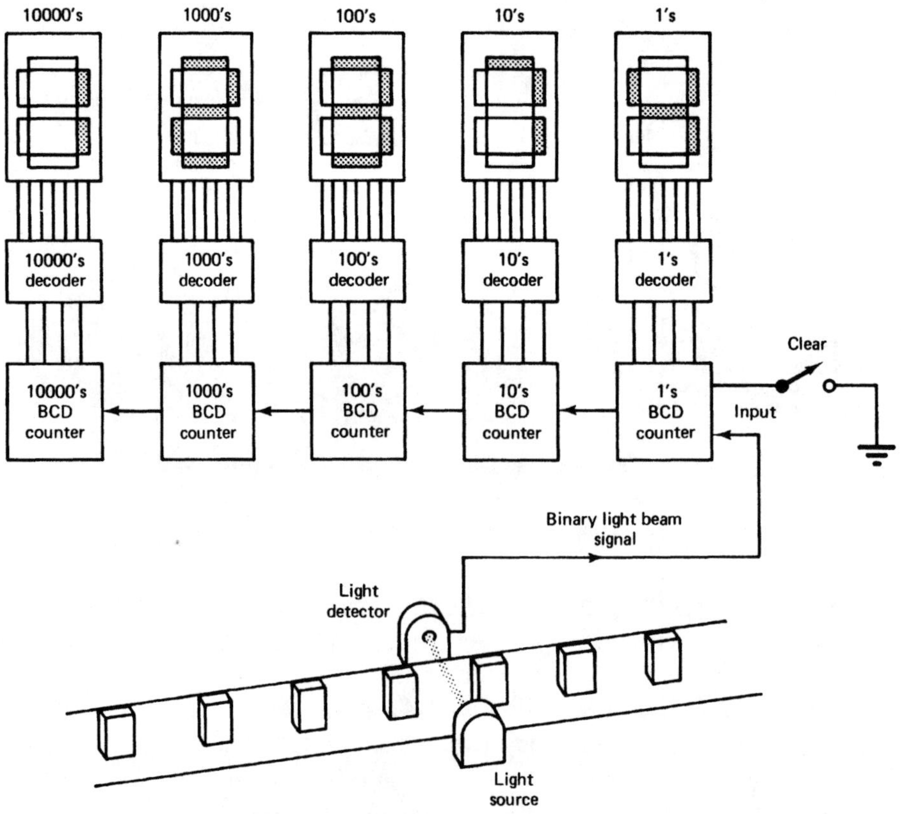

Figure 5-31. Production line counter.

Chapter 6

Computers and Microprocessors

INTRODUCTION

This chapter provides a basic introduction to computers and microprocessors and recent innovations in computer electronics technology. Computer technology has brought about significant changes in electronics. Computers have been used for a number of years to perform many operations. Calculations can be performed quickly, data can be stored and retrieved rapidly, and decisions can be made from data.

In the mid-1960s minicomputers were developed to provide computer technology for applications not requiring the capacity of large-scale computers. These units were significantly less costly than large computers. These computers demonstrated their usefulness with a wide range of applications. CRT displays were used with minicomputers to present an immediate information display for the user. Information is usually placed into these computers through keyboards.

OBJECTIVES

Upon completion of this chapter you will be able to:

1. Become familiar with the functional components of a computer.
2. Become familiar with the functional role of a microprocessor.
3. Describe the functional blocks of a microprocessor.
4. Distinguish between different types of memory used in computers and microprocessor systems.
5. Become familiar with microcomputer functions.
6. Recognize basic instructional sets used in computer systems.

MICROCOMPUTER BASICS

Microcomputers were introduced by several manufacturers in the early 1970s. These computers were an extension of the technology employed by the minicomputer. Large-scale integration (LSI) technology was used to place thousands of discrete solid-state components on a single IC chip. The entire *central processing unit* (CPU) of a microcomputer is achieved today by a single IC chip called a microprocessor unit (MPU). A microprocessor with memory, input-output interface chips, and a power source forms a microcomputer system. This combination of components can be built on a single printed circuit board and can be easily housed in a small cabinet. Microcomputer technology has revolutionized electronics by providing inexpensive small-capacity computers that can be used for many applications.

COMPUTER BASICS

The term *computer is* a general term that is used to cover a number of functions. It refers primarily to a system that will perform automatic computations when provided the appropriate information. Computers range from pocket calculators to complex centralized units that are used to serve an entire industry. Electric energy is needed to energize the unit and operation is performed through electronic components. Information is provided in two states—digital or analog.

Digital computers use numbers represented by the presence or absence of voltage, as discussed in Chapter 4. A high voltage level or pulse usually represents a 1 state, whereas no voltage indicates a 0 state. A voltage value or single pulse is described as a bit (binary digit) of information. A binary number has two states, or conditions, of operation. A group of pulses or voltage level changes produces a *word. A byte* consists of 8 bits.

Computer systems have certain fundamental parts, which may be arranged in a variety of different ways and still achieve the same function. The internal organization and design of each circuit differ considerably between manufacturers. Figure 6-1 shows a block diagram of a digital computer.

A digital computer is an electronic unit that consists of input and output devices, arithmetic logic and

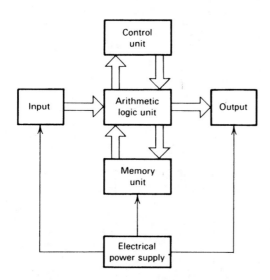

Figure 6-1. Digital computer system.

control circuitry, and some form of memory. Before the computer is placed into operation, a special set of instructions called a program must be supplied. These programs must be written in a *language* that the computer understands. This information is then supplied to the input unit by punched cards, punched tape, magnetic tape, or typewriter keyboard.

Input data supplied to the computer are translated into some type of number code, or *machine language*, before operation progresses. The arithmetic logic control unit then manipulates the input data according to its programmed instructions. After all the internal operations are complete, the coding process is reversed. The machine language is then translated back into another language. This information is used to actuate the output device.

The *memory* of a computer serves as a place to store the operating instructions that direct the CPU. Coded data in the form of 1s and 0s are "written" into memory through directions provided by a program. The CPU then "reads" these instructions from memory in a logical sequence and uses them to initiate processing operations. When the program is logical, processing proceeds in the correct manner with useful results.

Computers in operation today have one or more *output ports* that permit the CPU to communicate with the operator. Video display terminals (VDTs) for monitoring, line printers, and magnetic tape storage are typical output devices. Operational speed is an important characteristic of the output device.

The parts of a computer are basically the same for large-scale computers, minicomputers, and microcom-

puters. The differences between these units are in physical size. *Large-scale computers* usually employ thousands of discrete components and logic gates. This type of unit usually has an enormous memory capacity, high-speed operation, and simplified program routines. Typically, 32 or more bits of information are used to make programming words.

Minicomputer systems are made from logic gates that are arranged on numerous printed circuit boards. This type of unit has reduced memory capacity, requires a longer cycle time, and is less expensive. They are primarily designed for applications not requiring the capacity of a large-scale computer. Program word size may be 8, 16, or 32 bits.

Microcomputer systems are small units built around a single IC chip. The *microprocessor* of such a system is a single-chip IC package that performs the arithmetic logic and control functions. These units are relatively inexpensive and can be used for many applications. Microcomputer systems are generally used to serve a specific need, solve a specific type of problem, or handle one application. Such computers are often classified as dedicated systems, since they are designed for specific applications.

MICROCOMPUTER SYSTEMS

Microcomputer systems are very significant developments in the field of electronics. The capabilities of the microcomputer are complex when viewed in their entirety. The microcomputer has a microprocessor, memory, an interface adapter, and several distribution paths called buses. To clarify the system, look at a block diagram of its physical makeup, as shown in Figure 6-2.

This simplified diagram of the microcomputer is similar to the basic computer diagram of Figure 6-1, The microprocessor has replaced the arithmetic logic unit and control unit in the microcomputer diagram. Information within this system is of two types: instructions and data. In a simple addition problem such as 9 + 2 = 11, the numbers 9, 2, and 11 are data and the plus sign is an instruction. The data is distributed by the *data bus* to all parts of the system. Instructions are distributed by the *control bus* through a separate path. The address bus of the unit forms an alternate distribution path for the distribution of address data. It is normally used to place information into memory at the correct address. By an appropriate command from the microprocessor, data may be removed from memory and distributed to the output.

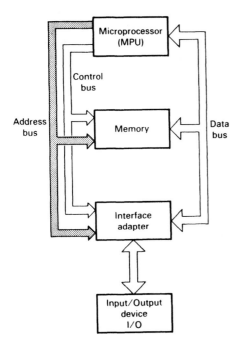

Figure 6-2. Simplified block diagram of microcomputer.

THE MICROPROCESSOR

The MPU is the arithmetic logic unit (ALU) and control section of a microcomputer; it fits into a single IC chip. Typical chip sizes are approximately 1/2 cm² and contain thousands of transistors, resistors, and diodes. Many companies are now manufacturing these chips in different designs and prices. Applications of this device are increasing at a remarkable rate.

A microprocessor is a digital device that is designed to receive data in the form of 1s and 0s. It may then store this data for future processing, perform arithmetic and logic operations in accordance with stored instructions, and deliver the results to an output device. In a sense, a microprocessor is a computer on a chip. The block diagram of a typical microprocessor, shown in Figure 6-3, contains a number of components connected in a rather unusual manner. Included in its construction are the ALU, an accumulator, a data register, address registers, program counter, instruction decoder, and a sequence controller.

ALU

All microprocessors contain an ALU. The ALU is a calculator chip that performs mathematical and logic operations on the data words supplied to it. It is made to work automatically by control signals developed in the instruction decoder. The ALU combines the contents of its two inputs, which are called the *data register* and the *accumulator*. Addition, subtraction, and logic comparisons

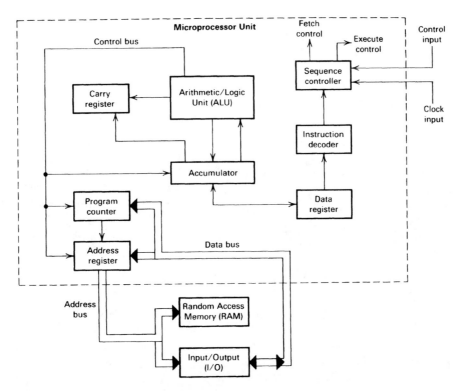

Figure 6-3. Simplified microprocessor block diagram.

are the primary operations performed by the ALU. The specific operation to be performed is determined by the control signal supplied by the instruction decoder.

The data supplied to the inputs of an ALU may be in the form of 8-bit binary numbers. They are combined by the ALU in accordance with the logic of binary arithmetic. Since mathematical operations are performed on the two data inputs, they are often called operands.

To demonstrate the operation of the ALU, assume now that two binary numbers are to be added—6_{10} and 8_{10}. Initially the binary number 00000110_2(6_{10}) is placed in the accumulator. The second operand, 00001000_2(8_{10}), is then placed into the data register. When a proper control line to the ALU is activated, binary addition is performed, producing an output 00001110_2, or which is the sum of the two operands This value is then stored in the accumulator. It replaces the operand that appeared there originally. The ALU only responds to binary numbers.

Accumulators

The *accumulators* of a microprocessor are temporary registers designed to store operands that are to be processed by the ALU. Before the ALU can perform, it must first receive data from an accumulator. After the data register input and accumulator input are combined, the logical answer, or output of the ALU, appears in the accumulator.

A typical instruction for a microprocessor is load accumulator. This instruction enables the contents of a particular memory location to be placed into the accumulator. A similar instruction is store accumulator. This instruction causes the contents of the accumulator to be placed in a selected memory location. The accumulator serves as an input source for the ALU and as a destination for its output.

Data Registers

The *data register* of a microprocessor serves as a temporary storage location for information applied to the data bus. Typically, this register accommodates an 8-bit data word. An example of a function of this register is operand storage for the ALU input. It may also be called upon to hold an instruction while the instruction is being decoded. Another function is to hold data temporarily prior to the data being placed in memory.

Address Registers

Address registers are used in microprocessors to store temporarily the address of a memory location that is to be accessed for data. In some units this register may be programmable. This means that it permits instructions to alter its contents. The program can also be used to build an address in the register prior to executing a memory reference instruction.

Program Counter

The *program counter* of a microprocessor is a memory device that holds the address of the next instruction to be executed in a program. This unit counts the instructions of a program in sequential order. When the MPU has fetched instructions addressed by the program counter, the count advances to the next location. At any given point during the sequence the counter indicates the location in memory from which the next information will be derived.

The numbering sequence of the program counter may be modified so that the next count may not follow a numerical order. Through this procedure, the counter may be programmed to jump from one point to another in a routine. This permits the MPU to have branch programming capabilities.

Instruction Decoder

Each specific operation that the MPU can perform is identified by binary numbers known as instruction codes. Eight-bit words are commonly used for this code. There are 2^8 (256_{10}) separate or alternative operations that can be represented by this code. After a typical instruction code is pulled from memory and placed in the data register, it must be decoded. The *instruction decoder* examines the coded word and decides which operation is to be performed by the ALU. The output of the decoder is first applied to the sequence controller.

Sequence Controllers

The *sequence controller* performs several functions in the operation of microprocessor. Using clock inputs, it maintains the proper sequence of events required to perform a processing task. After instructions are received and decoded, the sequence controller issues a control sign that initiates the proper processing action. In most units the controller has the capability of responding to external control signals.

Buses

The registers and other functional circuits of most microprocessors are connected together by a common *bus* network. A bus is defined as a group of conductor paths that are used to connect data words to various registers. A diagram of registers connected by a common bus line is shown in Figure 6-4.

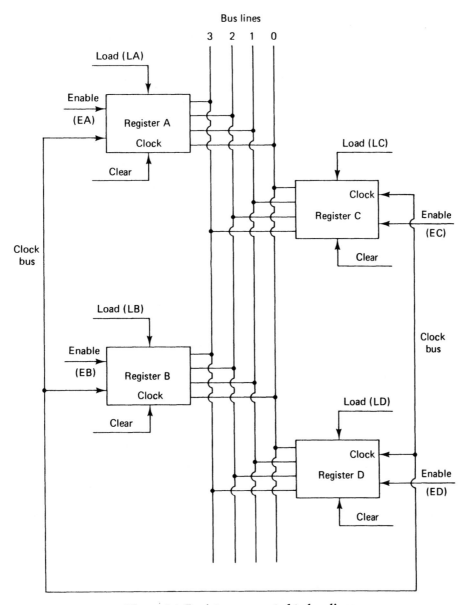

Figure 6-4. Registers connected to bus lines.

An advantage of bus-connected components is the ease with which a data word can be transferred or loaded into registers. Each register has inputs labeled clock, enable, load, and clear. When the load and enable input lines are at 0, each register is isolated from the common bus lines.

To transfer a word from one register to another, it is necessary to make the appropriate inputs convert to the 1 state. To transfer the data of register A to register D, enable A (EA) and D (LD) inputs both are placed in the 1 state. This causes the data of register A to appear on the common bus line. When a clock pulse arrives at the common inputs, the transfer is completed.

The word length of a bus is based upon the number of conductor paths. Buses for 4, 8, and 16 bits are used in microprocessors. MPUs use an 8-bit data bus and a 16-bit address bus.

Microprocessor Architecture

The physical layout, or *architecture*, of a microprocessor is rather complex.

The MC6800 is one type of microprocessor. It is housed in a 40-pin dual-in-line package shown in Figure 6-5. Two accumulators, an index register, a program counter register, a stack pointer register, and a condition code register are included in its physical makeup.

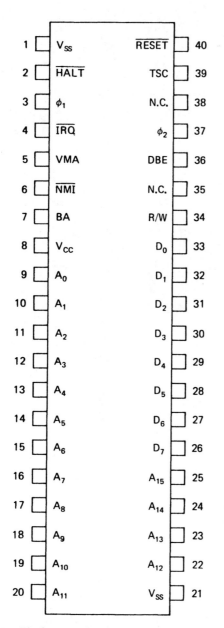

Figure 6-5. Pin layout of MC6 800 microprocessor chip.

MC6800 MICROPROCESSOR

Accumulators

The MC 6800 MPU has two *accumulators* labeled *AA* and *AB*. Each accumulator is 8 bits wide. Accumulators are designed to hold operands or data from the ALU.

Program instructions for each accumulator includes a mnemonic of its name and the operation to be performed. The loading of accumulator A is LDAA, and LDAB is the mnemonic for loading accumulator B. Storing data in accumulator A is STAA, with STAB achieving the same operation for accumulator B. With two accumulators in the MPU, arithmetic and logic operations can be performed

on two different numbers at the same time without shifting to memory.

Index Register

The *index register* of the MC6800 has a 16-bit, or 2-byte, capacity that is used to store memory addresses. This register has the capability of being loaded from two adjacent memory bytes. This allows its contents to be stored in two adjacent memory locations. Through this feature, data can be moved in 2-byte groups. The contents value increases by 1 when given the increment index register instruction or INX. A DEX instruction applied to the unit causes it to decrease by 1. This latter instruction is called decrementing the index register.

Program Counter

The *program counter (PC)* of the MC6800 is a 16-bit register that holds the address of the next byte to be fetched from memory. It can accommodate 216, or 65,536, different memory addresses. Two 8-bit bytes are used for obtaining a specific address location.

Stack Pointer

The *stack pointer (SP)* of the MC6800 is a special 16-bit (2-byte) register. It uses a section of memory that has a last-in, first-out capability. With this capability the status of the MPU registers may be stored when branch or interrupt program subroutines are being performed.

An address in the SP is the starting point of sequential memory locations in memory where the status of MPU registers is stored. After the register status has been placed into the SP, it is decremented. When the SP is accessed, the status of the last byte placed on the stack serves as the first byte to be restored.

MEMORY

Microcomputer applications range from single-chip microprocessor units to complex systems that employ several chips.

The primary difference in this range of applications is in the memory capabilities of the system. Single-chip microprocessors are limited in the amount of memory they possess. Additional memory is achieved economically through the use of auxiliary chips. The capabilities of a microcomputer system are extended by the range of memory they have.

Memory refers to the capability of a device to store logical data in such a way that a single bit or a group of bits can be easily accessed or retrieved. Memory is achieved in a variety of different ways. Microcomputer systems usually have read-write memory and read-only memory. These two types of memory are ordinarily accomplished by semiconductor circuits on a single IC chip.

Read-Write Memory

Read-write memory is widely used in microcomputer systems today. Read-write chips of the large-scale integration (LSI) type are capable of storing 16,384, or 16K, bits of data in an area less than $1/2$ cm². The structure of this chip includes a number of discrete circuits, each having the ability to store binary data in an organized rectangular configuration. Access to each memory location is provided by coded information from the microprocessor address bus. The read-write function indicates that data can be placed into memory or retrieved at the same rate.

A simplified representation of the memory process is illustrated by the 8 × 8 memory unit of Figure 6-6. Memory ICs are usually organized in a rectangular pattern of rows and columns. Figure 6-6 shows eight rows that can store 8-bit words, or 64-single bits of memory. To select a particular memory address, a 3-bit binary number is used to designate a specific row location, and 3 additional bits are used to indicate the column location. In this example the row address is 3_{10} or 011_2, and the column address is 5_{10} or 101_2. The selected memory cell address is at location 30.

Eight-bit word storage is achieved in the memory of Figure 6-6 by energizing one row and all eight columns simultaneously. The row and column decoders perform this operation. Some of the read-write memory units available today have capacities of 8 × 8, 32 × 8, 128 × 8, 1024 × 8, and 4096 × 8.

To write a word into memory, a specific address is first selected according to the data supplied by the address bus. Reading data from memory does not destroy the data at each cell. Data reading can take place as long as the memory unit is energized electrically. A loss of electrical power or turning the unit off destroys the data placed at each memory location. Solid-state read-write memory is classified as *volatile* because of this characteristic.

Read-Only Memory

Most microcomputer systems require memories that contain permanently stored or rarely altered data. Examples of this are math tables and permanent program data. Storage of this type is provided by *read-only memory* (ROM). Information is often placed in this type of memory unit when the chip is manufactured. ROM data are *nonvolatile*, which means that data are not lost when the power source is turned off.

A variation of the ROM is the *read-mostly memory*. This unit is used where read operations are needed more frequently than write operations. *Programmable read-only memory* (PROM) is another type that can be optically erased by exposure to an ultraviolet light source. After light exposure each memory cell of the unit goes to a 0 state. Writing data into the chip again initiates a new program.

An alternative to the optically erasable read-mostly memory is the *electrically altered read-only memory* (EAROM) or the *erasable programma-*

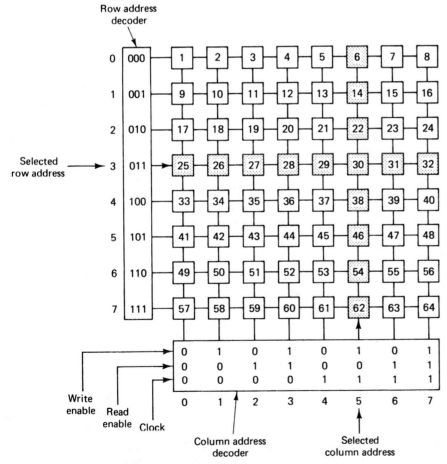

Figure 6-6. 8 × 8 simplified memory unit.

ble read-only memory (EPROM). This type of chip permits erasures of individual cells or word locations instead of the entire chip. Memory cell structure is very similar to the optically erasable ROM. The potential usefulness of the EAROM and EPROM is considerable for several applications.

MICROCOMPUTER FUNCTIONS

There are certain functions that are basic to most microcomputer systems. Among these functions are timing, fetch and execution, read memory, write memory, input-output transfer, and interrupts. A knowledge of these basic functions is important for understanding the operation of microcomputers.

Timing

The operation of a microcomputer system is achieved by a sequence of cycling instructions. The MPU fetches an instruction, executes it, and continues to operate in a cycling pattern. This means that all actions occur at or during a precisely defined time interval. An orderly sequence of operations like this requires some type of precision clock mechanism. A free-running electronic oscillator or clock is responsible for this function. In some systems the clock may be built into the MPU, whereas in others timing units feed the system through a separate clock control bus. The sequence of operations is controlled by a timing signal.

Sequential operations such as fetch and execute are achieved within a period called the MPU cycle. The fetch portion of this cycle consists of the same series of instructions. It therefore takes the same amount of time for each instruction. The execute phase of an operation may consist of many events and sequences, depending upon the specific instruction being performed. This portion of the cycle varies a great deal with the instruction being performed.

The total interval taken for a timing pulse to pass through a complete cycle from beginning to end is called a period. One or more clock periods may be needed to complete an operational instruction.

Fetch and Execute Operations

After programmed information is placed into the memory of a microcomputer, its action is directed by a series of fetch and execute operations. This sequence of operations is repeated until the entire program has cycled to its conclusion. This program tells the MPU specifically what operations it must perform.

The operation begins when the start function is initiated. This signal actuates the control section of the MPU, which automatically starts the machine cycle. The first instruction that the MPU receives is to fetch the next instruction from memory. The MPU may then issue a read operational code instruction. The contents of the program counter are then sent to the program memory, which returns the next instruction word. The first word of this instruction is then placed into the instruction register. If more than one word is included in an instruction, a longer cycling time is needed. After the complete instruction word is in the MPU, the program counter is incremented by one count. The instruction is then decoded, which prepares the unit for the next fetch instruction.

The execution phase of operation is based upon which instruction is to be performed by the MPU. This instruction may be read memory, write in memory, read the input signal, transfer to output, or an MPU operation such as add registers, subtract, or register-to-register transfer. The time of an operation is dependent upon the programmed information placed in memory.

The 8080 microprocessor takes a number of the clock periods to perform its operations. This particular chip can be operated at a clock rate of 2 MHz. A single cycle of the clock has a period $1/2,000,000$, or 0.0000005, s. Periods this small are best expressed in microseconds (0.5 μs) or nanoseconds (500 ns). The fastest instruction change that can be achieved by this chip requires 4 clock periods, or 4×500 ns = 2 μs. Its slowest instruction requires 18 periods, or 18×500 ns = 9 μs. The operational time of an MPU is a good measure of its effectiveness and how powerful it is as a functional device.

Read Memory Operation

The *read memory operation* is an instruction that calls for data to be read from a specific memory location and applied to the MPU. To initiate this operation, the MPU issues a read operation code and sends it to the proper memory address. The read-write memory unit then sends the data stored at the selected address into the data bus. This 8-bit number is fed to the MPU, where it is placed in the accumulator after timing pulse has been initiated.

Write Memory Operation

The *write memory operation* is an instruction that calls for data to be placed into or stored at a specific memory location. This function is initiated when the MPU issues a write operation code and sends it to a selected read-write memory unit. Data from the data bus are then placed into the selected memory location for storage. In practice, 8-bit numbers are usually stored as words in the memory unit.

Input-Output Transfer Operation

In a microcomputer system, *input-output (I/O) transfer operations* are very similar to read-write operations. The major difference is the op-code data number used to call up the operation. When the MPU issues an input-output op-code, it actuates the appropriate I/O port which either receives data from the input or sends it to the output device according to the coded instruction.

A simplified microcomputer system with a read-write memory, read-only memory, output, and input connected to a common address bus and data bus is shown in Figure 6-7. In this type of system data are processed by the MPU. Data move in either direction to the read-write memory and flow from the ROM or input into the data bus. The output flows from the data bus to the output device.

Interrupt Operation

Interrupt operations are often used to improve the efficiency of a microcomputer system. Interrupt signals are generated by peripheral equipment such as keyboards, displays, printers, or process control devices. These signals are generated by peripheral equipment and applied to the MPU. They inform the MPU that a peripheral device needs some type of attention.

Assume that a microcomputer system is designed to process a large volume of data and the output is a line printer. The MPU can output data at a very high rate of speed compared to the time needed to actuate a character on the line printer. This means that the MPU has to remain idle while waiting for the printer to complete its operation.

If an interrupt is used by the MPU of a microcomputer system, it can output a data byte and then return to processing data while the printer completes its operation. When the printer is ready for the next data byte, it requests an interrupt.

Upon acknowledging the interrupt, the MPU stops program execution and automatically branches to a subroutine that will output the next data byte to the printer. The MPU then continues the main program execution where it was interrupted. Through this procedure, high-speed MPU operation is not restricted by slow-speed peripheral equipment.

PROGRAMMING

The term *program* refers to a series of instructions developed for a computer that permit it to perform a prescribed operation. In microcomputer systems some programs are hard-wired, some are in read-only memories called firmware, and others are described as software. A *dedicated system (hard-wired)* designed for a specific function is usually hardware programmed so that it cannot be adapted for other tasks without a circuit change.

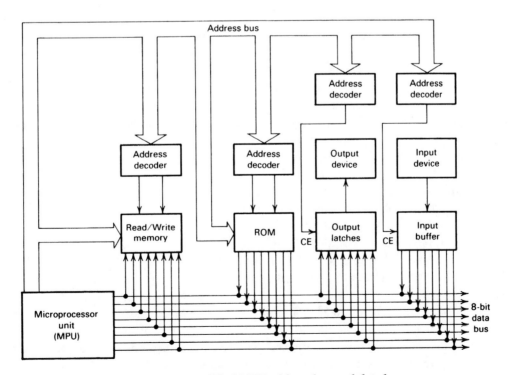

Figure 6-7. Simplified MPU address bus and data bus.

Firmware systems have programmed material placed on ROM chips. Program changes for this type of system can be achieved for changing the ROMs. A *software* type of program has the greatest flexibility. This type of program is created on paper and transferred to the system by a keyboard, punched cards, or magnetic tape. Instructions are stored at read-write memory locations and performed by calling for these instructions from memory.

The design of a microcomputer system and its purpose usually dictate the type of programming method used. Combined firmware and software programmed systems with a keyboard are very common.

Instructions

The *instructions* for a microcomputer system normally appear as a set of characters or symbols that are used to define a specific operation. These symbols are similar to those that appear on a typewriter. Included are decimal digits 0-9, the letters A-Z, and, in some cases, punctuation marks and specialized keyboard characters. Symbolized instructions may appear as binary numbers, hexadecimal numbers, or as mnemonic codes.

Microcomputers may employ any one of several different codes in a program according to its design. Each type of MPU has an instructional set it is designed to understand and obey. These instructions appear as binary data symbols and words. Machine instructions of this type are normally held in a ROM unit that is address selected and connected to the MPU through the common data bus. Instructional sets are an example of firmware because they are fixed at a specific memory location and cannot be changed.

Microcomputer instructions usually consist of 1, 2, or 3 bytes of data. This type of data must follow the instruction commands in successive memory locations. These instructions are usually called *addressing modes.*

One-byte instructions are often called inherent-mode instructions. These instructions are designed to manipulate data to the accumulator registers of the ALU. No address code is needed for this type of instruction because it is an implied machine instruction. The instruction CLA, for example, is a 1-byte op-code that clears the contents of accumulator A. No specific definition of data is needed, nor is an address needed for further data manipulation. This instruction simply clears the accumulator register of its data.

Inherent mode instructions differ a great deal between manufacturers. Some representative op-code instructions are shown in Figure 6-8 Note that the meaning of the instruction, its code in hexadecimal form, and a mnemonic are given for each instruction. Each microcom-

puter system has a number of instructions of the 1-byte type that contain only an op-code.

Immediate Addressing

Immediate addressing is accomplished by a 2-byte instruction that contains an op-code and an operand. In this addressing mode the op-code appears in the first 8 bytes and it is followed by an 8-bit operand. A common practice is to place intermediate addressing instructions in the first 256 memory locations. These instructions can be retrieved very quickly, since this is the fastest mode of operation.

Relative Addressing

Relative addressing instructions are designed to transfer program control to a location other than the next consecutive memory address. In this type of addressing, two 8-bit bytes are used for the instruction. Transfer operations of this type are often limited to a specific number of memory locations in front or in back of its present location. The 2-byte instruction contains an op-code in the first byte and an 8-bit memory location in the second byte. The second byte points to the location of the next instruction that is to be executed. This type of instruction is used for branching programming operations.

Indexed Addressing

Indexed addressing is achieved by a 2-byte instruction and is very similar to the relative addressing mode of instruction. In this type of addressing, the second byte of the instruction is added to the contents of the index register to form a new, or "effective," address. This address is obtained during program execution rather than being held at a predetermined location. A newly created effective address is held in a temporary memory address register so that it will not be altered or destroyed during processing.

Direct Addressing

Direct addressing is a common mode of instruction. In this type of instruction the address is located in the next byte of memory following the op-code. This permits the first 256 bytes of memory, from 0000 to $00FF_{16}$, to be addressed.

Extended Addressing

Extended addressing is a method of increasing the capability of direct addressing so that it can accommodate more data. This mode of addressing is used for memory locations above $00FF_{16}$ and requires 3 bytes of data for the instruction. The first byte is a standard 8-bit op-code.

Mnemonic	Opcode	MEANING
ABA	1B	Add the contents of accumulators A and B. The result is stored in accumulator A. The contents of B are not altered.
CLA	4F	Clear accumulator A to all zeros.
CLB	5F	Clear accumulator B.
CBA	11	Compare accumulators: Subtract the contents of ACCB from ACCA. The ALU is involved but the contents of the accumulators are not altered. The comparison is reflected in the condition register.
COMA	43	Find the ones complement of the data in accumulator A, and replace its contents with its ones complement. (The ones complement is simple inversion of all bits.)
COMB	53	Replace the contents of ACCB with its ones complement.
DAA	19	Adjust the two hexadecimal digits in accumulator A to valid BCD digits. Set the carry bit in the condition register when appropriate. The correction is accomplished by adding 06, 60, or 66 to the contents of ACCA.
DECA	4A	Decrement accumulator A. Subtract 1 from the contents of accumulator A. Store result in ACCA.
DECB	5A	Decrement accumulator B. Store result in accumulator B.
LSRA	44	Logic shift right, accumulator A or B.
LSRB	54	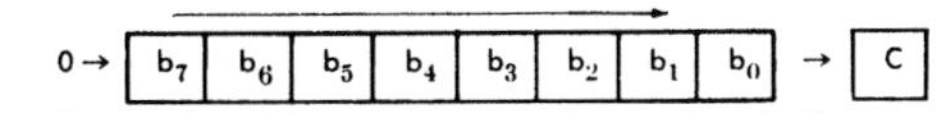 $0 \rightarrow b_7\ b_6\ b_5\ b_4\ b_3\ b_2\ b_1\ b_0 \rightarrow C$
SBA	10	Subtract the contents of accumulator B from the contents of accumulator A. Store results in accumulator A.
TAB	16	Transfer the contents of ACCA to accumulator B. The contents of register A are unchanged.
TBA	17	Transfer the contents of ACCB to accumulator A. The contents of ACCA are unchanged.
NEGA	40	Replace the contents of ACCA with its twos complement. This operation generates a negative number.
NEGB	50	Replace the contents of ACCB with its twos complement. This operation generates a negative number.
INCA	4C	Increment accumulator A. Add 1 to the contents of ACCA and store in ACCA.
INCB	5C	Increment accumulator B. Store results in AACB.
ROLA	49	Rotate left, accumulator A or B.
ROLB	59	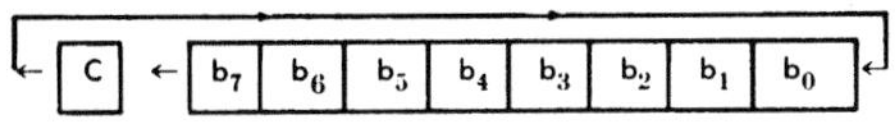 $C \leftarrow b_7\ b_6\ b_5\ b_4\ b_3\ b_2\ b_1\ b_0 \leftarrow$ Carry/borrow in condition register.
RORA	46	Rotate right, accumulator A or B.
RORB	56	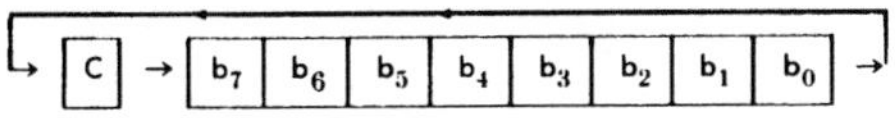 $C \rightarrow b_7\ b_6\ b_5\ b_4\ b_3\ b_2\ b_1\ b_0 \rightarrow$
ASLA	48	Shift left, accumulator A or B (arithmetic).
ASLB	58	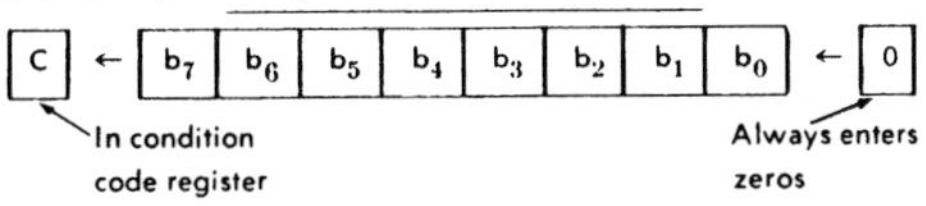 $C \leftarrow b_7\ b_6\ b_5\ b_4\ b_3\ b_2\ b_1\ b_0 \leftarrow 0$ In condition code register / Always enters zeros
ASRA	47	Shift right, accumulator A or B (arithmetic).
ASRB	57	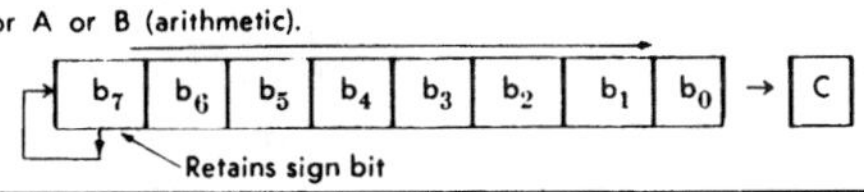 $b_7\ b_6\ b_5\ b_4\ b_3\ b_2\ b_1\ b_0 \rightarrow C$ Retains sign bit

Figure 6-8. Op-code instructions.

The second byte is an address location for the most significant, or highest, order of the data in 8 bits. This is followed by the third byte, which holds the address of the least significant, or lowest, 8 bits of the data number being processed.

Program Planning

A microcomputer system could not solve even the simplest type of problem without the help of a well-defined program. After the program has been developed, the system simply follows this procedure to accomplish the task. Programming is an essential part of nearly all computer system applications.

Before a program can be effectively prepared for a microcomputer system, the programmer must be fully aware of the specific instructions that are performed by the system. Microcomputer systems have a list of instructions that are used to control their operation. The *instruction set* of a microcomputer is the basis of all programs.

A programmer should be familiar with the instruction set of the system being used on a microcomputer. He or she should be able to decide what specific instructions are needed to solve a problem. A limited number of operations can usually be developed without the aid of a diagrammed plan of procedure. Complex problems require a specific plan in order to reduce confusion or to avoid the loss of an important operational step. Flow charts are commonly used to aid the programmer in this type of planning. Figure 6-9 shows some of the flowchart symbols that are used in program planning.

When programming a microcomputer system, the programmer must be aware of the specific instructions. He or she must decide upon what instructions are needed to solve a problem, plan a flowchart, develop a programming sheet, initiate the program, correct it if needed, and execute its operation. Microcomputers may be used for practically unlimited applications.

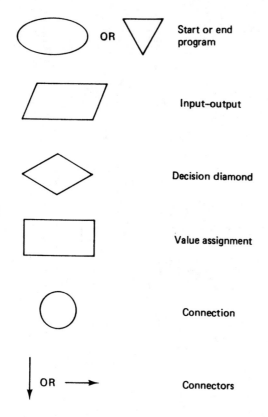

Figure 6-9. Programming flowchart symbols.

A PROGRAMMING EXAMPLE

To demonstrate the potential capabilities of a micro-computer system, its ability to solve a straight-line computation problem will be illustrated. In this problem the system will be used simply to add two numbers and then indicate the resulting sum. This type of problem is obviously quite simple and could be easily solved without the help of a microcomputer system. It is used in this situation to demonstrate a principle of operation and to show a plan of procedure.

In practice, a microcomputer system could not solve even the simplest type of problem without the help of a well-defined program that works out everything right down to the smallest detail. After the program has been developed, the system simply follows this procedure to accomplish the task. Programming is an essential part of nearly all computer system applications.

Before a program can be effectively prepared for a microcomputer system, the programmer must be fully aware of the specific instructions that can be performed by the system. In general, each microcomputer system has a unique list of instructions that are used to control its operation. The instruction set of a microcomputer is the basis of all program construction. Figure 6-10 shows an instruction

set for the Motorola MC6800 microprocessor system.

Assume now that the programmer is familiar with the instruction set of the system being used to solve this straight-line computation problem. The next step in this procedure is to decide what specific instructions are needed to solve the problem. A limited number of operations to perform can generally be developed without the aid of a diagrammed plan of procedure. Complex problems, by comparison, usually require a specific plan in order to reduce confusion or to avoid the loss of an important operation step. Flow charts are commonly used to aid the programmer in this type of planning. Figure 6-10 shows the flow-chart symbols that are commonly used in program planning.

The first step in preparing a program to solve our problem is to make a flow chart that shows the general plan of procedure to be followed. In this case we will set up the program to accomplish the following:

Find: the number N (a decimal value)

Use equation: $x + y = N$
Let: $\quad x = 0A$ (operand)
$\quad y = 07$ (operand)

Figure 6-11 shows a simple flow-chart plan of procedure to be followed by our program to solve the problem. The first step, numbered 0, is a simple statement of the problem. Steps 1 and 2 are then used as the operands of the problem and indicate the values of x and y. After these values have been obtained, each is placed into a specific accumulator. Step 3 is then responsible for the addition operation. Since this system deals with binary numbers, conversion to decimal values is also necessary. Step 4 is a conversion operation that changes binary numbers to BCD values. This value can then be used to energize an output so that it would produce a decimal readout. The fifth and final step of the program is an implied-halt opcode that stops the program.

A programming sheet for the problem example is shown in Figure 6-12. Notice that this sheet indicates the step number, a representative memory address in hexadecimal values, instruction bytes in hexadecimal values, op-code mnemonics, binary equivalents, addressing mode, and a description of each function. The memory address locations employed in the program begin with 6616. In practice, it is a common procedure to reserve the first 100 memory locations, or 00 to 6316, for branch instructions. We have arbitrarily selected location 6616 or 102,0 in order to avoid those addresses being reserved for branching operations.

ACCUMULATOR AND MEMORY INSTRUCTIONS

OPERATIONS	MNEMONIC	IMMED OP	~	=	DIRECT OP	~	=	INDEX OP	~	=	EXTND OP	~	=	IMPLIED OP	~	=	BOOLEAN/ARITHMETIC OPERATION (All register labels refer to contents)	H 5	I 4	N 3	Z 2	V 1	C 0
Add	ADDA	8B	2	2	9B	3	2	AB	5	2	BB	4	3				A + M → A	↕	•	↕	↕	↕	↕
	ADDB	CB	2	2	DB	3	2	EB	5	2	FB	4	3				B + M → B	↕	•	↕	↕	↕	↕
Add Acmltrs	ABA													1B	2	1	A + B → A	↕	•	↕	↕	↕	↕
Add with Carry	ADCA	89	2	2	99	3	2	A9	5	2	B9	4	3				A + M + C → A	↕	•	↕	↕	↕	↕
	ADCB	C9	2	2	D9	3	2	E9	5	2	F9	4	3				B + M + C → B	↕	•	↕	↕	↕	↕
And	ANDA	84	2	2	94	3	2	A4	5	2	B4	4	3				A · M → A	•	•	↕	↕	R	•
	ANDB	C4	2	2	D4	3	2	E4	5	2	F4	4	3				B · M → B	•	•	↕	↕	R	•
Bit Test	BITA	85	2	2	95	3	2	A5	5	2	B5	4	3				A · M	•	•	↕	↕	R	•
	BITB	C5	2	2	D5	3	2	E5	5	2	F5	4	3				B · M	•	•	↕	↕	R	•
Clear	CLR							6F	7	2	7F	6	3				00 → M	•	•	R	S	R	R
	CLRA													4F	2	1	00 → A	•	•	R	S	R	R
	CLRB													5F	2	1	00 → B	•	•	R	S	R	R
Compare	CMPA	81	2	2	91	3	2	A1	5	2	B1	4	3				A − M	•	•	↕	↕	↕	↕
	CMPB	C1	2	2	D1	3	2	E1	5	2	F1	4	3				B − M	•	•	↕	↕	↕	↕
Compare Acmltrs	CBA													11	2	1	A − B	•	•	↕	↕	↕	↕
Complement, 1's	COM							63	7	2	73	6	3				M̄ → M	•	•	↕	↕	R	S
	COMA													43	2	1	Ā → A	•	•	↕	↕	R	S
	COMB													53	2	1	B̄ → B	•	•	↕	↕	R	S
Complement, 2's	NEG							60	7	2	70	6	3				00 − M → M	•	•	↕	↕	①	②
(Negate)	NEGA													40	2	1	00 − A → A	•	•	↕	↕	①	②
	NEGB													50	2	1	00 − B → B	•	•	↕	↕	①	②
Decimal Adjust, A	DAA													19	2	1	Converts Binary Add. of BCD Characters into BCD Format	•	•	↕	↕	↕	③
Decrement	DEC							6A	7	2	7A	6	3				M − 1 → M	•	•	↕	↕	④	•
	DECA													4A	2	1	A − 1 → A	•	•	↕	↕	④	•
	DECB													5A	2	1	B − 1 → B	•	•	↕	↕	④	•
Exclusive OR	EORA	88	2	2	98	3	2	A8	5	2	B8	4	3				A ⊕ M → A	•	•	↕	↕	R	•
	EORB	C8	2	2	D8	3	2	E8	5	2	F8	4	3				B ⊕ M → B	•	•	↕	↕	R	•
Increment	INC							6C	7	2	7C	6	3				M + 1 → M	•	•	↕	↕	⑤	•
	INCA													4C	2	1	A + 1 → A	•	•	↕	↕	⑤	•
	INCB													5C	2	1	B + 1 → B	•	•	↕	↕	⑤	•
Load Acmltr	LDAA	86	2	2	96	3	2	A6	5	2	B6	4	3				M → A	•	•	↕	↕	R	•
	LDAB	C6	2	2	D6	3	2	E6	5	2	F6	4	3				M → B	•	•	↕	↕	R	•
Or, Inclusive	ORAA	8A	2	2	9A	3	2	AA	5	2	BA	4	3				A + M → A	•	•	↕	↕	R	•
	ORAB	CA	2	2	DA	3	2	EA	5	2	FA	4	3				B + M → B	•	•	↕	↕	R	•
Push Data	PSHA													36	4	1	A → M$_{SP}$, SP − 1 → SP	•	•	•	•	•	•
	PSHB													37	4	1	B → M$_{SP}$, SP − 1 → SP	•	•	•	•	•	•
Pull Data	PULA													32	4	1	SP + 1 → SP, M$_{SP}$ → A	•	•	•	•	•	•
	PULB													33	4	1	SP + 1 → SP, M$_{SP}$ → B	•	•	•	•	•	•
Rotate Left	ROL							69	7	2	79	6	3				M	•	•	↕	↕	⑥	↕
	ROLA													49	2	1	A	•	•	↕	↕	⑥	↕
	ROLB													59	2	1	B	•	•	↕	↕	⑥	↕
Rotate Right	ROR							66	7	2	76	6	3				M	•	•	↕	↕	⑥	↕
	RORA													46	2	1	A	•	•	↕	↕	⑥	↕
	RORB													56	2	1	B	•	•	↕	↕	⑥	↕
Shift Left, Arithmetic	ASL							68	7	2	78	6	3				M	•	•	↕	↕	⑥	↕
	ASLA													48	2	1	A	•	•	↕	↕	⑥	↕
	ASLB													58	2	1	B	•	•	↕	↕	⑥	↕
Shift Right, Arithmetic	ASR							67	7	2	77	6	3				M	•	•	↕	↕	⑥	↕
	ASRA													47	2	1	A	•	•	↕	↕	⑥	↕
	ASRB													57	2	1	B	•	•	↕	↕	⑥	↕
Shift Right, Logic	LSR							64	7	2	74	6	3				M	•	•	R	↕	⑥	↕
	LSRA													44	2	1	A	•	•	R	↕	⑥	↕
	LSRB													54	2	1	B	•	•	R	↕	⑥	↕
Store Acmltr	STAA				97	4	2	A7	6	2	B7	5	3				A → M	•	•	↕	↕	R	•
	STAB				D7	4	2	E7	6	2	F7	5	3				B → M	•	•	↕	↕	R	•
Subtract	SUBA	80	2	2	90	3	2	A0	5	2	B0	4	3				A − M → A	•	•	↕	↕	↕	↕
	SUBB	C0	2	2	D0	3	2	E0	5	2	F0	4	3				B − M → B	•	•	↕	↕	↕	↕
Subtract Acmltrs	SBA													10	2	1	A − B → A	•	•	↕	↕	↕	↕
Subtr with Carry	SBCA	82	2	2	92	3	2	A2	5	2	B2	4	3				A − M − C → A	•	•	↕	↕	↕	↕
	SBCB	C2	2	2	D2	3	2	E2	5	2	F2	4	3				B − M − C → B	•	•	↕	↕	↕	↕
Transfer Acmltrs	TAB													16	2	1	A → B	•	•	↕	↕	R	•
	TBA													17	2	1	B → A	•	•	↕	↕	R	•
Test, Zero or Minus	TST							6D	7	2	7D	6	3				M − 00	•	•	↕	↕	R	R
	TSTA													4D	2	1	A − 00	•	•	↕	↕	R	R
	TSTB													5D	2	1	B − 00	•	•	↕	↕	R	R

Condition code register header: | H | I | N | Z | V | C |

LEGEND

OP	Operation Code (Hexadecimal)	+	Boolean Inclusive OR.
~	Number of MPU Cycles.	⊙	Boolean Exclusive OR.
=	Number of Program Bytes.	M̄	Complement of M.
+	Arithmetic Plus.	→	Transfer Into.
−	Arithmetic Minus.	0	Bit Zero.
·	Boolean AND.	00	Byte Zero.
M$_{SP}$	Contents of memory location pointed to be Stack Pointer		

Note   Accumulator addressing mode instructions are included in the column for IMPLIED addressing

CONDITION CODE SYMBOLS

H	Half carry from bit 3.
I	Interrupt mask
N	Negative (sign bit)
Z	Zero (byte)
V	Overflow, 2's complement
C	Carry from bit 7
R	Reset Always
S	Set Always
↕	Test and set if true, cleared otherwise
•	Not Affected

Figure 6-10. MC6800 Instruction set. (Courtesy of Motorola Semiconductor Products, Inc.)

The program example used here is only one of literally thousands of programs that can be employed by a microcomputer to perform useful industrial operations. This problem showed only one simple problem that could be achieved by a microcomputer. The potential capabilities of this type of system are virtually unlimited. The type of microprocessor employed by a system and its unique instructional set are the primary factors that govern its operation in program planning.

Title: X + Y = N							**Date:** _____	
Purpose: Find N							**Time:** _____	
Steps	Memory Address (Hexadecimal)	Instruction			Opcode	Addressing Mode	Operand (Binary)	Description
		Byte 1	Byte 2	Byte 3				
1	66	86	–	–	LDA A	Immediate		Load accumulator A
	67	–	0A	–			0000 1010	Use data $0A_{16}$
2	68	C6	–	–	LDA B	Immediate		Load accumulator B
	69	–	08	–			0000 1000	Use data 08_{16}
3	6A	1B	–	–	ABA	Inherent		Add the contents of accumulators A and B
4	6B	19	–	–	DAA	Inherent		Correct for bcd output
5	6C	3E	–	–	HLT	Inherent		Stop all operations

Figure 6-11. Flow-chart plan of a problem.

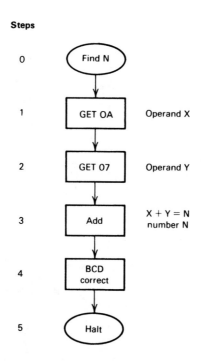

Figure 6-12. Programming sheet.

Chapter 7

Electronic Communications

INTRODUCTION

Sound is made louder by using electronic amplifiers. High-level sound amplification made it possible to communicate over a rather substantial distance. A public address amplifier system, for example, permits an announcer to communicate with a large number of people in a stadium or an arena. Even the most sophisticated sound system, however, has some limitations. Sound waves moving away from the source have a tendency to become somewhat weaker the farther they travel. An increase in signal strength does not, therefore, necessarily solve this problem. People near large speakers usually become uncomfortable when the sound is increased to a high level. It is possible, however, to communicate with people over long distances without increasing sound levels. Electromagnetic waves make this type of communication possible. Radio, television, long-distance telephone, and cellular phone communication are achieved by this process.

Electromagnetic wave communications systems play an important role in our daily lives. We listen to radio receivers, watch television, talk to friends on cellular telephones, and even communicate with astronauts by electromagnetic waves. Because this type of communication is widely used today, it is important that we have some basic understanding of it.

In this chapter we investigate how sound is transmitted through the air by electromagnetic waves. This type of communication uses high-frequency alternating current or radio frequency energy for its operation. These signals travel through the air at 186,000 miles per second. Electromagnetic communication permits sound to travel long distances al-most instantaneously.

OBJECTIVES

Upon completion of this chapter, you will be able to:

1. Explain the advantages, disadvantages, and characteristics of amplitude modulation and frequency modulation.

2. Explain the operation of AM and FM detectors.
3. Draw block diagrams of basic AM, CW, and FM transmitters and receivers.
4. Discuss the effects of heterodyning.
5. Discuss how intelligence is added to and removed from an RF carrier.
6. Calculate percentage modulation.
7. Explain the operation of a cathode-ray tube (CRT).
8. Describe how a picture is produced on the face of a picture tube.
9. Draw block diagrams of a black-and-white and a color TV set, and explain the function of each block.
10. Explain the functions of sweep signals and how they are synchronized.
11. Explain how high voltage is produced in a TV receiver.
12. Describe how video amplifiers obtain wide bandwidth, and list ways of increasing bandwidth.
13. Describe fiber optic systems.

COMMUNICATIONS SYSTEMS

A communication system is similar to any other electrical system. It has an energy source, transmission path, control, load device, and one or more indicators. These individual parts are all essential to the operation of the system. The physical layout of the communication system is one of its most distinguishing features.

Figure 7-1 shows a block diagram of a radio-frequency communication system. The signal source of the system is a radio-frequency (RF) transmitter. The transmitter is the center or focal point of the system. The RF signal is sent to the remaining parts of the system through space. Air is the transmission path of the system. RF finds air to be an excellent signal path. The control function of the system is directly related to the signal path. The distance that the RF signal must travel has much to do with its strength. The load of the system is an infinite number of radio receivers. Each receiver picks the signal out of the air and uses it to do work. Any number of receivers can be used without directly influencing the output of the transmitter. System indicators may be found at several locations. Meters, indicating lamps, and waveform monitoring oscilloscopes are typical indicators. These basic functions apply in general to all RF communication systems.

Many different RF communication systems are in operation today. As a general rule, these systems are classified according to the method by which signal information or intelligence is applied to the transmitted signal. Three common communication systems are continuous wave (CW), amplitude modulation (AM), and frequency modulation (FM). Each system has several unique features that distinguish it from others. The transmitter and receiver of the system are quite different. They are all similar to the extent that they use an RF signal energy source as a source of operation.

A more realistic way to look at an RF communication system is to divide the transmitter and receiver into separate systems. The transmitter then has all its system parts in one discrete unit. Its primary function is to generate an RF signal that contains intelligence and radiate this signal into space. The receiver can also be viewed as an independent system. It has a complete set of system parts. Its primary function is to pick up the RF transmitted signal, extract the intelligence, and develop it into a usable output. A one-way communication system has one transmitter and an infinite number of receivers. Commercial AM, FM, and TV communication is achieved by this method. Two-way mobile communication systems have a transmitter and a receiver at each location. This type of system permits direct communication between each location. A citizen's band (CB) radio and cellular telephones are considered to be two-way communication.

Figure 7-2 shows a simplification of the transmitter and receiver as independent systems. The transmitter has an RF oscillator, amplifiers, and a power source. The antenna ground at the output circuit serves as the load device for the system. Intelligence is applied according to the design of the system, which may be CW, AM, or FM.

The RF receiver function is represented as an independent system. It will not be operational unless a signal is sent out by the transmitter. The receiver has tuned RF amplification, detection, reproduction, and a power supply. The detection function picks out the intelligence from the received signal. The reproduction unit, which is usually a speaker, serves as the load device. The receiver will only respond to the RF signal sent out by the transmitter.

ELECTROMAGNETIC WAVES

Electromagnetic communication systems rely on the radiation of high-frequency energy from an antenna-ground net-work. This particular principle was discovered by Heinrich Hertz around 1885. He found that high-frequency waves produced by an electrical spark caused electrical energy to be induced in a coil of wire some distance away. He also discovered that current passing through a coil of wire produces a strong electromagnetic field. The fundamental unit of frequency is expressed in hertz, which is in recognition of his electromagnetic discoveries.

When direct current is applied to a coil of wire, it causes the field to remain stationary as long as current is flowing. When dc is first turned on, the field expands. When it is turned off, the field collapses and cuts across the coil. As a general rule, dc electromagnetic field de-

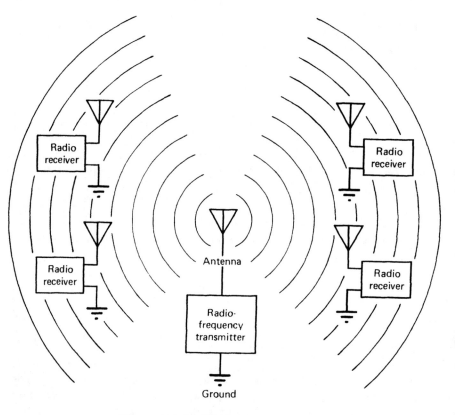

Figure 7-1. Radio-frequency system.

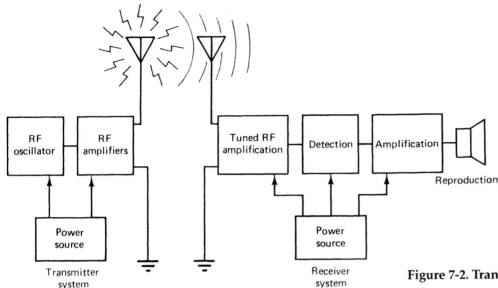

Figure 7-2. Transmitter-receiver system.

velopment is of no significant value in radio communication.

When low-frequency ac is applied to an inductor, it produces an electromagnetic field. Because ac is in a constant state of change, the resulting field is also in a state of change. It changes polarity twice during each operational cycle.

Figure 7-3 shows how the electromagnetic field of a coil changes when ac is applied. In Figure 7-3(a), the field is expanding during the first half of the positive alternation. At the peak of the alternation, coil current begins to decrease in value. Figure 7-3(b) shows the field collapsing. The current value decreases to the zero baseline at this time and completes the positive alternation or first 180° of the sine wave. Figure 7-3(c) shows the field expanding for the negative alternation. The polarity of this field is now reversed. At the peak of the negative alternation, the field begins to collapse. Figure 7-3(d) shows the field change for the last part of the alternation. The cycle is complete at this point. The sequence is repeated for each new sine wave.

When high-frequency ac is applied to an inductor, it also produces an electromagnetic field. The directional change of high-frequency ac, however, occurs quickly. The change is so rapid that there is not enough time for the collapsing field to return to the coil. As a result, a portion of the field becomes detached and is forced away from the coil. In a sense, the field "radiates out" from the coil. The word *radio* was developed from these two terms.

The electromagnetic wave of an RF communication system radiates out from the antenna of the transmitter. This wave is the end result of an interaction between electric and magnetic fields. The resultant wave is invisible,

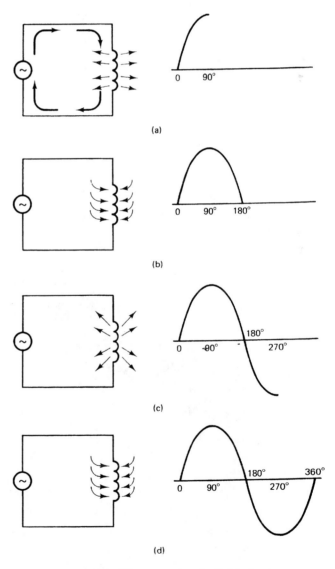

Figure 7-3. Electromagnetic field charges.

inaudible, and travels at the speed of light. These waves become weaker after leaving the antenna. Stronger electromagnetic waves can be effectively used in a radio communication system. These waves do not affect humans like high-powered sound waves. RF waves are also easier to work with in the receiver circuit.

Electromagnetic wave radiation is directly related to its frequency. Low-frequency (LF) waves of 30 to 300 kHz and medium-frequency (MF) waves of 300 to 3000 kHz tend to follow the curvature of the earth. These waves are commonly called *ground waves*. They are usable during the day or night for distances up to 100 miles. Daytime reception of commercial AM radio is largely through ground waves.

A different part of the electromagnetic wave radiated from an antenna is called a *sky wave*. Depending on their frequency, sky waves are reflected back to the earth by a layer of ionized particles called the *ionosphere*. These waves permit signal reception well beyond the range of the ground wave. Sky-wave signal patterns change according to ion density and the position of the ionized layer. During the daylight hours, the ionosphere is dense and near the surface of the earth. Signal reflection, or skip, of this type is not suitable for long-distance communication. In the evening hours the ionosphere is less dense and moves to higher altitudes. Sky-wave reflection patterns have larger angles and travel greater distances. Figure 7-4 shows some sky-wave patterns and the ground wave.

Very-high-frequency (VHF) signals of 30 to 300 MHz tend to move in a straight line. The height of the antenna and its radiation angle have much to do with the direction of the signal path. Line-of-sight transmission describes this type of radiation pattern. FM, TV, and satellite communication are achieved by VHF signal radiation. The physical size of the wave, or its wavelength, decreases with an increase in frequency. High-frequency signals have a short wavelength. These signals pass readily through the ionosphere without being reflected or distorted. Communication between an FM or TV transmitter and a receiver on earth is limited to a few hundred miles. Communication between an earth station and a satellite can involve thousands of miles.

CONTINUOUS-WAVE COMMUNICATION

Continuous-wave (CW) communication is the simplest of all communication systems. It consists of a transmitter and a receiver. The transmitter employs an oscillator for the generation of an RF signal. The oscillator generates a continuous sine wave. In most systems the CW signal is amplified to a desired power level by an RF power amplifier. The output of the final power amplifier is then connected to the antenna-ground network. The antenna radiates the signal into space. Information or intelligence is applied to the CW signal by turning it on and off, which causes the signal to be broken into a series of pulses or short bursts of RF energy. The pulses conform to an intelligible code. The international Morse code is in common usage today. Figure 7-5 shows a simplification of the Morse code. A short burst of RF represents a dot. A burst three times longer represents a dash. Signals of this type are called *keyed* or *coded continuous waves*. The Federal Communications Commission (FCC) classifies

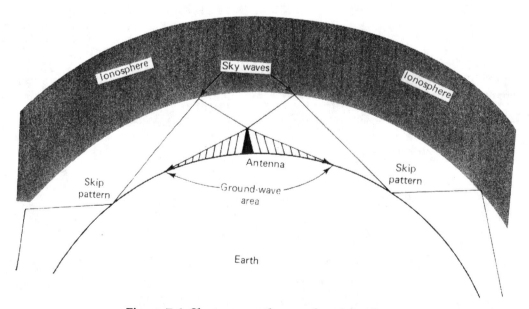

Figure 7-4. Sky-wave and ground-wave patterns.

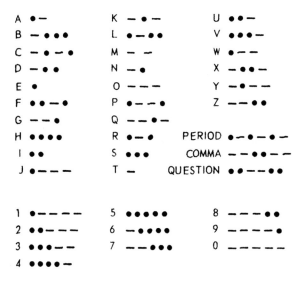

A	•—	K	—•—	U	••—
B	—•••	L	•—••	V	•••—
C	—•—•	M	——	W	•——
D	—••	N	—•	X	—••—
E	•	O	———	Y	—•——
F	••—•	P	•——•	Z	——••
G	——•	Q	——•—		
H	••••	R	•—•	PERIOD	•—•—•—
I	••	S	•••	COMMA	——••——
J	•———	T	—	QUESTION	••——••

1	•————	5	•••••	8	———••
2	••———	6	—••••	9	————•
3	•••——	7	——•••	0	—————
4	••••—				

Figure 7-5. Morse code.

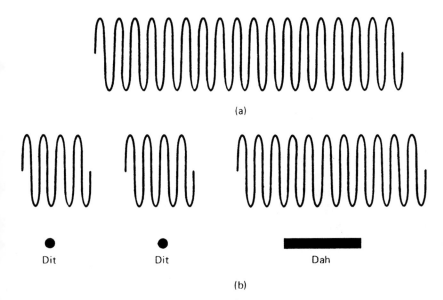

(a)

Dit   Dit   Dah

(b)

Figure 7-6. Comparison of CW signals: (a) continuous-wave signal; (b) coded or keyed continuous wave, letter *U*.

keyed CW signals as type A1. This is considered to be radio *telegraphy* with on-off keying. Figure 7-6 shows a comparison between a CW signal and a keyed CW signal.

The receiver function of a CW communication system is somewhat more complex than the transmitter function. It has an antenna-ground network, a tuning circuit, an RF amplifier, a heterodyne detector, a beat frequency oscillator, an audio amplifier, and a speaker. The receiver is designed to pick up a CW signal and convert it into a sound signal that drives the speaker. Figure 7-7 shows a block diagram of the CW communication system.

CW Transmitter

A CW transmitter in its simplest form has an oscillator, a power supply, an antenna-ground network, and a keying circuit. The power output developed by the circuit is usually quite small. In this type of system, the active device serves jointly as an oscillator and a power amplifier to reduce the number of circuit stages in front of the antenna. An oscillator used in this manner generally has reduced frequency stability. As a rule, frequency stability is an important operational consideration. Single-stage transmitters are rarely used today because of their instability.

A single-stage CW transmitter is shown in Figure 7-8. This particular circuit is a tuned base Hartley oscillator. The output frequency is adjustable up to 3.5 MHz. The power output is approximately 0.5 W. When the output of the oscillator is connected to a load, it usually causes a shift in frequency. The load should therefore be of a constant impedance value for this type of transmitter to be effective.

Before a CW transmitter can be placed into operation, it is necessary to close the code key. This operation supplies forward-bias voltage to the base through the divider network of R_1 and R_2. With the transistor properly biased, a feedback signal is supplied to the T_1-C_1 tank circuit which charges C_1 and causes the circuit to oscillate. Opening the key causes the oscillator to stop functioning. Intelligence can be injected into the CW signal in the form of code. The code key simply responds as a fast-acting on-off switch.

The output of the transmitter is developed across a coil wound around T_1. Inductive coupling of this type is common in RF transmitter circuits. A low-impedance output is needed to match the antenna-ground impedance, which, when accomplished, provides a maximum transfer of energy from the oscillator to the antenna.

A single-stage CW transmitter has two shortcomings—the power output is usually held to a low level and the frequency stability is rather poor. These problems can be overcome to some extent by adding a power amplifier after the oscillator. Systems of this type are called *master oscillator power amplifier (MOPA) transmitters.*

Figure 7-9 shows the circuitry of a low-power MOPA transmitter system. Q_1 is a modified Pierce oscillator. The crystal (Y_1) provides feedback from the collector to the

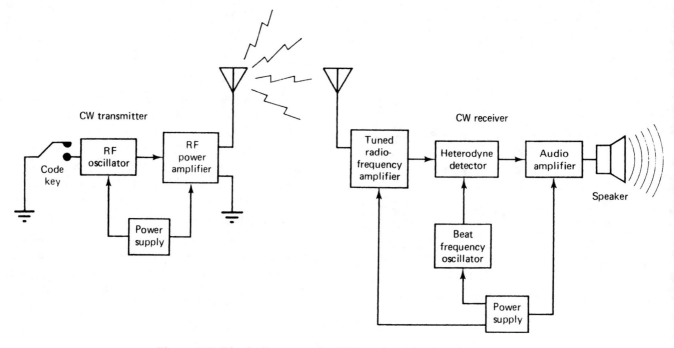

Figure 7-7. Block diagram of a CW communication system.

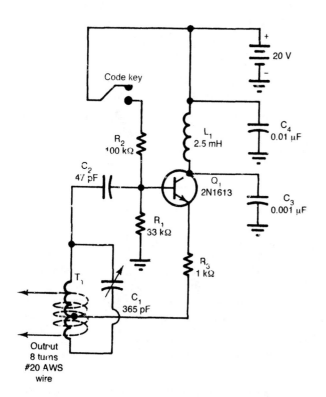

Figure 7-8. Single-stage CW transmitter.

base. Transformer T_1 is used to match the collector impedance of Q_1 to the base of transistor, Q_2. Q_2 is the power amplifier. It operates as a class C amplifier. The oscillator is therefore isolated from the antenna through Q_2, which provides a fixed load for the oscillator. Operational

stability is improved and power output is increased with this modification.

CW Receiver

The receiver of a CW communication system is responsible for intercepting an RF signal and recovering the information it contains. Information of this type is in the form of a radiotelegraph code. The signal is an RF carrier wave that is interrupted by a coded message. A CW receiver must perform several functions to achieve this operation, including signal reception, selection, RF amplification, detection, AF amplification, and sound reproduction. These functions are also used in the reception of other RF signals. In fact, CW reception is usually only one of a number of operations performed by a communication receiver. This function is generally achieved by placing the receiver in its CW mode of operation. A switch is usually needed to perform this operation.

Figure 7-10 shows the functions of a CW communication receiver in a block diagram. Note that electrical power is needed to energize the active components of the system. Not all blocks of the diagram are supplied operating power, meaning that some of the functions can be achieved without a transistor or a vacuum tube. The antenna, tuning circuit, and detector do not require power supply energy for operation. These functions are achieved by signal energy. A radio receiver circuit has a signal energy source and an operational energy source. The power supply provides operational energy directly to the circuit.

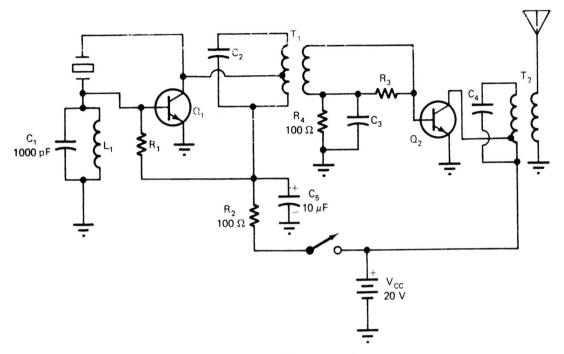

Figure 7-9. MOPA transmitter.

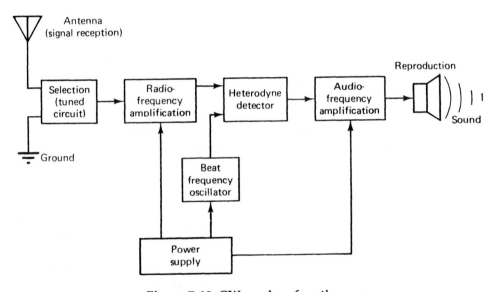

Figure 7-10. CW receiver functions.

The signal source for RF energy is the transmitter. This signal must be intercepted from the air and processed by the receiver during its operation.

Antennas

The antenna of a CW receiver is responsible for the reception function. It intercepts a small portion of the RF signal sent out by the transmitter. Ordinarily, the receiving antenna only develops a few microwatts of power. In a strict sense, the antenna is a transducer. It is de-signed to convert electromagnetic wave energy into RF signal voltage. The signal voltage then causes a corresponding current flow in the antenna-ground network. RF antenna power is the product of signal voltage and current.

Receiving antennas come in a variety of styles and types. In two-way communication systems, the transmitter and receiving antenna are of the same unit. The antenna is switched back and forth between the transmitter and receiver. In one-way communication systems, antenna construction is not particularly critical. A long piece of wire can respond as a receiving antenna. In weak signal areas, antennas may be tuned to resonate at the particular frequency being received. Portable radio receivers use small antenna coils attached directly to the circuit.

When the electromagnetic wave of a transmitter passes over the receiving antenna, ac voltage is induced into it. This voltage causes current to flow as shown in Figure 7-11. Starting at point A of the waveform, the electrons are at a standstill. The induced signal then causes the voltage to rise to its positive peak. The resultant current flow is from the antenna to the ground. It rises to a peak, then drops to zero at point C on the curve. The polarity of the induced voltage changes at this point. Between points C and E it rises to the peak of the negative alternation and returns again to zero. The resultant current flow is from the ground into the antenna. This induced signal is repeated for each succeeding alternation.

With radio frequency induced into the antenna, a corresponding RF is brought into the receiver by the antenna coil. The antenna coil serves as an impedance-matching transformer. It matches the impedance of the antenna to the impedance of the receiver input circuit. When outside antennas are used, the primary winding is an extension of the antenna. The loop antenna coil of a portable receiver is attached directly to the input circuit.

Signal Selection

The signal selection function of a radio receiver refers to its ability to pick out a desired RF signal. In receiver circuits, the antenna is generally designed to intercept a band or range of different frequencies. The receiver must then select the desired signal from all those intercepted by the antenna. Signal selection is primarily achieved by

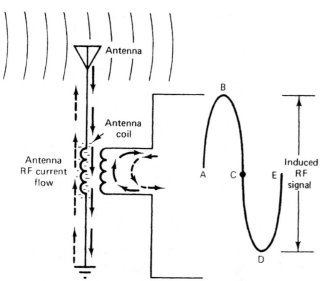

Figure 7-11. Antenna signal voltage and current.

an *LC* resonant circuit. This circuit provides a low-impedance path for its resonant frequency. Nonresonant frequencies see a high impedance. In effect, the resonant frequency signal is permitted to pass through the tuner without opposition.

Figure 7-12 shows the input tuner of a CW receiver. In this circuit, L_2 is the secondary winding of the antenna transformer; C1, is a variable capacitor. By changing the value of C1, the resonant frequency of L_2-C1 is tuned to the desired frequency. The selected frequency is then permitted to pass into the remainder of the receiver. Signal selection for AM, FM, and TV receivers all respond to a similar tuning circuit.

In communication receiving circuits, several tuning stages are needed for good signal selection. Each stage of tuning permits the receiver to be more selective of the desired frequency. The tuning response curve of one tuned stage is shown in Figure 7-12. In a crowded radio-frequency band, several stations operating near the selected frequency might be received at the same time. Additional tuning improves the selection function by narrowing the response curve. Figure 7-13 shows some representative loop antenna coils.

RF Amplification

Most CW receivers employ at least one stage of radio-frequency amplification. This specific function is designed to amplify the weak RF signal intercepted by the antenna. Some degree of amplification is generally needed to boost the signal to a level where its intelligence can be recovered. RF amplification can be achieved by a variety of active devices. Vacuum tubes have been used for a number of years. Solid-state circuits can employ bipolar transistors and MOSFETs in their design. Several manufacturers have now developed IC RF amplifiers. Any of these active devices can be used to achieve RF signal amplification.

A representative RF amplifier circuit is shown in Figure 7-14. The primary function of the circuit is to determine the level of amplification developed. Note that the amplifier has both input and output tuning circuits. This circuit is called a *tuned radio frequency (TRF) amplifier*. The broken line between the two variable capacitors indicates that they are "ganged" together. When one capacitor is tuned, the second capacitor is adjusted to a corresponding value.

Heterodyne Detection

Heterodyne detection is an essential function of the CW receiver. Heterodyning is simply a process of mixing two ac signals in a nonlinear de-vice. In the CW receiver,

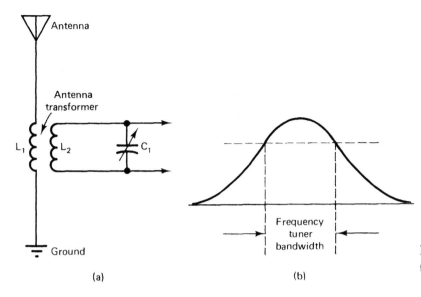

(a)   (b)

Figure 7-12. (a) Input tuning circuit; (b) tuner frequency response curve.

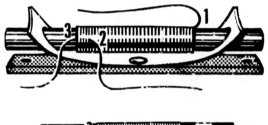

Figure 7-13. Antenna coils.

the nonlinear device is a diode. The applied signals are called the *fundamental frequencies*. The resultant output of this circuit has four frequencies. Two of these are the original fundamental frequencies. The other two are beat frequencies. Beat frequencies are the addition and the difference in the fundamental frequencies. Heterodyne detectors used in other receiver circuits are commonly called *mixers*.

Figure 7-15 shows the signal frequencies applied to a heterodyne detector. Fundamental frequency F_1, is representative of the incoming RF signal. This signal has been keyed on and off with a telegraphic code. Fundamental frequency F_2 comes from the beat frequency oscillator

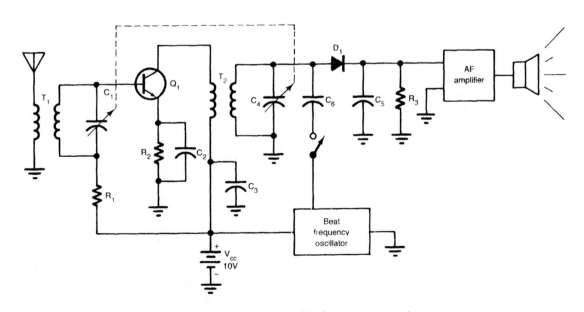

Figure 7-14. Tuned radio-frequency receiver.

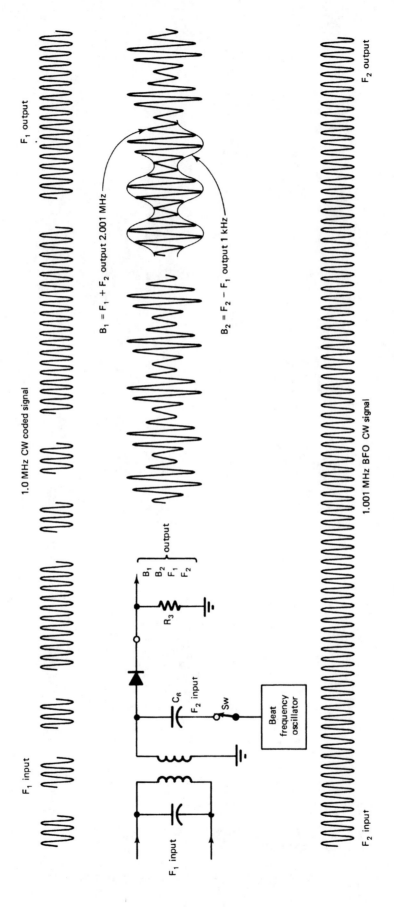

Figure 7-15. Heterodyne detector frequencies.

(BFO) of the receiver. It is a CW signal of 1.001 MHz. Both RF signals are applied to the receiver's diode.

The resulting output of a heterodyne detector is F_1, F_2, $F_1 + F_2$, and $F_2 - F_1$. $F1$ is 1.0 MHz and F_2 is 1.001 MHz. Beat frequency B_1 is 2.001 MHz and beat frequency B_2 is 1000 Hz. The output of the detector contains all four frequencies. The difference beat frequency of 1000 Hz is in the range of human hearing. F_1 F_2, and F_3 are RF signals.

When two RF signals are applied to a diode, the resultant output will be the sum and difference signals. If the two signals are identical, the output frequency will be the same as the input, and is generally called *zero beating*. There is no developed beat frequency when this occurs.

When the two RF input signals are slightly different, beat frequencies will occur. At one instant, the signals will move in the same direction. The resultant output will be the sum of the two frequencies. When one signal is rising, and one is falling at a different rate, at times the signals will cancel each other. The sum and difference of the two signals are therefore frequency dependent. B_1 and B_2 are combined in a single RF wave. The amplitude of this

wave varies according to frequency difference in the two signals. In a sense, the mixing process causes the amplitude of the RF wave to vary at an audio rate.

The diode of a heterodyne detector is responsible for rectification of the applied signal frequencies. The rectified output of the added beat frequency is of particular importance. It contains both RF and AF. Figure 7-16 shows how the diode responds. Initially, the RF signals are rectified. The output of the diode would be a series of RF pulses varying in amplitude at an AF rate. C_5 added to the output of the diode responds as a C-input filter. Each RF pulse will charge C_5 to its peak value.

During the off period C_5 will discharge through R_3. The output will be a low-frequency AF signal of 1000 Hz, which will occur only when the incoming CW signal has been keyed. The AF signal is representative of the keyed information imposed on the RF signal at the transmitter.

Beat Frequency Oscillator

Nearly any basic oscillator circuit could be used as the BFO of a CW receiver. Figure 7-17 shows a Hartley oscillator. Feedback is provided by a tapped coil in the base

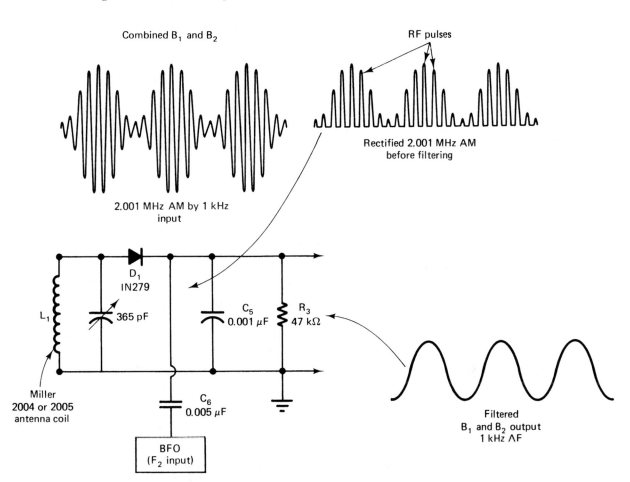

Figure 7-16. Heterodyne diode output.

circuit. C_1 and T_1 are the frequency-determining components. Note that this particular oscillator has variable-frequency capabilities. C_1 is usually connected to the front panel of the receiver. Adjustment of C_1, is used to alter the tone of the beat frequency signal. When the BFO signal is equal to the coded CW signal, no sound output will occur. This is where zero beating takes place. When the frequency of the BFO is slightly above or below the incoming signal frequency, a low AF tone is produced. Increasing the BFO frequency causes the pitch of the AF tone to increase. Adjustment of the AF output is a matter of personal preference. In practice, an AF tone of 400 Hz to I kHz is common.

Output of the BFO is coupled to the diode through capacitor C_6. In a communication receiver, BFO output is con-trolled by a switch. With the switch on, the receiver will produce an output for CW signals. With the switch off, the receiver will respond to AM and possibly FM signals. The BFO is only needed to receive coded CW signals.

Audio-frequency Amplification

The audio-frequency (AF) amplifier of a CW receiver is responsible for increasing the level of the developed sound signal. The type and amount of signal amplification varies widely among different receivers. Typically, a small-signal amplifier and a power amplifier are used.

The small-signal amplifier responds as a voltage amplifier. This amplifier is designed to increase the signal voltage to a level that will drive the power amplifier. Power amplification is needed to drive the speaker. The power output of a communication receiver rarely ever exceeds 5 W. A number of the AF amplifier circuits in Chapter 14 could be used in a CW receiver.

AMPLITUDE MODULATION COMMUNICATION

Amplitude modulation or AM is an extremely important form of communication. It is achieved by changing the physical size or amplitude of the RF wave by the intelligence signal. Voice, music, data, and picture intelligence can be transmitted by this method. The intelligence signal must first be changed into electrical energy. Transducers such as microphones, phonograph cartridges, tape heads, and photo electric devices are designed to achieve this function. The developed signal is called the *modulating component*. In an AM communication system, the RF transmitted signal is much higher in frequency than the modulating component. The RF component is an uninterrupted CW wave. In practice, this part of the radiated signal is called the *carrier wave.*

An example of the signal components of an AM system is shown in Figure 7–18. The unmodulated RF carrier

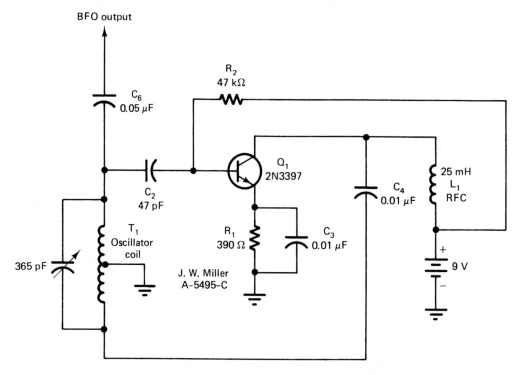

Figure 7-17. Hartley oscillator used as a BFO.

is a CW signal that is generated by an oscillator. In this example, the carrier is 1000 kHz or 1.0 MHz. A signal of this frequency would be in the standard AM broadcast band of 535 to 1605 kHz. When listening to this station, a receiver would be tuned to the carrier frequency. Assume now that the RF signal is modulated by the indicated 1000-Hz tone. The RF component will change 1000 cycles for each AF sine wave. The amplitude of the RF signal will vary according to the frequency of the modulating signal. The resulting wave is called an *amplitude-modulated RF carrier*.

To achieve amplitude modulation, the RF and AF components are applied to a nonlinear device. A vacuum-tube or solid-state device operating in its nonlinear region can be used to produce modulation. In a sense, the two signals are mixed or heterodyned together. This operation causes beat frequencies to be developed. The signals of Figure 7-18 will have two beat frequencies. Beat frequency 1 is the sum of the two signal frequencies. B_1 is 1,000,000 Hz plus 1000 Hz or 1,001,000 Hz. B_2 is 1,000,000 Hz minus 1000 Hz or 999,000 Hz. The resulting AM signal therefore contains three RF signals. These are the 1.0-MHz carrier, 0.999 MHz, and 1.001 MHz.

When the modulating component of an AM signal is music, the resultant beat frequencies become quite complex. As a rule, this involves a range or a band of frequencies. In AM systems, these are called *sidebands*. B_1 is the upper sideband and B_2 is the lower sideband.

The space that an AM signal occupies with its frequency is called *a channel*. The bandwidth of an AM channel is twice the highest modulating frequency. For our 1-kHz modulating component, a 2-kHz bandwidth is needed, which would be 1 kHz above and below the carrier frequency of 1 MHz. In commercial AM broadcasting, a station is assigned a 10-kHz channel which limits the AM modulation component to a frequency of 5 kHz. Figure 7-19 shows the side-bands produced by a standard AM station.

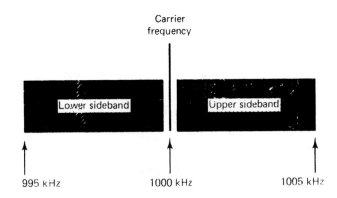

Figure 7-19. Sideband frequencies of an AM signal.

Interestingly, the carrier wave of an AM signal contains no modulation. All the modulation appears in the sidebands. If the modulating component is removed, the sidebands will disappear. Only the carrier will be transmitted. The sidebands are directly related to the carrier and the modulating component. The carrier has a constant frequency and amplitude. The sidebands vary in frequency and amplitude according to the modulation component. In AM radio, the receiver is tuned to the carrier wave.

Percentage of Modulation

In AM radio, it is not permitted by law to exceed 100% modulation. Essentially, this means that the modulating component cannot cause the RF component to vary over 100% of its unmodulated value. Figure 7-20 shows an AM signal with three different levels of modulation.

When the peak amplitude of a modulating signal is less than the peak amplitude of the carrier, modulation is less than 100%. If the modulation component and carrier amplitudes are equal, 100% modulation is achieved. A modulating component greater than the carrier will cause

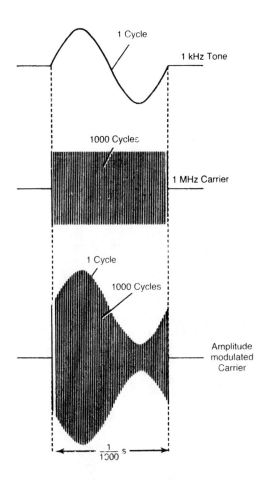

Figure 7-18. AM signal components.

overmodulation. An overmodulated wave has an interrupted spot in the carrier wave. *Overmodulation* causes increased signal bandwidth and additional sidebands to be generated, thus causing interference with adjacent channels.

Modulation percentage can be calculated or observed on an indicator. When operating voltage values are known, the percentage of modulation can be calculated. The formula is

$$\% \text{ modulation} = V_{max} - V_{min}/2V \text{ carrier} \times 100\%$$

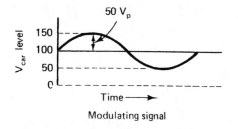

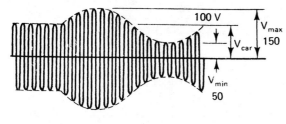

Modulated carrier

(a)

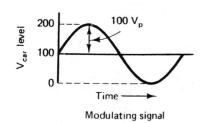

Using the values of Figure 7-20(a), compute the percentage of modulation. Compute the percentage of modulation for Figure. 7-20(b). If the V_{min} value of the overmodulated signal is considered to be 0 V, compute the modulation percentage.

AM Communication System

A block diagram of an AM communication system is shown in Figure 7-21. The transmitter and receiver respond as independent systems. In a one-way communication system, there is one transmitter and an infinite number of receivers. Commercial AM radio is an example of one-way communication. Two-way communication systems have a transmitter and a receiver at each location. CB radio systems are of the two-way type. The operating principles are basically the same for each system. The *transmitter* of an AM system is responsible for signal generation. The RF section of the transmitter is primarily the same as that of a CW system. The modulating signal component is, however, a unique part of the transmitter. Essentially, this function is achieved by an AF amplifier. A variety of amplifier circuits can be used. Typically, one or two small-signal amplifiers and a power amplifier are suitable for low-power transmitters. The developed

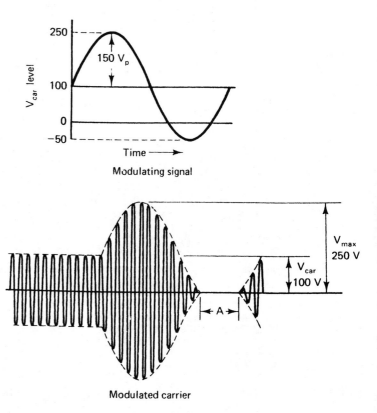

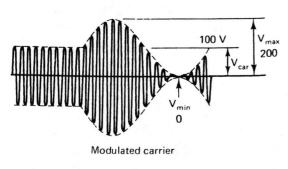

Modulated carrier

(b)

Modulated carrier

(c)

Figure 7-20. AM modulation levels: (a) undermodulation; (b) 100% modulation; (c) overmodulation.

modulation component power must equal the power level of the RF output. Modulation of the two signals can be achieved at several places. High-level modulation occurs in the RF power amplifier. Low-level modulation is achieved after the RF oscillator. All amplification following the point of modulation must be linear. Only high-level modulation will be discussed in this chapter.

The *receiver* of an AM system is responsible for signal interception, selection, demodulation, and reproduction. Most AM receivers employ the heterodyne principle in their operation. This type of receiver is known as a *superheterodyne circuit.* It is somewhat different from the het-

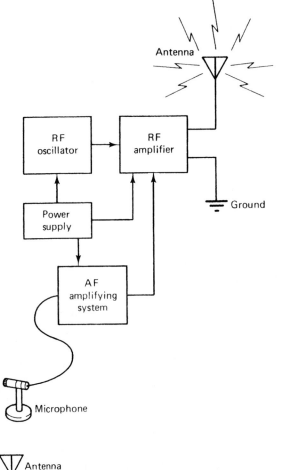

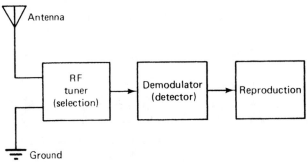

Figure 7-21. AM communication system.

erodyne detector of the CW receiver. A large part of the circuit is the same as the CW receiver. We will only discuss the new functions of the AM receiver.

AM Transmitters

Many types of AM transmitters are available today. Toy "walkie-talkie" units with an output of less than 100 MW are popular. Citizens band (CB) transmitters are designed to operate at 27 MHz with a 5-W output. AM amateur radio transmitters are available with a power output of up to 1 kW of power. Commercial AM transmitters are assigned power output levels from 250 W to 50 kW. These stations operate between 535 and 1605 kHz. In addition, there are 2-meter mobile communication systems, military transmitters, and public service radio systems that all use AM. As a general rule, frequency allocations and power levels are assigned by the Federal Communications Commission.

An AM transmitter has a number of fundamental parts regardless of its operational frequency or power rating. It must have an oscillator, an audio signal component, an antenna, and a power supply. The oscillator is responsible for the RF carrier signal. The audio component is responsible for the intelligence being transmitted. The modulation function can be achieved in a variety of ways. The signal is radiated from the antenna.

A simplified AM transmitter with a minimum of components is shown in Figure 7-22. Transistor Q_1 is an AF signal amplifier. The sound signal being amplified is developed by a crystal microphone. It is then applied to transistor Q_2. This transistor is a Colpitts oscillator. L_1, C_4, and C_5 determine the RF frequency of the oscillator. The amplitude of the oscillator varies according to the AF component. Resistor R_3 is used to adjust the amplitude level of the AF signal. The developed AM output signal is applied to the transmitting antenna. With proper design of the transmitter, it can be tuned to the standard AM broadcast band. The frequency of the system is adjusted by capacitor C_4. The operating range of this unit is several hundred feet.

A schematic of an improved low-power AM transmitter is shown in Figure 7-23. Transistors Q_1, Q_2, and Q_3 of this circuit are responsible for the RF signal component. The AF component is developed by Q_4, Q_5, and Q_6. High-level modulation is achieved by this transmitter. The AF signal modulates the RF power amplifier Q_3.

Transistor Q_1 is the active device of a Hartley oscillator. This particular oscillator has a variable-frequency output of 1 to 3.5 MHz. The frequency-determining components are C_1 and L_{T1}. The output of the oscillator is coupled to the base of Q_2 by capacitor C_3. The emitter-follow-

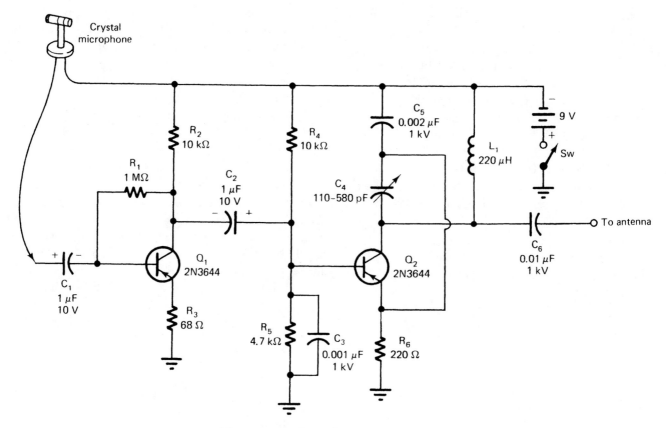

Figure 7-22. Simplified AM transmitter.

er output of the oscillator has a low impedance.

Transistor Q_2 is an RF signal amplifier. This transistor is primarily responsible for increasing the signal level of the oscillator. It is also used to isolate the RF load from the oscillator, which is needed to improve oscillator frequency stability. When Q_2 is used in this regard, it is called a *buffer amplifier.*

The signal output of Q_2 must be capable of driving the power amplifier. In high-power transmitters, several RF signal amplifiers may be found between the oscillator and the power amplifier. Each stage is responsible for increasing the signal level to a suitable level. RF amplifiers of this type are often called *drivers*. Q_3 is an RF power amplifier. It is designed to increase the power level of the RF signal applied to its input. The output is used to drive the transmitting antenna. The load of Q_3 is a tuned circuit composed of C_{10}, L_5, and C_{11}. This is a pi-section filter. A filter of this type is used to remove signals other than those of the resonant frequency. C_{11} is the output capacitor of the filter. It is adjusted to match the impedance of the antenna. C_{10} is the input capacitor. It is used to resonate the filter to the applied carrier frequency. Resonance of the tuning circuit occurs when the collector current meter dips to its lowest value. In the broadcasting field an

adjustment of this type is called "dipping the final." Q_3 is operated as a class C power amplifier.

The modulating component of the transmitter is developed by an AF amplifier. Q_4 and Q_5 are push-pull AF power amplifiers. The developed AF signal is applied to the modulation transformer T_2. This signal causes the collector volt-age of Q_3 to vary at an AF rate instead of being dc, which causes the RF output to vary in amplitude according to the AF component. The output signal has a carrier and two sidebands. The power output is approximately 5 W.

Do not connect the output of this transmitter to an out-side antenna unless you hold a valid radio-telephone operator's license. A load lamp is used for operational testing. The intensity of the load lamp is a good indication of the RF power developed by the transmitter.

Simple AM Receiver

An AM receiver has four primary functions that must be achieved for it to be operational. No matter how complex or involved the receiver, it must accomplish these functions: signal interception, selection, detection, and reproduction. Figure 7–24 shows a diagram of an AM receiver that accomplishes these functions.

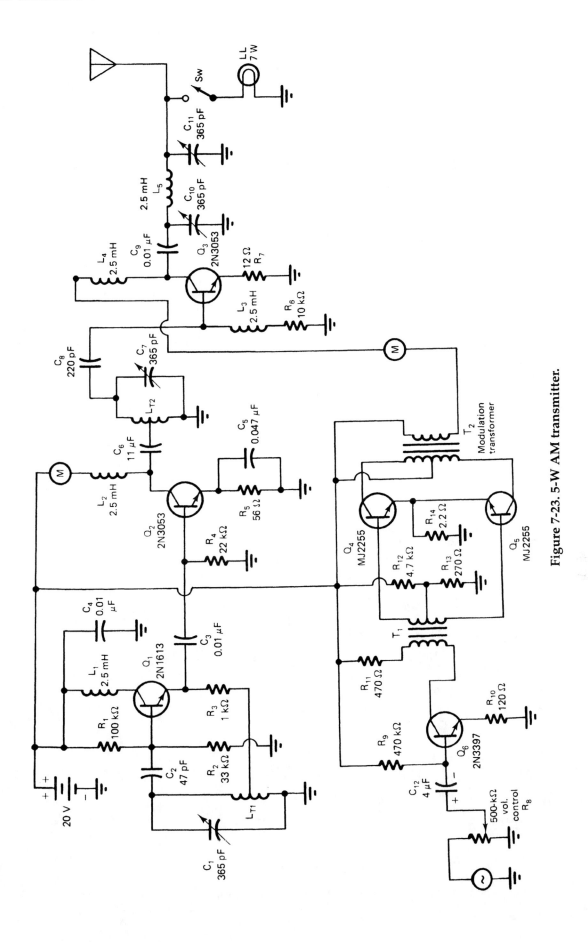

Figure 7-23. 5-W AM transmitter.

Note that the circuit does not employ an amplifying device. No electrical power source is needed to make this receiver operational. The signal source is intercepted by the antenna-ground network. The receiver is energized by the intercepted RF signal energy. A receiver of this type is generally called a *crystal radio.* The detector is a crystal diode.

The functional operation of our crystal diode radio receiver is similar to the CW receiver. This particular circuit does not employ an RF amplifier, a beat frequency oscillator, or an AF amplifier. A strong AM signal is needed to make this receiver operational. Signal interception and selection are achieved in the same way as in the CW receiver.

The detection or demodulation function of an AM receiver is responsible for removing the AF component from the RF signal. The detector is essentially a half-wave rectifier for the RF signal. Germanium diodes are commonly used as detectors. They are more sensitive to RF signals than a silicon diode.

The waveforms of Figure 7-24 show how the crystal diode responds. The selected AM signal is applied to the diode. Detection is accomplished by rectification and filtering. The detected wave is a half-wave rectification version of the input. C_2 responds as an RF filter. It charges to the peak value of each RF pulse. It then discharges

through the resistance of the earphone when the diode is reverse biased. The average value of the RF component appears across C_2 and is representative of the AF signal component. It energizes a small coil in the earphone. A thin metal disk in the earphone fluctuates according to coil energy. The earphone therefore changes electrical energy into sound waves. AF signal re-production is achieved by the earphone.

SUPERHETERODYNE RECEIVERS.

Practically all AM radio receivers in operation today are of the super-heterodyne type. This type of system accomplishes all the basic receiver functions. It has a number of circuit modifications which provide improved reception capabilities. It has excellent selectivity and is sensitive to long-distance signal reception. Figure 7-25(a) shows a block diagram of an AM superheterodyne receiver. This particular receiver is de-signed for signal reception between 535 and 1605 kHz.

Superheterodyne operation is somewhat unusual compared with other methods of radio reception. Special RF amplifiers that are tuned to a fixed frequency are used in this circuit. These amplifiers are called *intermediate frequency (IF) amplifiers.* Each incoming station frequency is

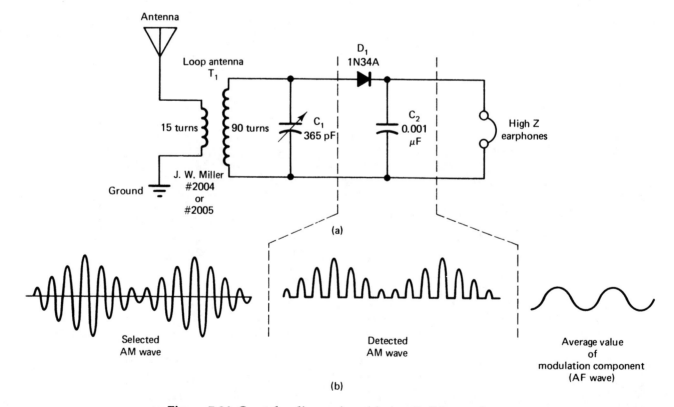

Figure 7-24. Crystal radio receiver (a) circuit; (b) waveforms.

changed into the IF. The IF amplifier responds in the same manner to all incoming signals. Each receiver has a tunable CW oscillator. This local oscillator generates an unmodulated RF signal of constant amplitude. This signal and the selected station are then mixed together. Mixing or heterodyning these two signals produces the IF signal. The IF contains all the modulation characteristics of the incoming RF signal. The IF signal is the difference beat frequency. After suitable amplification, the IF signal is applied to the detector. The resultant AF output signal is then amplified and applied to the speaker for sound reproduction.

Figure 7-26 shows a schematic diagram of a representative AM superheterodyne receiver. Transistor Q_1 is a tuned RF amplifier. Adjustment of capacitor C_1 selects the desired RF signal frequency. Q_2 is the mixer stage. The selected RF signal and the local oscillator are mixed together in Q_2. Q_3 is the local oscillator. The oscillator signal is fed to the base Q_2 through C_8. Transistor Q_4 is the IF amplifier. The output of Q_2 is coupled to Q_4 by transformer T_4. This trans-former is tuned to pass only an IF of 455 kHz. T_5 is the output IF transformer. The IF signal is coupled to the diode detector D_1 through this transformer. The detector rectifies the IF signal and develops the AF modulating component. C_{15} is the RF bypass filter capacitor and R_{12} is the volume control. Q_5 is the first AF signal amplifier. Push-pull AF power amplification is achieved by transistors Q_6 and Q_7.

The AF output signal is changed into sound by the speaker. The entire circuit is energized by a 9-V battery. Switch SW-1 is connected to the volume control and turns the circuit on and off.

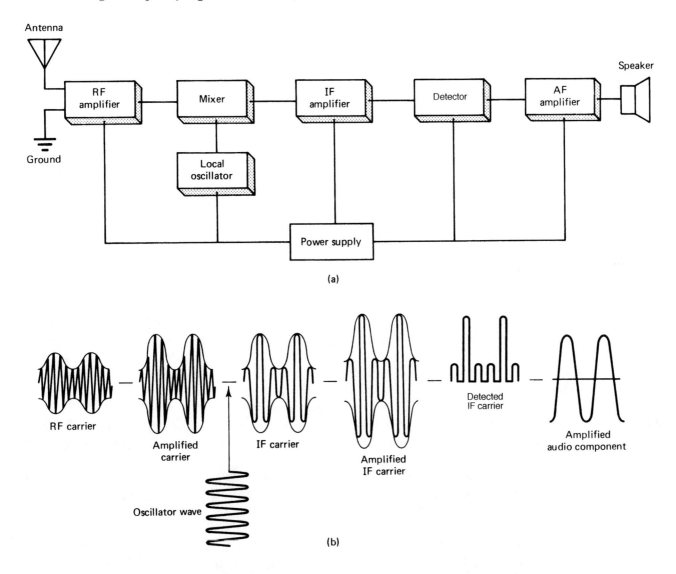

Figure 7-25. Syperheterodyne receiver: (a) block diagram; (b) waveforms.

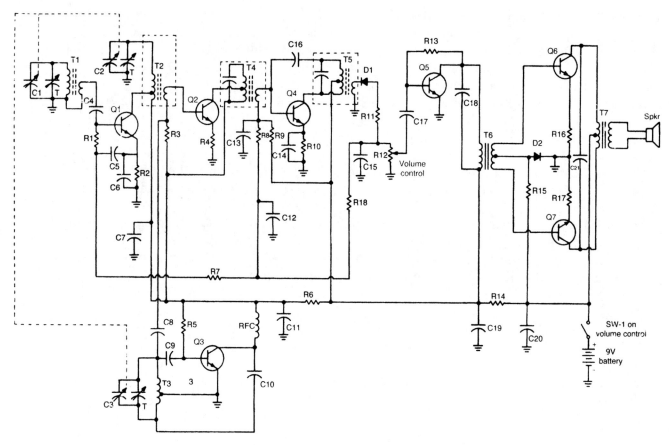

Figure 7-26. Transistor AM receiver.

Practically all the superheterodyne receiver circuits have been discussed in conjunction with other system functions. The local oscillator, for example, is a variable-frequency Hartley oscillator. Mixing of the oscillator signal and selected RF signal has been described as "heterodyning." Its output contains two input frequencies and two beat frequencies. The detector is essentially a crystal diode. An AF amplifying system is used to develop the audio signal component. The speaker was previously described as a transducer. All these operations remain the same when used in the superheterodyne receiver. Operational frequencies and circuit performance are generally somewhat different than previously described. We will therefore direct our attention to specific circuit performance.

Using Figures 7-25 and 7-26, we will now see how the superheterodyne responds when tuned to a specific frequency. Assume now that a number of AM signals are intercepted by the antenna-ground network. The desired station to be received occupies a frequency of 1000 kHz. The tuner dial is adjusted to 1000 kHz. The AF signal component then produces sound from the speaker. The volume control is adjusted to a desired signal level. The receiver is operational and performing its intended function.

For our receiver to develop a sound signal, a number of operational steps must be performed. Adjustment of the tuning dial changes the *LC* circuit of the RF amplifier. Capacitors C_1, C_2, and C_3 are all ganged together. One tuner adjustment alters the three tuned circuits at the same time. C_1 and C_2 tune the input and output of the RF amplifier to 1000 kHz. C_2 adjusts the frequency of the oscillator to 1455 kHz. The *LC* components of the oscillator are designed to generate a frequency that will be 455 kHz higher than the selected RF signal.

The 1000-kHz signal now passes through the input tuner and is applied to the base of Q_1. This common-emitter amplifier increases the voltage level of the applied signal. C_2 and T_2 are also tuned to pass the signal to the base of Q_2. The oscillator signal is coupled to the base of Q_2 through capacitor C_8. The incoming signal is AM and the oscillator signal is CW. By heterodyning action the collector of Q_2 has four signals. The *fundamental frequencies* are 1000-kHz AM and 1455-kHz CW. The *beat frequencies* are 2455-kHz AM and 455-kHz AM.

The IF amplifier has fixed tuning in its input and

output circuits. T_4 is the input IF transformer. It is tuned to resonate at 455 kHz. This frequency will pass into the base of Q_4 with a minimum of opposition. The other three signals will encounter a high impedance. 455 kHz will be amplified by Q_4. The signal is then coupled to the detector through transformer T_5. The detector recovers the AF component and applies it to the AF amplifier system for sound reproduction.

Assume now that the receiver is tuned to select a station at 1340 kHz. C_1, C_2, and C_3 all change to the new signal frequency. The oscillator now develops a CW signal of 1795 kHz. The mixer has 1340 kHz and 1795 kHz applied to its input. The collector of Q_2 has these two fundamental frequencies plus 3135 kHz and 455 kHz. The IF amplifier section will process the 455-kHz signal and apply it to the detector. The AF component is then recovered and processed for re-production.

Suppose now that the receiver is tuned to a station located at 600 kHz. This frequency would be amplified and applied to the mixer. The oscillator would now send a CW signal of 1055 kHz to the mixer. The collector of Q_2 will have signals of 600, 1055, 1655, and 455 kHz. The IF amplifier will again only pass the 455-kHz signal for reproduction.

The basic operation of an AM superheterodyne receiver should be obvious by this time. The oscillator is designed so that it will always be higher in frequency than the incoming station frequency. In effect, its frequency is the incoming station frequency plus the IF. The standard IF for AM receivers is 455 kHz. Through this type of circuit design, the IF can be adjusted to a specific frequency. Figure 7-27 shows two IF transformers that are used in transistor AM receivers. The metal can surrounding the transformer shields it from stray electromagnetic fields. This type of transformer is designed for printed circuit board (PCB) installations.

FREQUENCY MODULATION COMMUNICATION

Frequency modulation (FM) has several unique advantages over other communication systems. It uses low audio signal levels for modulation, has excellent frequency reproduction capabilities, and is nearly immune to noise and interference. FM is commonly used in the VHF range of 30 to 300 MHz and extends into the UHF range of 300 to 3000 MHz. The commercial FM band is 88 to 108 MHz. FM is the dominant form of communication for private two-way mobile communication services.

In an FM communication system, intelligence is superimposed on the carrier by variations in frequency. An

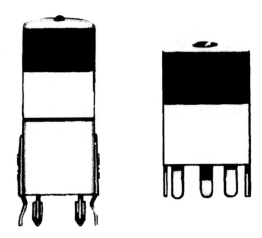

Figure 7-27. AM receiver IF transformers.

FM signal does not effectively change in amplitude. The modulating component does, however, cause the carrier to shift above and below its center frequency. Each FM station has an assigned center frequency. Receivers are tuned to this frequency for signal reception. When the carrier is unmodulated, it rests at the center frequency. Frequency allocations for FM are made by the FCC.

Examine the FM signal of Figure 7-28. Note that the modulating component is low-frequency AF and the carrier is RF. In commercial FM, the modulating component could be an audio signal of 20 Hz to 15 kHz. The carrier would be of some RF value between 88 and 108 MHz. As the modulating component changes from 0° to 90°, it causes an increase in the carrier frequency. The carrier rises above the center frequency during this time. Between 90° and 180° of the audio component, the carrier decreases in frequency. At 180° the carrier re-turns to the center frequency. As the audio signal changes from 180° to 270°, it causes a decrease in the carrier frequency. The carrier drops below the center frequency during this period. Between 270° and 360° the carrier rises again to the center frequency, showing that without modulation applied the carrier rests at the center frequency. With modulation, the carrier will shift above and below the center frequency.

We have seen that the modulating component of an FM system causes the carrier to shift above and below its center frequency. The amplitude of the modulating component is also extremely important in this operation. The larger the modulating signal, the greater the frequency shift. For example, if the center frequency of an FM signal is 100 MHz, a weak audio signal may cause a 1-kHz change in frequency. The carrier would change from 100.001 to 99.999 MHz. A strong audio signal could cause the carrier to change as much as 75 kHz. This ac-

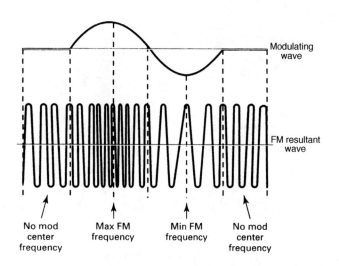

Figure 7-28. Frequency modulation waves.

rier changes during the same period for each modulating component.

FM Communication System

A block diagram of an FM communication system is shown in Figure 7—31. The transmitter and receiver respond as independent systems. In commercial FM, each station has a transmitter. There can be an infinite number of FM receivers.

Two-way mobile communication systems have a transmitter and receiver at each location. FM mobile communication systems are classified as narrowband FM. This type of system only transmits voice signals. The carrier of a narrowband FM system only deviates ±5 kHz. Commercial FM is considered to be wideband FM. The carrier of a commercial FM system deviates ± 75 kHz at 100% modulation. The operational principles of narrowband and wideband FM are primarily the same for each system.

The transmitter of an FM system is primarily responsible for signal generation [see Figure 7-31(a)]. Note that it employs an oscillator, an RF signal amplifier, and a power amplifier. The modulating component of an FM system is applied directly to the oscillator. AF changes in the applied signal cause the oscillator to deviate. Amplifiers following the oscillator are classified as linear devices.

The receiver of an FM system is responsible for signal interception, selection, demodulation, and reproduction [see Figure 7-31(b)]. Most FM receivers are of the superheterodyne type. These receivers are similar in operation to the AM superheterodyne circuit. The selection frequency is much higher (88 to 108 MHz) for commercial FM. A standard IF of 10.7 MHz is used in an FM receiver

tion would cause a shift of 100.075 to 9.925 MHz. It is important to note that the amplitude of the carrier does not change. Figure 7-29 shows how the amplitude of the audio component changes the frequency of the carrier.

The frequency of the modulation component is another important consideration in FM. Modulating frequency deter-mines the rate at which the frequency change takes place. With a 1-kHz modulating component, the carrier will swing back and forth 1000 times per second. A modulating component of 10 kHz will cause the carrier to shift 10,000 times per second.

Figure 7-30 shows how the FM carrier responds to two different frequencies. Here $t1$ has two cycles of modulation in the same time that t_2 has three cycles; $t2$ is of slightly higher frequency than $t1$ Note how the car-

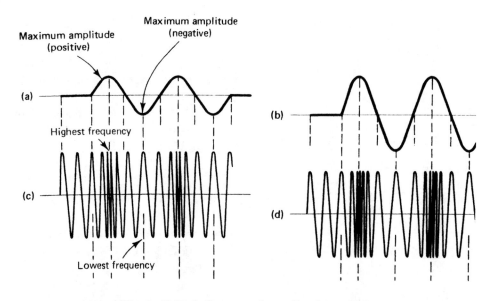

Figure 7-29. Influence of amplitude on FM.

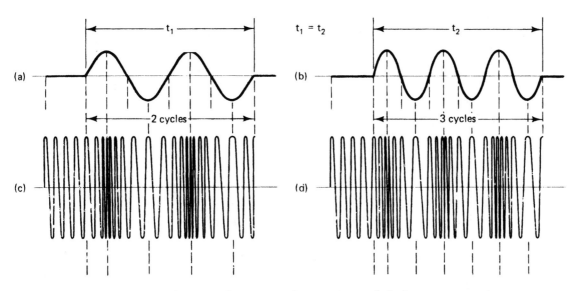

Figure 7-30. FM due to frequency changes in modulating component.

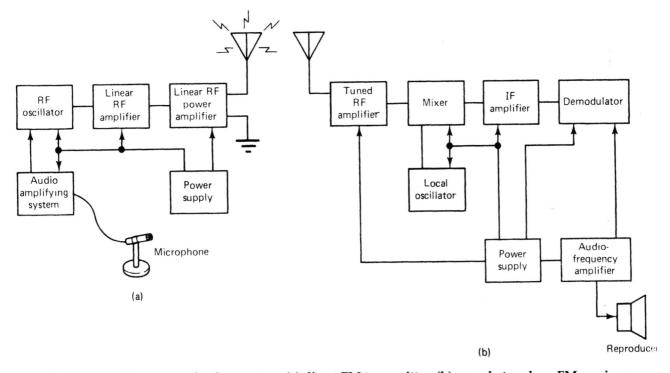

Figure 7-31. FM communication system: (a) direct FM transmitter; (b) superheterodyne FM receiver.

today. The demodulator of an FM receiver is uniquely different from an AM detector. The FM demodulator is designed to respond to changes in the carrier frequency. The modulating component is recovered from these frequency changes. The remainder of the circuit is the same as the AM superheterodyne.

FM Transmitters

There is a rather extensive list of different FM transmitters available today. Wireless microphones have an output of approximately 100 MW. Maritime mobile FM communications systems operate with a few watts of power. Two-way land mobile communications employs low-power transmitters. Educational FM transmitters can operate with a power of several hundred thousand watts. Commercial FM transmitters have a similar power rating. In addition to these types are public service radio and military FM. The frequency allocation and power capabilities of an FM transmitter are allocated and regulated by the FCC.

An FM transmitter regardless of its power rating or operational frequency has a number of fundamental parts. The oscillator is responsible for the development of the RF carrier. Many FM transmitters employ direct oscillator modulation, meaning that the frequency of the oscillator is shifted by direct application of the modulation component. The RF signal is then processed by a number of amplifiers. The power amplifier ultimately feeds the FM carrier into the antenna for radiation.

A simplified direct FM transmitter is shown in Figure 7-32. A basic Hartley oscillator is used as the RF signal source. With power applied, the oscillator will generate the RF carrier. The frequency of the oscillator is determined by the value of C_4 and T_1. Without modulation, the oscillator would generate a constant frequency, which is representative of the unmodulated carrier wave.

Modulation of the oscillator is achieved by voltage changes across a varicap diode. The capacitance of this diode changes with the value of the reverse-bias voltage. Bias volt-age is used to establish a dc operating level for this diode. A change in audio signal voltage causes the capacitance of the diode to change at an audio rate. Note that the series network of C_1, D_1, and C_2 is connected in parallel with capacitor C_4 of the oscillator tank circuit. A

change in audio signal level will cause a corresponding change in tank circuit capacitance. This, in turn, causes the oscillator frequency to change according to the modulating component. The modulating component is applied directly to the oscillator's frequency-determining components.

Direct FM can be achieved in a number of ways. The varicap diode method is commonly found in new transmitter equipment. This application is only one of a large number of circuit variations of the varicap diode. Circuits of this type are particularly well suited for voice communication in narrowband FM.

The power output of an FM transmitter is dependent primarily on its application. In some systems the oscillator output can be applied directly to the antenna for radiation. In other systems the power level must be raised to a higher level. The output of the oscillator would be applied to linear amplifiers for processing. The amplification function would be primarily the same as that achieved by CW and AM systems. We will not repeat the discussion of RF power amplification for FM transmitters. The primary difference in FM power amplification is the resonant frequency of the tuned circuits. FM usually operates in the VHF band which generally calls for smaller capacitance

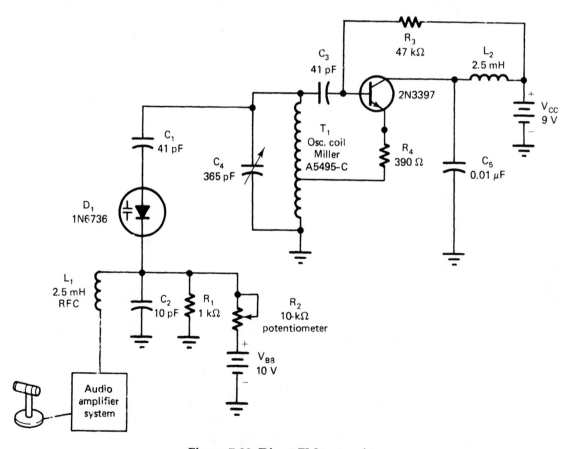

Figure 7-32. Direct FM transmitter.

and inductance values in the resonant circuits.

FM Receivers

FM signal reception is achieved in basically the same way as AM and CW. An FM superheterodyne circuit is designed to respond to frequencies in the VHF band. Commercial FM signal reception is in the 88- to 108-MHz range. Higher-frequency operation generally necessitates some change in the design of antenna, RF amplifier, and mixer circuits. These differences are due primarily to the in-creased frequency rather than the FM signal. The RF and IF sections of an FM receiver are somewhat different. They must be capable of passing a 200-kHz bandwidth signal in-stead of the 10-kHz AM signal. The most significant difference in FM reception is in the demodulator. This part of the receiver must pick out the modulating component from a signal that changes in frequency. In general, this circuit is more complicated than the AM detector. The AF amplifier section of an FM receiver is generally better than an AM receiver. It must be capable of amplifying frequencies of 30 Hz to 15 kHz. Figure 7-33 shows a block diagram of an FM super-heterodyne receiver. Some differences in FM reception are indicated by each functional block.

FM Demodulation

The demodulator of an FM receiver is the primary difference between AM and FM superheterodyne receivers. A number of demodulators have been used to achieve this receiver function. Today the ratiodetector seems to be the most widely used. The circuit is an outgrowth of earlier FM discriminators. Solid-state diodes and transformer design have made the ratio detector a practical circuit.

A basic ratio detector is shown in Figure 7-34. The operation of this detector is based on the phase relationship of an FM signal that is applied to the input. This signal comes from the IF amplifier section of the receiver (see Figure 7–33). It is 10.7 MHz and deviates in frequency. Design of the circuit causes two IF signal components to appear in the secondary winding of the transformer. One part of this component is coupled by capacitor C_3 to the center connection of the transformer. This resulting voltage is considered to be V_I. Note the location of V_1 in the circuit and voltage-line diagrams.

The second IF signal component is developed inductively by transformer action. Design of the secondary winding causes two voltage values to be developed. These voltages are labeled V_2 and V_3 in the circuit and voltage-line diagrams—they are of equal amplitude and 180° out of phase with each other. The center-tap connection of the secondary winding is used as the common reference point for these voltage values.

The resultant secondary voltages are based on the combined component voltage values of V_1, V_2, and V_3. V_{D1} and V_{D2} are the resulting secondary voltages. These two voltage values will appear across diodes D_1 and D_2. The developed voltage for each diode is based on the phase relationship of the two IF components. Note the location of V_{D1} and V_{D2} in the circuit and the voltage line diagram.

Operation of the ratio detector is based on voltage developed by the transformer for diodes D_1 and D_2. With no modulation applied to the FM carrier, the transformer will have a 10.7-MHz IF applied to the transformer. The developed voltage values are shown by the center-frequency voltage-line diagram. This method of voltage display is generally called a *vector diagram*. A vector shows the relationship of voltage values (line length) and their

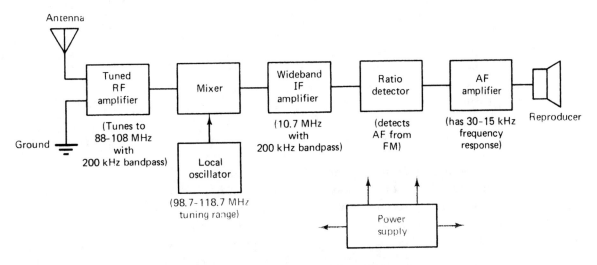

Figure 7-33. Block diagram of an FM receiver.

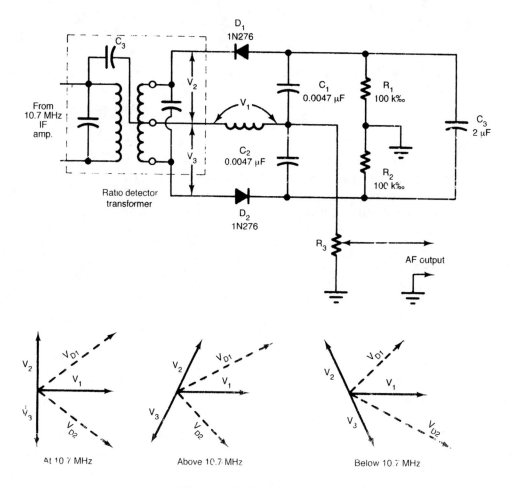

Figure 7-34. Ratio detector.

phase relationship (direction). At the center frequency, note that the resultant diode voltage values (V_{D1} and V_{D2}) are of the same length, meaning that each diode will receive the same voltage value. D_1 and D_2 will conduct an equal amount of current.

When the carrier swings above the center frequency, it causes the IF to swing above its resonant frequency. Note the resultant vector for this change in frequency. V_{D1} is longer than V_{D2}, meaning that D_1 will conduct more current than D_2. In the same regard, note how the vector changes when the frequency swings below resonance. This condition will cause D_1 to be less conductive and D_2 to be more conductive. Essentially, this means that the IF signal is translated into different diode voltage values.

Let us now see how the input RF voltage is used to develop an AF signal. The two diodes of the ratio detector are connected in series with the secondary winding and capacitors C_1 and C_2. For one alternation of the input signal both diodes are reverse biased. No conduction occurs during this alternation [see Figure 7-35(a)]. For the next alternation both diodes are forward biased.

The input signal voltage is then rectified [see Figure 7-35(b)]. Essentially, this means that the incoming signal is changed into a pulsating waveform for one alternation.

Figure 7-36 shows how the ratio detector responds when the input signal deviates above and below the center frequency. Keep in mind that conduction occurs only for one alternation of the input. Figure 7-36(a) shows how the circuit will respond when the input is at its 10.7-MHz center frequency. Each diode has the same input voltage value for this condition of operation. Capacitors C_1 and C_2 will charge to equal voltages as indicated. With respect to ground, the output voltage will be 0 V. Note this point on the output volt-age waveform in Figure 7-36(d).

Assume now that the input IF signal swings to 10.8 MHz as shown in Figure 7-36(b). This condition causes D_1 to receive more voltage than D_2. C_1, charges to 12 V while C_2 charges to – 1 V. With respect to ground, the output voltage rises to + 1 V. Note this point on the output voltage waveform in Figure 7-36(d).

The input IF signal then swings to 10.6 MHz as shown in Figure 7-36(c). This condition causes D_2 to receive more voltage than D_1. C_2 charges to – 2 V while C_1

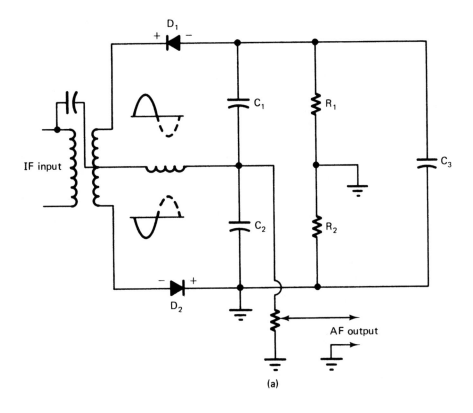

(a)

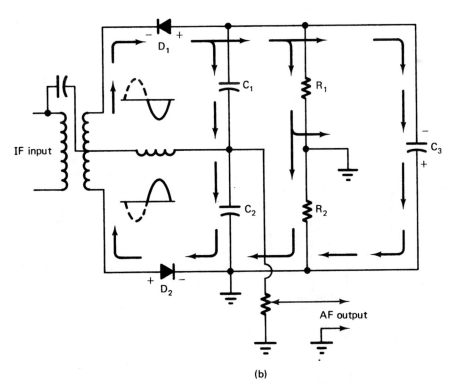

(b)

Figure 7-35. Diode conduction of a ratio detector: (a) nonconduction; (b) conduction.

charges to +1 V. With respect to ground, the output voltage drops to –1 V. See this point on the output voltage waveform of Figure 7-36(d).

The output of the ratio detector is an AF signal of 2-V p-p. This signal corresponds to the frequency changes placed on the carrier at the transmitter. In effect, we have recovered the AF component from the FM carrier signal. This signal can then be amplified by the AF section for reproduction.

TELEVISION COMMUNICATION

Nearly everyone has had an opportunity to view a television communication system in operation. This communication process plays an important role in our lives. Few people spend a day without watching television. It is probably the most significant application of

electronics today.

The signal of a television system is quite complex. It is made up of a number of unique parts or components. Basically, the transmitted signal has a picture carrier and a sound carrier. These two signals are transmitted simultaneously from a single antenna. The picture carrier is amplitude modulated. The sound carrier is frequency modulated. A television receiver intercepts these two signals from the air. They are tuned, amplified, demodulated, and ultimately reproduced. Sound is reproduced by a speaker. Color and picture information are reproduced by a picture tube. Television is primarily a one-way communication system with a central transmitting station and an infinite number of receivers. A simplification of the television communication system is shown in Figure 7-37.

The television camera of our system is basically a transducer. It changes the light energy of a televised scene into electrical signal energy. Light energy falling on

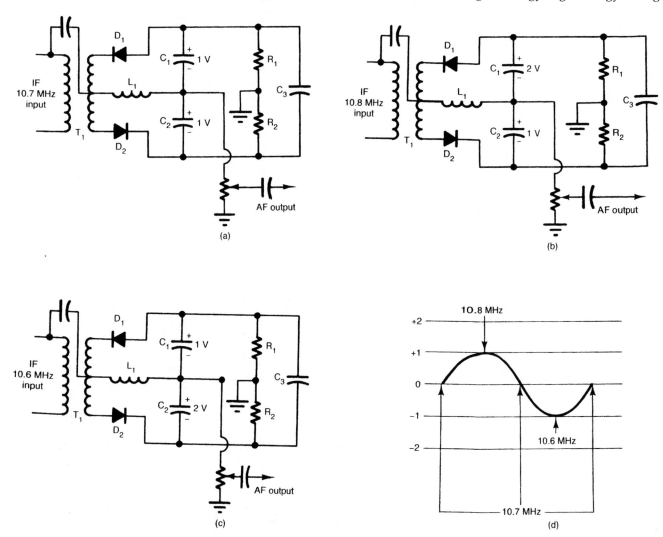

Figure 7-36. Ratio detector response to frequency: (a) center frequency; (b) above center; (c) below center; (d) output waveform.

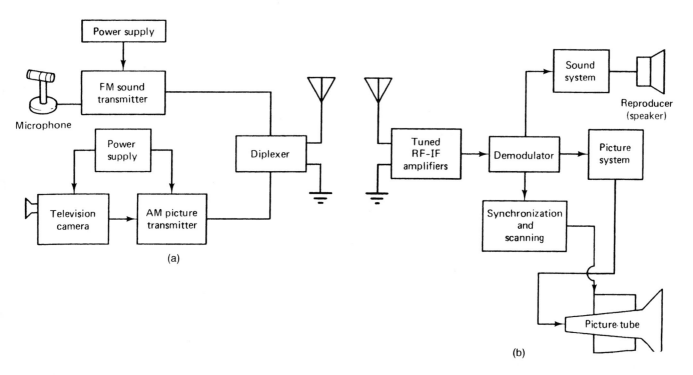

Figure 7-37. TV communication system: (a) transmitter; (b) receiver.

a highly sensitive surface varies the conduction of current through a resistive material. The resultant current flow is proportional to the brightness of the scene. An electron beam scans horizontally and vertically across the light-sensitive surface. Traditionally, the camera tube of a TV system has been a videcon. Videcons are were replaced with solid-state tubes called *charge-coupled devices* or *CCDs*. These devices are smaller, use less power, and are more sensitive to low light levels. CCDs are also used in portable video recorders or camcorders.

Figure 7-38 shows a simplification of the television camera tube. Note that a complete circuit exists between the cathode and power supply. Electrons are emitted from the heated cathode. These are formed into a thin beam and directed toward the back side of the photoconductive layer. Conduction through the layer is based on the intensity of light from the scene being televised. A bright or intense light area becomes low resistant. Dark areas have a higher resistance. As the electron beam scans across the back of the photoconductive layer, it sees different resistance values. Conduction through the layer is based on this resistance. A discrete area with low resistance causes a large current through the layer. Dark areas cause less current flow. Current flow is directly related to the light intensity of the televised scene. Output current flow appears across the load resistor (R_L). Voltage developed across R_L is amplified and ultimately used to modulate the picture carrier. In practice, the developed camera tube voltage is called a video signal. *Vide* means "to see" in Latin. A camera tube sees things electronically.

The scene being televised by a camera tube must be broken into small parts called *picture elements* or *pixels*. For this to occur, it is necessary to scan the light-sensitive surface of a camera tube with a stream of electrons, similar to reading a printed page. Letters, words, and sentences are placed on the page by printing. We do not determine what is on a printed page at one instant. Our eyes must scan the page starting at the upper left-hand corner one line at a time. They move left to right, drop down one line and quickly return to the left, then scan right again for the next line. The process continues until all lines are scanned.

In a similar way the electron beam of a camera tube scans the back surface of the photoconductive layer. The electron beam is deflected horizontally and vertically by an electromagnetic field. A coil fixture known as the *deflection yoke is* placed around the neck of the camera tube. This coil deflects the electron beam. Current flow in the deflection yoke is varied so that the field rises to a peak value, then drops to its starting value. Figure 7-39 shows the deflection yoke current and the resultant electron beam scanning action. Notice that each line has a trace and a retrace time. During the trace period, the line is scanned from left to right, which takes a rather large por-

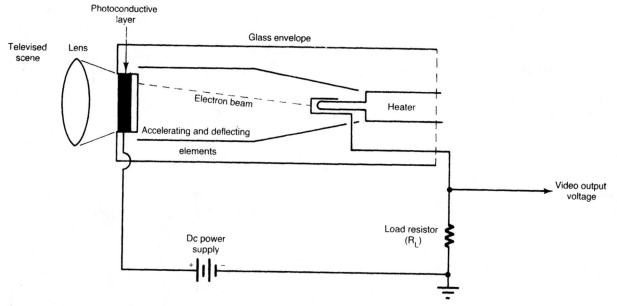

Figure 7-38. Simplification of a TV camera tube.

tion of the complete saw tooth wave. Retrace occurs when the beam returns from right to left. Notice that this takes only a small portion of the total waveform. The same condition applies to the vertical sweep waveform.

Figure 7-39 shows another unique difference in the scanning lines. During the trace time, the scanning line is solid which indicates that the electron beam is conducting during this time. It also shows a broken line during the retrace period which indicates that the electron beam is nonconductive during this period. In effect, conduction occurs during the trace time and no conduction occurs during retrace. This same condition applies to the vertical trace and retrace time.

The scanning operation of a camera tube requires two complete sets of deflection coils. One set is for horizontal deflection and one is for vertical. The current needed to produce deflection comes from two saw tooth oscillators—a vertical blocking oscillator and a horizontal multivibrator could be used for this operation. In U.S. television, the horizontal sweep fre-

quency is 15,750 Hz and the vertical frequency is 60 Hz.

To produce a moving television scene, there must be at least 30 complete pictures produced per second.

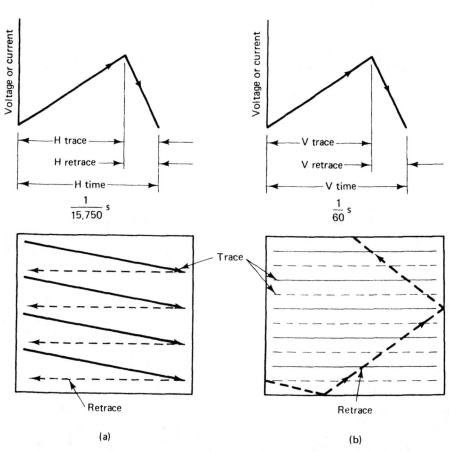

Figure 7-39. Scanning and sweep signals: (a) horizontal; (b) vertical.

In a TV system a complete picture is called *a frame*. The U.S. television system has 525 horizontal lines in a frame. These lines are not scanned progressively, such as 1, 2, 3, 4, 5, and so on. They are, however, divided into two fields. One field contains 262.5 odd-numbered lines. Lines 1, 3, 5, 7, 9, 11, and so on, make up the odd field. The other field has 262.5 even-numbered lines. Lines 2, 4, 6, 8, 10, and so on, are included in the even field. A complete picture or frame has one odd-lined field and one even-lined field.

Picture production in a TV system employs interlace scanning. To produce one frame, the odd-line field is scanned first. After scanning all the odd-numbered lines, the electron beam is deflected from the bottom position to the top. The even-numbered-line field is then scanned. A complete frame has 262.5 odd-numbered lines and 262.5 even-numbered lines. The odd-line field starts with a complete line and ends with a half line. The even field starts with a half line and ends with a complete line.

The picture repetition rate of U.S. television is 30 frames per second. Because two fields are needed to produce one frame, the vertical frequency is 2 30, or 60 Hz. This particular frequency was chosen to coincide with the ac power line frequency. In some foreign countries the vertical frequency is 50 Hz.

In television signal production, the vertical and horizontal sweep circuits must be synchronized properly to produce a picture. The signal sent out by the transmitter must contain synchronization or sync information. This signal is used to keep the oscillator of the receiver in step with the correct signal frequency. Separate generators are used to develop the sync signal which is added to the video signal developed by the camera.

Picture Signal

The picture signal of a TV transmitter contains several important parts. Each part of the signal plays a specific role in the operation of the system. The video signal, for example, is developed by the camera tube. It represents instantaneous variations in scene brightness. Figure 7-40(a) shows the video signal for one horizontal line.

The video signal of a television system has negative picture phase, which means that the highest amplitude part of the signal corresponds to the darkest picture area. Bright picture areas have the lowest amplitude level. Signal levels that are 75% of the total amplitude range are considered to be in the black region. All light disappears in this region. Signal-level amplitude percentages are shown in Figure 7-40(a).

In addition to the video information, the picture signal must also provide some way of cutting off the electron beam at certain times. When scanning occurs, the elec-

tron beam is driven to cut off during the retrace period Horizontal retrace occurs at the end of one line and vertical retrace occurs at the end of each field. A rectangular pulse of sufficient amplitude is needed to reach cutoff. This condition permits the electron beam to retrace without producing unwanted lines. This part of the signal is called *blanking*. A composite signal has both horizontal and vertical blanking pulses. Figure 7-40(a) shows the location of a horizontal blanking pulse. Vertical blanking occurs after 262.5 horizontal lines. Figure 7-40(b) shows the vertical blanking time.

A composite TV picture signal also has vertical and horizontal synchronization pulses. These pulses ride on the top of the blanking pulses. Horizontal sync pulses are shown in both parts of Figure 7-40. The serrated pulses of Figure 7-40(b) provide continuous horizontal sync during the vertical retrace time. The vertical sync pulse is made up of six rectangular pulses near the center of the vertical blanking time. The width of a vertical sync pulse is much greater than that of the horizontal sync pulse. All these pulses, plus blanking and the video signal, are described as a compositive picture signal.

Television Transmitter

A television transmitter is divided into two separate sections or divisions that feed outputs into a common antenna. The video section is responsible for the picture part of the signal. A crystal oscillator is used for carrier-wave generation. As a general rule, the frequency is multiplied to bring it up to an allocated channel in the VHF or UHF band. An intermediate power amplifier and a final power amplifier follow the last multiplier. The modulating component is a composite picture signal. It contains video, blanking pulses, sync, and equalizing pulses. The composite signal is amplified and ultimately applied to the final power amplifier. The final power amplifier is amplitude modulated by the composite picture signal. This section of the transmitter is essentially the same as that of a commercial AM station. The obvious differences are frequency and power output. Figure 7-41 shows a block diagram of a black-and-white TV transmitter.

A unique difference in TV and commercial AM transmitter circuitry appears after the final power amplifier. The TV output signal is applied to a vestigial sideband filter. This filter is designed to remove all lower sideband frequencies 1.25 MHz below the carrier frequency. The entire upper sideband of the signal is transmitted. A large portion of the lower sideband is purposely suppressed to reduce the frequency occupied by a channel. With the lower sideband suppressed, the bandwidth of a TV channel is 6 MHz.

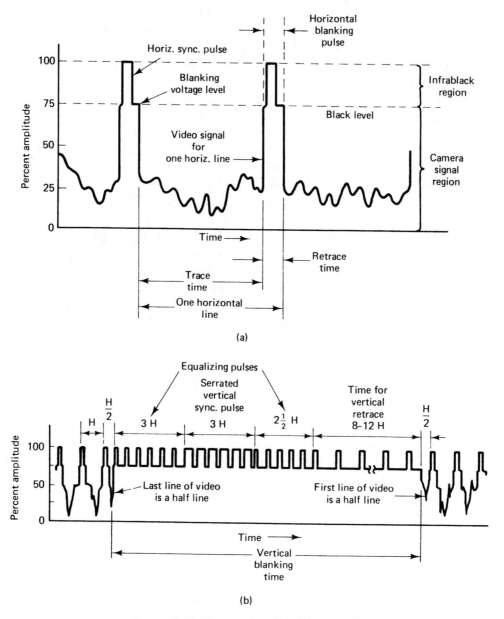

Figure 7-40. Composite TV picture signal.

The vestigial sideband signal is then applied to a diplexer, which is a filter circuit that isolates the picture and sound carriers. Essentially, the sound carrier will not pass into the picture section and the picture carrier will not pass into the sound section, thus preventing undesirable interaction between the two carrier signals.

The sound section of a TV system is primarily an FM transmitter similar to a commercial FM transmitter. The center frequency of the FM carrier is always 4.5 MHz above the picture carrier. Carrier deviation is ±25 kHz in a TV system. *Modulation* is normally applied to the oscillator of the FM sound system. The remainder of the sound section is similar to that of a commercial FM

transmitter. See the FM sound section of the transmitter in Figure 7-41.

Television Receiver

A television receiver is designed to intercept the electromagnetic waves sent out by the transmitter and use them to develop sound and a picture. The received signals are in the VHF or UHF band. The FCC has allocated a 6-MHz bandwidth for each TV channel. Channels 2 through 13 are in the VHF band. These frequencies are from 54 to 216 MHz. Channels 14 through 83 are in the UHF band which ranges from 470 to 890 MHz. All channels in the immediate area induce a signal into the anten-

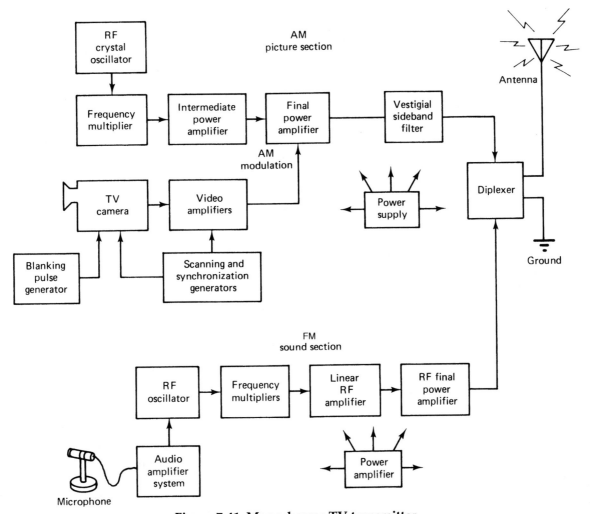

Figure 7-41. Monochrome TV transmitter.

na. A desired station is selected by altering a tuning circuit. This *LC* circuit passes only the selected channel and rejects the others.

A functional block diagram of a black-and-white television receiver is shown in Figure 7-42. Practically all of this circuitry has been used in other communication systems. The front end of a TV receiver, for example, is a super-heterodyne circuit. It has a tuned RF amplifier, a mixer, and an oscillator. This section of the receiver is called the *tuner.* It is housed in a shielded metal container to reduce interference. The output of the tuner is an IF signal. A standard IF for television is 41.25 MHz for the sound and 45.75 MHz for the picture. The IF must pass both sound and picture carriers and their sidebands which necessitates a 6-MHz band-pass for the IF amplifiers.

The *demodulation* function of a TV receiver is achieved by a diode. This circuit is primarily an AM detector. The output of the detector has all the picture in-

formation placed on the carrier at the transmitter. After demodulation the picture carrier is discarded. Three different kinds of signal information appear at the output of the detector. It recovers the video signal and sync signals for immediate use. The sound carrier passes into the sound IF section.

The video signal recovered by the detector is processed by the video amplifier. The output of this section is then used to control the brightness of the electron beam of the picture tube. A dark spot detected by the TV camera will cause a corresponding reduction in picture tube brightness. A bright spot or an intense picture element will cause the picture tube to conduct quite heavily, causing a corresponding bright spot to appear on the picture tube face. The video signal developed by the camera of the transmitter will be faithfully reproduced on the picture tube screen.

The sync signal of the video detector is used to synchronize the vertical and horizontal sweep oscillators.

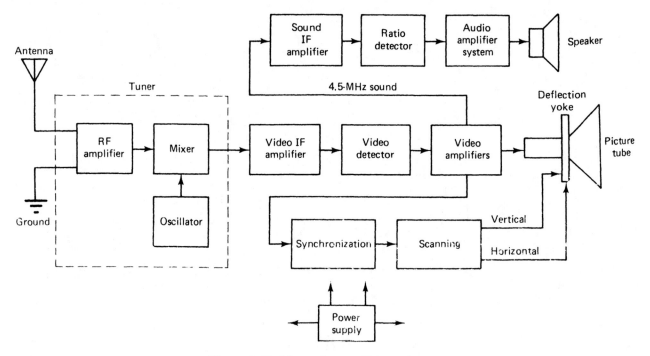

Figure 7-42. Monochrome TV receiver.

These oscillators develop the saw tooth waves that are used to deflect the electron beam. The electron beam must be in step with the transmitted signal for the picture to be stable. The transmitted sync signal is recovered and used to trigger the two receiver sweep oscillators.

An AM video detector does not respond effectively to frequency changes in the IF sound carrier. It does, however, heterodyne the two signals together. The difference between the 45.75- and 41.25-MHz IF signals is 4.5 MHz. This signal takes on the FM modulation characteristic of the sound carrier. A reading of 4.5 MHz is called the sound IF. The sound IF deviates ±25 kHz above and below 4.5 MHz.

A ratio detector can be used to demodulate the sound IF signal. The recovered audio component is then processed by an AF amplifier. It is ultimately used to drive a loud speaker for sound reproduction. The FM sound section of a TV receiver is similar to that of a standard FM broadcast receiver.

The picture tube of a TV receiver is responsible for changing an electrical signal into light energy. A picture tube is also called a *cathode-ray tube (CRT)*. The intensity of an electron beam changes as it scans across the face of the tube. The inside face of the tube is coated with phosphor. When the electron beam strikes tiny grains of phosphor, it produces light. A combination of different light and dark phosphor grains causes a picture to ap-

pear on the inside of the face area. The resultant picture can be observed by viewing the front of the face area.

Figure 7-43 shows a simplification of the CRT of a TV receiver. The tube is divided into three parts. The gun area is responsible for electron beam production. The coil fixture attached to the neck of the tube deflects the electron beam. The viewing area changes electrical energy into light energy.

A CRT is a thermionic emission device. Heat applied to the cathode causes it to emit electrons. These electrons form into a fine beam and move toward the face of the tube. A high positive charge is placed on a conductive coating on the inside of the glass housing. This coating serves as the an-ode. The electron beam is attracted to this area of the tube be-cause of its positive charge. Voltage is supplied to the coating by an external anode connection terminal. The anode voltage of a black-and-white CRT is normally 15 kV.

After leaving the cathode, electrons pass through the control grid. Voltage applied to the grid controls the flow of electrons. A strong negative voltage will cause the electron beam to be cut off. Varying values of grid voltage will cause the screen to be illuminated. The video signal developed by the receiver can be used to control the intensity level of the electron beam. Focusing of the electron beam is achieved by the next two electrodes. They shape the electron beam into a fine trace. These electrodes are generally called *focusing anodes*.

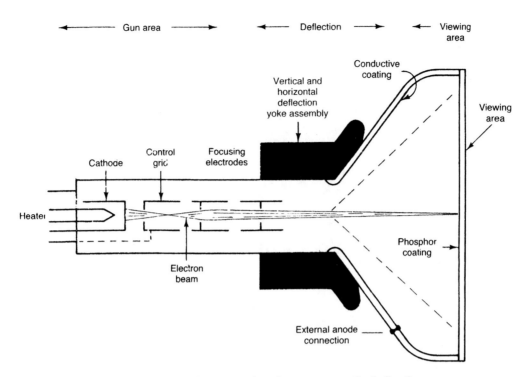

Figure 7-43. Electron tube-electromagnetic deflection.

Electromagnetic deflection of the electron beam is achieved by the yoke assembly. This coil fixture produces an electromagnetic field that causes deflection of the beam. Both vertical and horizontal deflection are needed to produce scanning.

Color Television

The transmission of a color picture by television complicates the system to some extent. The principles involved in transmission are primarily the same as those of monochrome, or black-and-white, television. One problem of color television is that it must be compatible, meaning that programs de-signed for color reception must also be received on a monochrome receiver. The transmitted signal must therefore contain color information as well as the monochrome signal. All this must fit into the 6-MHz channel allocation.

The picture portion of the color signal is basically the same as the monochrome signal. The video signal is, however, made up of three separate color signals. Each color signal is produced by an independent camera tube. One camera tube is used to develop a signal voltage that corresponds to the red content of the scene being televised. Blue and green camera tubes are used in a similar arrangement to produce the other two color signals. Red, green, and blue are considered to be the primary colors of the video signal. White is a mixture of all three colors. Black is the absence of all three primary colors.

Color Cameras

Figure 7-44 shows a simplified color television camera. The scene being televised is focused by a lens onto a special mirror. This mirror reflects one-third of the light and passes two-thirds of the light. The reflected image goes through a filter and is applied to the blue camera tube. Two-thirds of the image passing through the mirror is applied to a divider mirror. One-third of it is applied to the red camera and one-third to the green camera. Each camera tube provides an output signal that is proportional to the light level of the primary color. The brightness, or luminance, signal is a mixture of the three primary colors. These proportions of the color signal are 59% green, 30% red, and 11% blue. The luminance signal is generally called a Y signal. A monochrome receiver responds only to the Y signal and produces a standard black-and-white picture.

The human eye needs two stimuli to perceive color. One of these is called *hue*, which refers to a specific color. Green leaves have a green hue and a red cap has a red hue. The other consideration is called *saturation*, which refers to the amount or level of color present. A vivid color is highly saturated. Light or weak colors are diluted with white light. A scene being televised by a color camera has hue, saturation, and brightness.

Color Transmitters

Figure 7-45 shows a block diagram of a color trans-

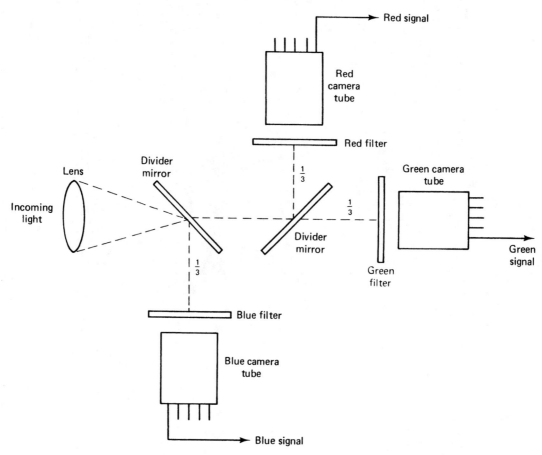

Figure 7-44. Simplified color TV camera.

mitter. The color section adds significantly to the transmitter. The remainder of the diagram has been reduced to a few blocks. This part of the transmitter is essentially the same for either a color or a black-and white system.

In a color transmitter, three separate color signals are developed by the TV camera. These colors are applied to a matrix and a mixer circuit. The mixer develops the luminance signal. Green, red, and blue are mixed in correct pro-portions. Luminance is the equivalent of a black-and-white signal. The output of the mixer (Y signal) goes directly to the carrier modulator. This signal is the same as the modulating component of a black-and-white transmitter.

The matrix is a rather specialized mixing circuit. It combines the three color signals into an I and a Q signal. These two signals can be used to represent any color developed by the system. The Q signal corresponds to green or purple information. The I signal refers to orange or cyan signals. These two signals are used to amplitude modulate a 3.58-MHz subcarrier signal. The carrier part of this signal is suppressed to prevent interference. The I signal is kept in phase with the 3.58-MHz subcarrier. The Q signal is one quadrant or 90° out of phase with the sub-

carrier. The I and Q signals are sideband frequencies of the subcarrier. Only this sideband information is used to modulate the transmitter picture carrier.

A synchronization signal is also generated by the color transmitter. This signal is applied to both the modulator and the camera scanning circuit. Scanning of the three camera tubes must be synchronized with the modulated carrier out-put. The sync signal is an essential part of the picture carrier modulation component. The modulating component contains I, Q, Y, and sync signals.

Color Receivers

A color television receiver picks up the transmitted signal and uses it to develop sound and a picture. It has all the basic parts of a monochrome receiver plus those needed to recover the original three color signals. Figure 7-46 shows a block diagram of the color TV receiver. The shaded blocks denote the color section of the receiver where the primary difference between color and monochrome TV receivers occurs.

Operation of the color receiver is primarily the same as that of a black-and-white receiver up to the video demodulator. After demodulation, the composite video sig-

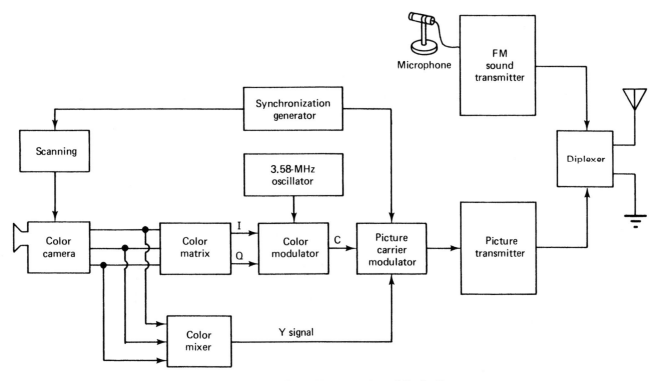

Figure 7-45. Color TV transmitter block diagram.

nal is then divided into chroma (C) and Y signals. The Y or luminance signal is coupled to a delay circuit which slows the Y signal. It will not get to the matrix at the same time as the chroma signal. The C signal is applied to the chroma amplifier. This signal must pass through a great deal more circuitry before reaching the matrix.

Remember that the chroma signal contains I and Q color and a 3.58-MHz suppressed carrier. To demodulate this signal, the 3.58-MHz signal must be reinserted. Notice that the 3.58-MHz oscillator is connected to the I and Q demodulators where the carrier reinsertion function takes place.

The I demodulator receives a 3.58-MHz signal directly from the oscillator where the in-phase signal is derived. The 3.58-MHz signal fed into the Q demodulator is shifted 90° where the quadrature signal is derived. I and Q color signals appear at the output of the demodulator.

The matrix of the receiver has three signals applied to its input. The Y signal contains the luminance information. The I and Q signals are the demodulated color signals. The matrix combines these three signals in proper proportions. Its output comprises the original red, green, and blue color signals. These signals are applied to the R, G, and B guns of the picture tube. The electron beam of each gun strikes closely spaced red, green, and blue phosphor spots on the face of the tube. These spots glow in differing amounts according to the signal level. Any color

can be reproduced on the face of the tube by the three primary colors.

FIBER OPTICS

Air is a common transmission medium for communications systems. Light travels through air rapidly. Typical indications of the speed of light are 300,000 kilometers per second or 186,000 miles per second. Light travels in this medium through waves. These waves are of the *transverse* type, which means that their motion is in a plane that is at right angles to the direction of travel. We believe that light waves travel in a straight line. This condition is called *rectilinear propagation*. Most optoelectric systems respond to light-wave propagation through air for only a limited distance.

Optical fibers are also frequently used in many systems as a transmission medium. These fibers are made from glass or plastic. The fiber material serves as the actual trans-mission medium. Glass fibers are used frequently and are available in diameters from 2 micrometers to more than 6 millimeters.

Each fiber has a central core that has its outside coated or clad with a reflective surface. The cladding material has a reflective index slightly lower than that of the core. To a light wave, the interface between the core and

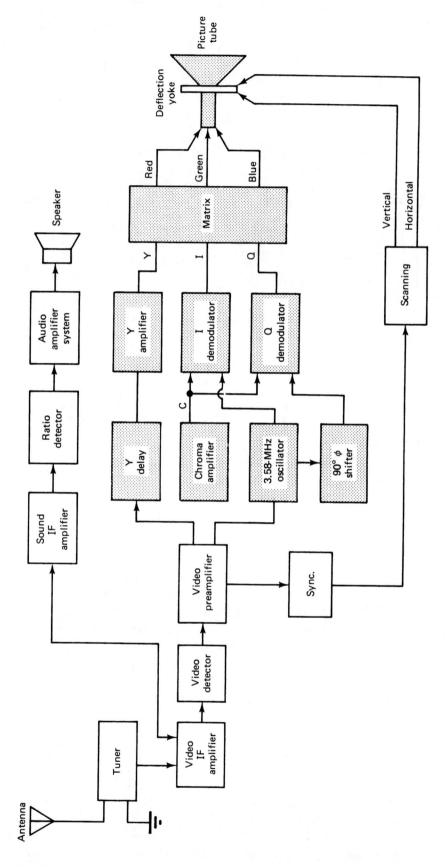

Figure 7-46. Color TV receiver block diagram.

the cladding material acts like a mirror. Light striking the interface is reflected back into the core. It moves and remains in the core until it strikes the core-cladding interface again. The process is then repeated. As many as 500 or more reflections can occur in 30 centimeters of the fiber core.

The principle of fiber optics, illustrated in Figure 7-47, utilizes optical fibers made of glass or plastic to transmit light from one point to another. Light can be transmitted in very unconventional ways, such as around corners, in limited space, or over long distances by using the fiber-optics principle. Light will transmit through the fiber-optic material regardless of how it is bent or shaped. Advances in fiber-optic design have made possible low-loss fiber lengths that are used for numerous applications. Optical communications are the most common use of fiber optics.

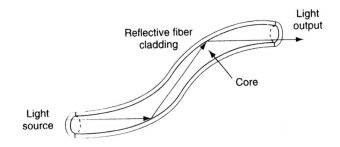

Figure 7-47. Cutaway view of a fiber-optic cable.

Chapter 8

Electronic Power Control

INTRODUCTION

Control is a basic function of all electronic systems. It is primarily responsible for altering the flow of electrical energy between the source and load. Anytime a system is placed into operation, some form of control is involved. The process of turning a system on and off is a prime example of control. This action is usually called *full* or *complete control*. All systems have some form of full control. In addition, control may involve some type of an adjustment. Partial control describes this function. Electronically, partial control can be achieved in a number of ways. Adjusting the volume of a stereo amplifier, the speed of a motor, and the brightness of a lamp are examples of partial control.

Partial control can be achieved in a variety of ways. Essentially, much of this book has been dealing in some way with partial control. In this regard, we have been concerned with control that alters the shape, amplitude, and frequency of an electronic signal. Rectification is used to alter the shape of a waveform. Amplification changes the amplitude of a signal. Frequency control is achieved by a tuning circuit or an oscillator. These control processes are all extremely important electronic functions.

Electronic power control often deals with devices that alter the conduction time of a waveform. This function is usually quite unique when compared with other methods of control. It may be used, for example, to change only one alternation of a sine wave. The conduction may be varied anywhere between 0% and 100% of the alternation. Both alternations of a sine wave may also be controlled. They may be changed between 0% and 100% of their conduction time. Control of this type is accomplished by several rather specialized electronic devices.

OBJECTIVES

Upon completion of this chapter, you will be able to:
1. Explain how electrical power is controlled by switching.
2. Identify different electronic power control devices.

3. Explain how an SCR is used to achieve partial control.
4. Identify the different parts of an *I-V* characteristic curve for an SCR, a triac, and a diac.
5. Distinguish among amplitude, phase, and dc control of an SCR.
6. Identify the different elements of an SCR, a triac, and a diac.
7. Distinguish among circuits that achieve amplitude, RC, and phase control of a triac.

ELECTRICAL POWER CONTROL SYSTEM

Electrical power is commonly used as an energy source for such things as welding, electric heat, and a variety of other operations. These items, in general, refer to the load of a power system. A certain amount of work is achieved by the load device when it is energized. Controlling the load of an electrical power system has been a rather significant problem through the years. Control is usually not achieved very efficiently.

Figure 8-1 shows how a rheostat is used to control an electrical power circuit. The rheostat in this case is used to change the brightness of an incandescent lamp. An increase in resistance will normally cause the lamp to be dim. A decrease in resistance will cause it to become brighter. With the rheostat set to zero resistance, the circuit will have 1 A of current flow, determined by the formula

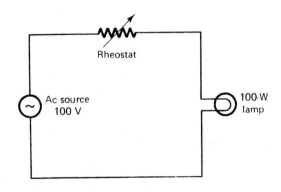

Figure 8-1. Rheostat-controlled lamp circuit.

261

current (*I*) = Power (*P*)/voltage (*V*)

or

$$I = P/V = 100\ \text{W}/100\ \text{V} = 1\ \text{A}$$

This formula shows when the lamp would be operating at its maximum output level. The resistance of the circuit in this condition is

resistance (*R*) = voltage (*V*)/current (*I*)

or

$$R = V/I = 100\ \text{V}/1\ \text{A}$$
$$= 100\ \Omega$$

We know that the current flow of the lamp circuit is directly related to resistance. An increase in resistance will reduce the total circuit current. Assume now that the rheostat is adjusted to produce a resistance of 100 fl. The total resistance of the circuit will now change to 2001 Z. This would be the original resistance of 100 SI plus the rheostat resistance of 100 SI. The resultant circuit current flow will decrease in value, which is shown to be

current (*I*) = voltage (*V*)/total resistance (*R_T*)

or

$$I = V/R_T = 100\ \text{V}/200\ \Omega$$
$$= 0.5\ \text{A}$$

Note that the circuit current has decreased to half of its original value.

A rather interesting change in the lamp circuit occurs when the total resistance increases. Some of the source voltage will appear across the rheostat and will be deducted from the lamp voltage. With the lamp and the rheostat both being 100 SI, there will now be 50 V across each. The lamp will obviously be operating with reduced power which is determined to be

power (*P*) = current (*I*) × voltage (*V*) or

$$P = I\ V = 0.5\ \text{A} \times 50\ \text{V}$$
$$= 25\ \text{W}$$

The lamp will therefore receive only 25 W of its original power. Its brightness will be reduced significantly. The rheostat in this case is consuming as much power as the load device. Power consumed by the rheostat appears as heat. Because the purpose of our circuit is to control light, heat is not a desirable feature. In effect, any heat dissipated by R is considered to be wasted energy. We must pay for this power to control lamp brightness. Control of a power system by changes in current is therefore not efficient.

As an alternative, electrical power can be controlled by variations in circuit voltage. Figure 8-2 shows a voltage-controlled lamp dimmer circuit. In this circuit, lamp brightness is controlled by a variable transformer. A reduction in voltage will produce a corresponding change in brightness. This type of control is more efficient than the rheostat circuit. In effect, the lamp is the primary resistive device. Power consumed by the circuit will appear only at the lamp. Some power is consumed by the primary winding of the transformer. This power is usually rather nominal compared with the rheostat circuit.

Power control by a variable transformer is more efficient than the rheostat method. A variable transformer is, however, a rather expensive item. When large amounts of power are being controlled, this device also becomes quite large. Control of this type is more efficient but has a number of limitations.

One of the most efficient methods of electrical power control that we have today is through circuit switching. When a switch is turned on, power is consumed by the load. When the switch is turned off, no power is consumed. The switching method of power control is shown in Figure 8-3. If the switch is low resistant, it consumes little power. When the switch is open, it consumes no power. By switching the circuit on and off rapidly, the average current flow can be reduced. The brightness of the lamp

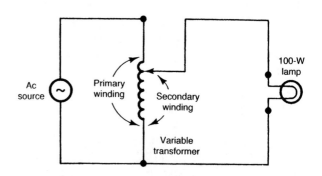

Figure 8-2. Voltage-controlled lamp circuit.

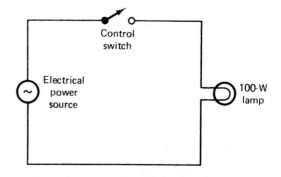

Figure 8-3. Switch-controlled lamp circuit.

will also be reduced. In effect, lamp brightness is controlled by the switching speed. Power in this method of control is not consumed by the control device. This type of control is both efficient and effective.

The switching method of electrical power control cannot be achieved effectively with a manual switch. The mechanical action of the switch will not permit it to be turned on and off quickly. A switch would soon wear out if used in this manner. Electronic switching can be used to achieve the same result. Switching action of this type can be accomplished in a lamp circuit without a noticeable flicker. Electronic power control deals with the switching method of controlling power.

SILICON-CONTROLLED RECTIFIERS

A silicon-controlled rectifier or SCR is probably the most popular electronic power control device today. The SCR is used primarily as a switching device. Power control is achieved by switching the SCR on and off during one alternation of the ac source voltage. For 60 Hz ac the SCR would be switched on and off 60 times per second. Control of electrical power is achieved by altering or delaying the turn-on time of an alternation.

An SCR, as the name implies, is a solid-state rectifier device. It conducts current in only one direction. It is similar in size to a comparable silicon power diode. SCRs are usually small, rather inexpensive, waste little power, and require practically no maintenance. The SCR is available today in a full range of types and sizes to meet nearly any power control application. Presently, they are available in current ratings from less than 1 A to over 1400 A. Voltage values range from 15 to 2600 V.

SCR Construction

An SCR is a solid-state device made of four alternate layers of *P*- and *N*-type silicon. Three *P-N* junctions are formed by the structure. Each SCR has three leads or terminals. The anode and cathode terminals are similar to those of a regular silicon diode. The third lead is called the *gate*. This lead determines when the device switches from its off to on state. An SCR will usually not go into conduction by simply for-ward biasing the anode and cathode. The gate must be forward biased at the same time. When these conditions occur, the SCR becomes conductive. The internal resistance of a conductive SCR is less than 1 Ω. Its reverse or off-state resistance is generally in excess of 1 MΩ.

A schematic symbol and the crystal structure of an SCR are shown in Figure 8-4. Note that the device has a *PNPN* structure from anode to cathode. Three distinct *P-*

N junctions are formed. When the anode is made positive and the cathode negative, junctions 1 and 3 are forward biased. J_2 is reverse biased. Reversing the polarity of the source alters this condition. J_1 and J_3 would be reverse biased and J_2 would be forward biased and would not permit conduction. Conduction will occur only when the anode, cathode, and gate are all forward biased at the same time.

Some representative SCRs are shown in Figure 8-5. Only a few of the more popular packages are shown here. As a general rule, the anode is connected to the largest electrode if there is a difference in their physical size. The gate is usually smaller than the other electrodes. Only a small gate current is needed to achieve control. In some packages, the SCR symbol is used for lead identification.

SCR Operation

Operation of an SCR is not easily explained by using the four-layer *PNPN* structure of Figure 8-4. A rather simplified method has been developed that describes the crystal structure as two interconnected transistors. Figure 8-6(a) shows this imaginary division of the crystal. Notice that the top three crystals form a *PNP* transistor. The lower three crystals form an *NPN* device. Two parts of each transistor are interconnected. The two-transistor circuit diagram of Figure 8-6(b) is an equivalent of the crystal structure division.

Figure 8-7 shows how the two-transistor equivalent circuit of an SCR will respond when voltage is applied. Note that the anode is made positive and the cathode negative. This condition forward biases the emitter of each transistor and reverse biases the collector. Without base voltage applied to the *NPN* transistor, the equivalent circuit is nonconductive. No current will flow through the load resistor. Reversing the polarity of the voltage source

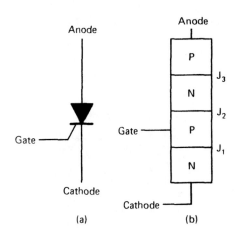

Figure 8-4. SCR crystal: (a) symbol; (b) structure.

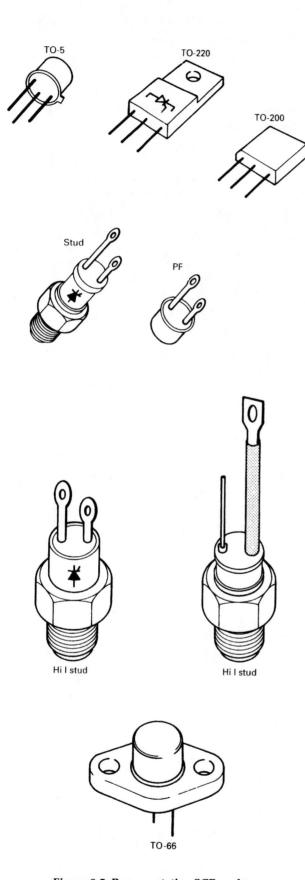

will not alter the conduction of the circuit. The emitter of each transistor will be reverse biased by this condition, meaning that no conduction will take place when the anode and cathode are reverse biased. With only anode-cathode voltage applied, no conduction will occur in either direction.

Assume now that the anode and cathode of an SCR are forward biased, as indicated in Figure 8-7. If the gate is momentarily made positive with respect to the cathode, the emitter-base junction of the *NPN* transistor becomes forward biased. This action will immediately cause collector current to flow in the *NPN* transistor. As a result, base current will also flow into the *PNP* transistor, which in turn will cause a corresponding collector current to flow. *PNP* collector current now causes base current to flow into the *NPN* transistor. The two transistors therefore latch or hold when in the conductive state. This

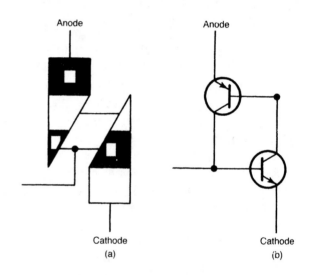

Figure 8-6. Equivalent SCR.

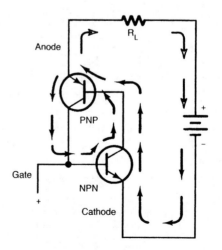

Figure 8-5. Representative SCR packages.

Figure 8-7. SCR response.

action continues even when the gate is disconnected. In effect, when conduction occurs, the device will latch in its "on" state. Current will continue to flow through the SCR as long as the anode-cathode voltage is of the correct polarity.

To turn off a conductive SCR, it is necessary to momentarily remove or reduce the anode-cathode voltage. This action will turn off the two transistors. The device will then remain in this state until the anode, cathode, and gate are forward biased again. With ac applied to an SCR, it will automatically turn off during one alternation of the input. Control is achieved by altering the turn-on time during the conductive or "on" alternation.

SCR *I-V* Characteristics

The current-voltage characteristics of an SCR tell much about its operation. The *I-V* characteristic curve of Figure 8-8 shows that an SCR has two conduction states. Quadrant I shows conduction in the forward direction, which shows how conduction occurs when the forward breakover voltage (V_{BO}) is exceeded. Note that the curve returns to approximately zero after the V_{BO} has been exceeded. When conduction occurs, the internal resistance of the SCR drops to a minute value similar to that of a for-

ward-biased silicon diode. The conduction current (I_{AK}) must be limited by an external resistor. This current, however, must be great enough to maintain conduction when it starts. The holding current or I_H level must be exceeded for this to take place. Note that the I_H level is just above the knee of the I_{AK} curve after it returns to the center.

Quadrant III of the *I-V* characteristic curve shows the reverse breakdown condition of operation. This characteristic of an SCR is similar to that of a silicon diode. Conduction occurs when the peak reverse voltage (PRV) value is reached. Normally, an SCR would be permanently damaged if the PRV is exceeded. Today, SCRs have PRV ratings of 25 to 2000 V.

For an SCR to be used as a power control device, the forward V_{BO} must be altered. Changes in gate current will cause a decrease in the V_{BO}. This occurs when the gate is forward biased. An increase in I_G will cause a large reduction in the forward V_{BO}. An enlargement of quadrant I of the *I-V* characteristics is shown in Figure 8-9, which also shows how different values of I_G change the V_{BO}. With 0 I_G it takes a V_{BO} o of 400 V to produce conduction. An increase in I_G reduces this quite significantly. With 7 mA of I_G the SCR conducts as a forward-biased silicon diode. Lesser values of I_G will cause an increase in

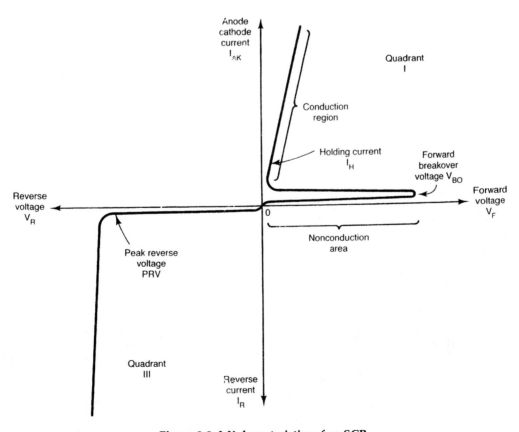

Figure 8-8. *I-V* characteristics of an SCR.

the V_{BO} needed to produce conduction.

The gate current characteristic of an SCR shows an important electrical operating condition. For any value of I_G there is a specific V_{BO} that must be reached before conduction can occur, which means that an SCR can be turned on when a proper combination of I_G and V_{BO} is achieved. This characteristic is used to control conduction when the SCR is used as a power control device.

DC Power Control with SCRs

When an SCR is used as a power control device, it responds primarily as a switch. When the applied source voltage is below the forward breakdown voltage, control is achieved by increasing the gate current. Gate current is usually made large enough to ensure that the SCR will turn on at the proper time. Gate current is generally applied for only a short time. In many applications this may be in the form of a short-duration pulse. Continuous I_G is not needed to trigger an SCR into conduction. After conduction occurs, the SCR will not turn off until the I_{AK} drops to zero.

Figure 8-10 shows an SCR used as a dc power control switch. In this type of circuit, a rather high load current is controlled by a small gate current. Note that the electrical power source (V_S) is controlled by the SCR. The polarity of V_S must forward bias the SCR, which is achieved by making the anode positive and the cathode negative.

When the circuit switch is turned on initially, the load is not energized. In this situation the $V_B o$ is in excess of the V_S voltage. Power control is achieved by turning on SW-1 which forward biases the gate. If a suitable value of I_G occurs, it will lower the V_{BO} and turn on the SCR. The I_G can be removed and the SCR will remain in conduction. To turn the circuit off, momentarily open the circuit switch. With the circuit switch on again, the SCR will remain in the off state. It will go into conduction again by closing SW-1.

Dc power control applications of the SCR require two switches to achieve control, but this application of the SCR is not practical. The circuit switch would need to be capable of handling the load current. The gate switch could be rated at an extremely small value. If several switches were needed to control the load from different locations, this circuit would be more practical. More practical dc power circuits can be achieved by adding a number

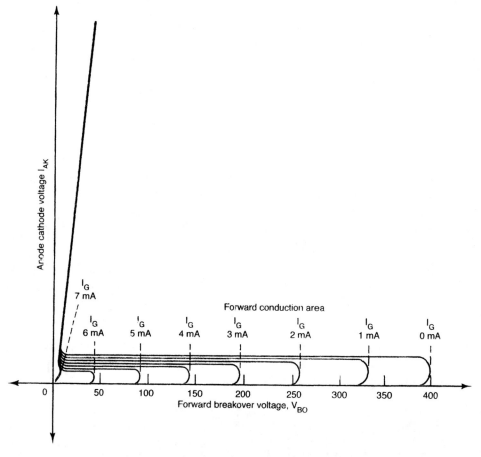

Figure 8-9. Gate current characteristics of an SCR.

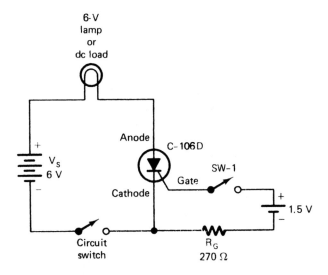

Figure 8-10. Dc power control switch.

of additional components. Figure 8-11 shows a dc power control circuit with one SCR being controlled by a second SCR. SCR_1 would control the dc load current. SCR_2 controls the conduction of SCR_1. In this circuit, switching of a high-current load is achieved with two small, low-current switches. SCR_1 would be rated to handle the load current. SCR_2 could have a rather small current-handling capacity. Control of this type could probably be achieved for less than a circuit employing a large electrical contactor switch.

Operation of the dc control circuit of Figure 8-11 is based on the conduction of SCR_1 and SCR_2. To turn on the load, the "start" pushbutton is momentarily closed, and thus forward biases the gate of SCR_1. The V_{BO} is reduced and SCR1 goes into conduction. The load current latches SCR_1 in its conduction state. This action also causes C_1 to charge to the indicated polarity. The load will remain energized as long as power is supplied to the circuit.

Turn-off of SCR_1 is achieved by pushing the stop button, which momentarily applies I_G to SCR_2 and causes it to be conductive. The charge on C_1 is momentarily applied to the anode and cathode of SCR_1, which reduces the I_{AK} of SCR_1 and causes it to turn off. The circuit will remain in the off state until it is energized by the start button. An SCR power circuit of this type can be controlled with two small pushbuttons. As a rule, control of this type would be more reliable and less expensive than a dc electrical contactor circuit.

Ac Power Control with SCRs

Ac electrical power control applications of an SCR are common. As a general rule, control is easy to achieve. The SCR automatically turns off during one alternation of the ac input and thus eliminates the turn-off problem with the dc circuit. The load of an ac circuit will see current only for one alternation of the input cycle. In effect, an SCR power control circuit has half-wave output. The conduction time of an alternation can be varied with an SCR circuit. We can have variable output through this method of control.

A simple SCR power control switch is shown in Figure 8-12. Connected in this manner, conduction of ac will only occur when the anode is positive and the cathode negative. Conduction will not occur until SW-1 is closed. When this takes place, there is gate current. The value of I_G lowers the V_{BO} to where the SCR becomes conductive. R_G of the gate circuit limits the peak value of I_G. Diode (D_1) prevents reverse voltage from being applied between the gate and cathode of the SCR. With SW-1 closed, the gate will be forward biased for only one alternation, which is the same alternation that forward biases the anode and cathode. With a suitable value of I_G and correct anode-cathode voltage (V_{AK}), the SCR will become conductive.

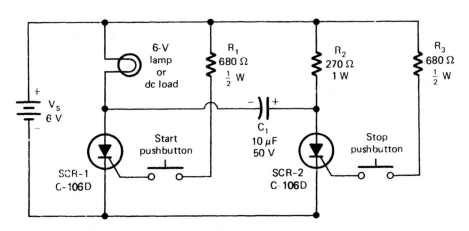

Figure 8-11. Dc power control circuit

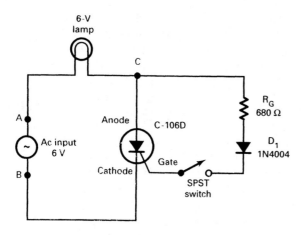

Figure 8-12. SCR power control switch.

The ac power control switch of Figure 8-12 is designed primarily to take the place of a mechanical switch. With a circuit of this type, it is possible to control a rather large amount of electrical power with a rather small switch. Control of this type is reliable. The switch does not have contacts that spark and arc when changes in load current occur. Control of this type, however, is only an on-off function.

SCRs are widely used to control the amount of electrical power supplied to a load device. Circuits of this type respond well to 60 Hz ac. This type of circuit is called a *half-wave SCR phase shifter*. Figure 8-13 shows a simplified version of the variable power control circuit.

Operation of the variable power control circuit is based on the charge and discharge of capacitor C_1. Assume now that the positive alternation occurs. Point A will be positive and point B negative, and will forward bias the SCR. It will not turn on immediately, however, because there is no gate current. Capacitor C_1 begins to charge through R_1 and R_2. In a specific time, the gate voltage builds to where the gate current is great enough to turn on the SCR. Diode D_1 is forward biased by the capacitor voltage. I_G then flows through D_1. Once the SCR goes into conduction, it continues for the remainder of the alternation. Current flows through the load. In effect, turn-on of the SCR is delayed by the value of R-C.

When the negative alternation occurs, point A becomes negative and point B goes positive, which reverse biases the SCR. No current flows through the load for this alternation. Diode D_2 is, however, forward biased by this voltage. Capacitor C_1 then discharges quickly through D_2 and bypasses the resistance of R_1. The capacitor is reset or in a ready state for the next alternation. The process repeats itself for each succeeding cycle.

A change in the resistance of R_1 will alter the con-

duction time of the SCR power control circuit. When R_1 is increased, C_1 cannot charge as quickly during the positive alternation. It therefore takes more time for C_1 to build its charge voltage, which means that the turn-on of I_G will be delayed for a longer time. As a result, conduction of the SCR will be delayed at the beginning of the alternation. See the waveform of Figure 8-13(c). The resultant load current for this condition will be somewhat less than that of the waveform in Figure 8-13(b). Variable control of the load is achieved by altering the conduction time of the SCR.

A further increase in the resistance of R_1 will cause more delay in the turn-on of the SCR during the positive alternation. Waveforms of Figures 8-13(d), (e), and (f) show the result of this change. Waveform (d) has a 90° delay of SCR conduction time. Waveform (e) shows a 135° delay in the conduction time. If the value of R_1 is great enough, the SCR will delay its turn-on for the full alternation. The waveform in Figure 8-13(f) shows a 180° turn-on delay. Note that the load does not receive any current for this condition. Control of this circuit is from full conduction as in waveform (b) to 180° delay as in waveform (f).

Power control with an SCR is highly efficient. If no delay of the conduction time occurs, the load develops full power. In this case, full power occurs only for one alternation. If conduction is delayed 90°, the load develops only half of its potential power. If the conduction delay is 135°, the load develops only 25% of its potential power. It is important to note that nearly all the power controlled by the circuit is applied to the load device—little is consumed by the SCR. In variable power control applications, it is extremely important that the load be supplied to its full value of power. With an SCR control circuit, no power is consumed by other components. By delaying the turn-on time of conduction, power can be controlled with the highest level of efficiency.

TRIAC POWER CONTROL

Ac power control can be achieved with a device that switches on and off during each alternation. Control of this type is accomplished with a special solid-state device known as a *triac*. This device is described as a three-terminal ac switch. Gate current is used to control the conduction time of either alternation of the ac waveform. In a sense, the triac is the equivalent of two reverse-connected SCRs feeding a common load device.

A triac is classified as a gate-controlled ac switch. For the positive alternation it responds as *a PNPN* device. An alternate crystal structure of the *NPNP* type is used

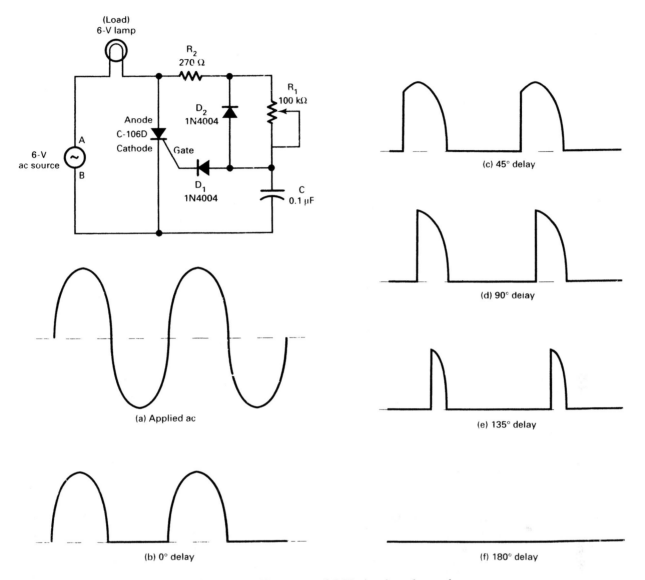

Figure 8-13. Phase control SCR circuit and waveforms.

for the negative alternation. Each crystal structure is triggered into conduction by the same gate connection. The gate has a dual-polarity triggering capability.

Triac Construction

A triac is a solid-state device made of two different four-layer crystal structures connected between two terminals. We do not generally use the terms *anode* and *cathode* to describe these terminals. For one alternation they would be the anode and cathode. For the other alternation they would respond as the cathode and anode. It is common practice therefore to use the terms *main 1* and *main 2* or *terminal 1* and *terminal 2* to describe these leads. The third connection is the gate. This lead determines when the device switches from its off to its on state. The gate G will normally go into conduction when it is forward bi-

ased, and is usually based on the polarity of terminal 1. If T_1 is negative, G must be positive. When T_1 is positive, the gate must be negative. This means that ac voltage must be applied to the gate to cause conduction during each alternation of the T_1-T_2 voltage.

The schematic symbol and the crystal structure of a triac are shown in Figure 8-14. Notice the junction of the crystal structure simplification. Looking from T_1 to T_2, the structure involves crystals N_1, P_1, N_2, and P_2. The gate is used to bias P_1. This is primarily the same as an SCR with T_1 serving as the cathode and T_2 the anode.

Looking at the crystal structure from T_2 to T_1, it is N_3, P_2, N_2, and P_1. The gate is used to bias N_4 for control in this direction, which is similar to the structure of an SCR in this direction. Notice that T_1, T_2, and G are all connected to two pieces of crystal. Conduction will take

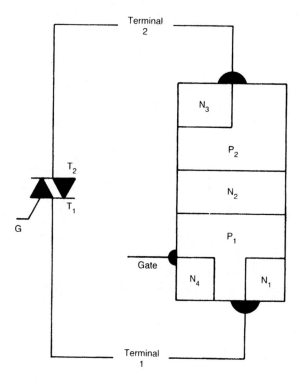

Figure 8-14. Triac symbol and crystal structure.

place only through the crystal polarity that is forward biased. When T_1 is negative, for example, N_1 is forward biased and P_1 is reverse biased. Terminal selection by bias
polarity is the same for all three terminals.

The schematic symbol of the triac is representative
of reverse-connected diodes. The gate is connected to
the same end as T_1, which is an important consideration
when connecting the triac into a circuit. The gate is normally forward biased with respect to T_1.

Triac Operation

When ac is applied to a triac, conduction can occur
for each alternation, but T_1 and T_2 must be properly biased with respect to the gate. Forward conduction occurs
when T_1 is negative, the gate (G) is positive, and T_2 is positive. Reverse conduction occurs when T_1 is positive, G is
negative, and T_2 is negative. Conduction in either direction is similar to that of the SCR. The two-transistor operation used with the SCR applies to the triac in both directions. Figure 8-15 shows an example of the operational
conduction polarities of a triac. Note that dc is used as a
power source. It is used here only to denote operational
polarities. In practice, dc would not be used as a power
source for a triac. It is an ac power control device.

The operation of a triac is primarily the same as that
of the SCR. When T_1 and T_2 are of the correct polarity, gate
current must occur to reduce the breakover voltage. Small

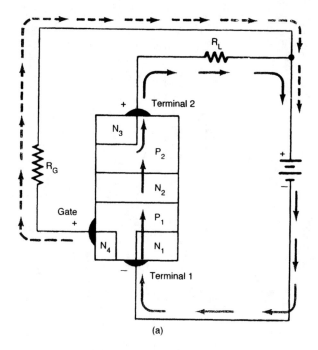

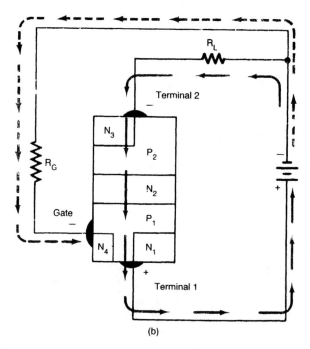

**Figure 8-15. Triac condition polarities: (a) forward conduction
$T_1 - T_2 + G$ +; (b) reverse condition.**

values of I_G will cause the breakover voltage to be quite
large. An increase in I_G decreases the breakover voltage.
This applies equally to both forward and reverse conduction. After conduction occurs, the gate loses its effectiveness.

When a triac goes into conduction, its internal resistance drops to an extremely low value. Typical resistance values are less than 1Ω. The device remains in this

state for the remainder of the alternation. It is essential that current flowing through the triac be great enough to maintain conduction, which is generally called the *holding current* or I_H. The resistance of the load device and the value of the source voltage determine conduction current. This current must be slightly greater than I_H to maintain conduction.

Operation of a triac has one rather unique problem that does not occur in the SCR which refers to its conduction commutation. Commutation is used to describe the turn-off of conduction. In triac operation, commutation time can be critical. With a resistive load, turn-off starts when the conduction current drops below the I_H level. It usually continues into the next alternation until the breakover voltage is reached. In most applications the triac has sufficient time to reach turn-off. With an inductive load, conduction time is extended somewhat due to the inductive voltage caused by the collapsing magnetic field. A special RC circuit called a *snubber* is placed across the triac to reduce this problem. A 0.1-μF capacitor with a 100-Ω series resistor is placed across the triac. The capacitor charges during the collapsing of the inductive field and generally reduces the commutation problem.

Triac I-V Characteristics

The *I-V* characteristic of a triac shows how it responds to forward and reverse voltages. Figure 8-16 is a typical triac *I-V* characteristic. Note that conduction occurs in quadrants I and III. The conduction in each quadrant is primarily the same. With 0 IG, the breakover voltage is usually quite high. When breakover occurs, the curve quickly returns to the center. This shows a drop in the internal resistance of the device when conduction occurs. Conduction current must be limited by an external resistor. The holding current or I_H of a triac occurs just above the knee of the I_T curve. I_H must be attained or the device will not latch during a specific alternation.

Quadrant III is normally the same as quadrant I and thus ensures that operation will be the same for each alternation. Because the triac is conductive during quadrant III, it does not have a peak reverse voltage rating. It does, however, have a maximum reverse conduction current value the same as the maximum forward conduction value. The conduction characteristics of quadrant III are mirror images of quadrant I.

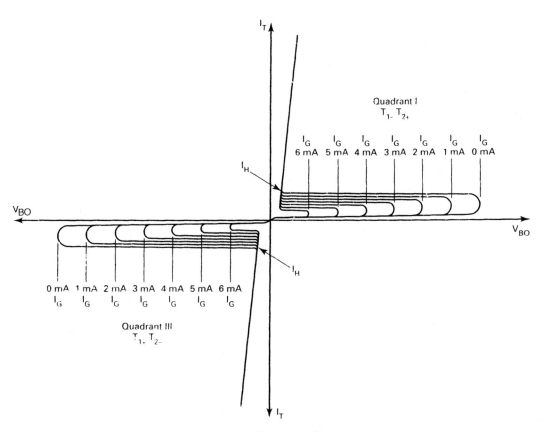

Figure 8-16. *I-V* characteristics of a triac.

Triac Applications

Triacs are used primarily to achieve ac power control. In this application the triac responds primarily as a switch. Through normal switching action, it is possible to control the ac energy source for a portion of each alternation. If conduction occurs for both alternations of a complete sine wave, 100% of the power is delivered to the load device. Conduction for half of each alternation permits 50% control. If conduction is for one fourth of each alternation, the load receives less than 25% of its normal power. It is possible through this device to control conduction for the entire sine wave, which means that a triac is capable of controlling from 0% to 100% of the electrical power supplied to a load device. Control of this type is efficient as practically no power is consumed by the triac while performing its control function.

Static Switching

The use of a triac as a static switch is primarily an on-off function. Control of this type has a number of advantages over mechanical load switching. A high-current energy source can be controlled with a very small switch. No contact bounce occurs with solid-state switching which generally reduces arcing and switch contact destruction. Control of this type is rather easy to achieve. Only a small number of parts are needed for a triac switch.

Two rather simple triac switching applications are shown in Figure 8-17. The circuit in Figure 8-17(a) shows the load being controlled by an SPST switch. When the switch is closed, ac is applied to the gate. Resistor R_1 limits the gate current to a reasonable operating value. With ac applied to the gate, conduction occurs for the entire sine wave. The gate of this circuit requires only a few milliamperes of current to turn on the triac. Practically any small switch could be used to control a rather large load current.

The circuit of Figure 8-17(b) is considered to be a three-position switch. In position 1, the gate is open and the power is off. In position 2, gate current flows for only one alternation. The load receives power during one alternation, which is the half-power operating position. In position 3, gate current flows for both alternations. The load receives full ac power in this position.

Start-stop Triac Control

Some electrical power circuits are controlled by two push-buttons or start-stop switches. Control of this type begins by momentarily pushing the start button. Operation then continues after releasing the depressed button. To turn off the circuit, a stop button is momentarily pushed. The circuit then resets itself in preparation for the next starting operation. Control of this type is widely used in motor control applications and for lighting circuits. A triac can be adapted for this type of power control.

A start-stop triac control circuit is shown in Figure 8-18. When electrical power is first applied to this circuit, the triac is in its nonconductive state. The load does not receive any power for operation. All the supply voltage appears across the triac because of its high resistance. No voltage appears across the RC circuit, the gate, or the load device initially.

To energize the load device, the start pushbutton is momentarily pressed. C_1 charges immediately through R_1 and R_2, which in turn causes I_G to flow into the gate. The V_{BO} of the triac is lowered and it goes into conduc-

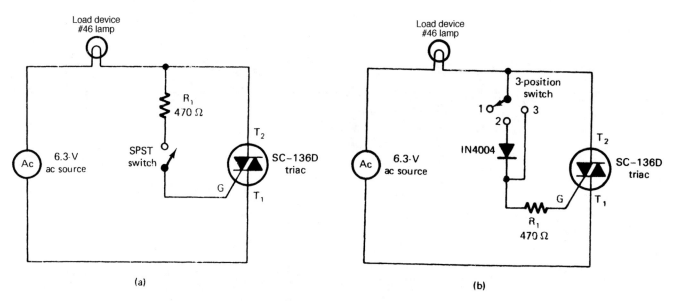

Figure 8-17. Triac switching circuits: (a) static triac switch; (b) three-position static switch.

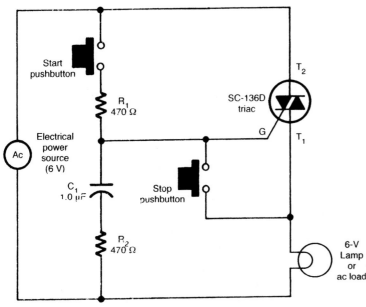

Figure 8-18. Start-stop control.

tion. Voltage now appears across the load, R_2-C_1, and the gate. The charging current of R_2-C_1 and the gate continue and are at peak value when the source voltage alternation changes. The gate then retriggers the triac for the next alternation. C_1 is recharged through the gate and R_2. The next alternation change causes I_G to again flow for retriggering of the triac. The load receives full power from the source. The process will continue into conduction as long as power is supplied by the source. To turn off the circuit, the stop button is momentarily depressed. This action immediately bypasses the gate current around the triac. With no gate current, the triac will not latch during the alternation change. As a result, C_1 cannot be recharged. The triac will then remain off for each succeeding alternation change. Conduction can be restored only by pressing the start button. This circuit is a triac equivalent of the ac motor electrical contactor.

Triac Variable Power Control

The triac is widely used as a variable ac power control device. Control of this type is generally called *full-wave control*. Full wave refers to the fact that both alternations of a sine wave are being controlled. Variable control of this type is achieved by delaying the start of each alternation. This process is similar to that of the SCR. The primary difference is that triac conduction applies to the entire sine wave. For this to be accomplished, ac must be applied to both the gate and the conduction terminals.

Variable ac power control can be achieved rather easily when the source is low voltage. Figure 8-19 shows

a simple low-voltage variable lamp control circuit. Note that the gate current of this circuit is controlled by a potentiometer. Connected in this manner, adjustment of R_1 determines the value of gate current for each alternation.

Conduction of a triac is controlled by the polarity of T_1 and T_2 with respect to the gate voltage. For the positive alternation, assume that point A is positive and B is negative, thus causing a $+T_2$, a $-T_1$, and a $+I_G$. The value of the circuit gate current is determined by the resistance of R_1 and R_2. For a high-resistance setting of R_1, the triac may not go into conduction at all. For a smaller resistance value, conduction can be delayed in varying amounts. Generally, conduction delay will occur only during the first 90° of the alternation. If conduction does not occur by this time, it will be off for the last 90° of the alternation.

Variable control of the same type also occurs during the negative alternation. For this alternation, point A is negative and point B is positive, thus causing a $-T_2$, a $+T_1$, and a $-I_G$. Gate current will flow and cause conduction during this alternation. The resistance setting of R_1 influences I_G in the same manner as it did for the positive alternation. Both alternations will therefore be controlled equally. Variable control of this type applies to only 50% of the source voltage. If conduction does not occur in the first 90% of an alternation, no control will be achieved.

Triac Phase Control

With ac phase control of a triac, it is possible to achieve nearly 100% control of the conduction time. For

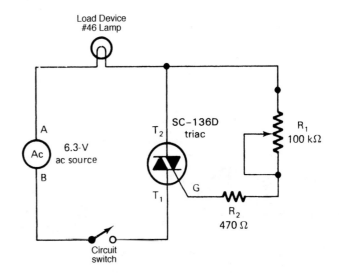

Figure 8-19. Variable lamp control.

this to be achieved, two distinct ac voltage values are applied to the triac. The ac source voltage is applied to terminals T_1 and T_2. This voltage must not be changed. A second ac signal is applied to the gate. Gate current is developed by this voltage. Control of the triac is achieved by shifting the phase of the gate voltage. Changes from 0° to 90° or 0° to 180° are possible depending on the circuit design. The gate voltage must lag behind the source voltage to initiate control.

Figure 8-20 shows five different operating states of a triac phase control circuit. In Figure 8-20(a), the T_1–T_2 source voltage and the gate voltage are in phase, which causes a 0° delay and full conduction of the triac. The load device receives 100% of the source voltage and operates at full power. Figure 8-20(b) shows where the gate voltage is shifted 45° behind the source voltage, causing power to be applied to the load for approximately 75% of the time. A 90° delay is shown in Figure 8-20(c). The gate voltage has shifted behind the source voltage by 90°, causing power to be applied to the load for approximately 50% of the time. In Figure 8-20(d), the load receives power for less than 25% of the time. Figure 8-20(e) shows a 180° delay and no power supplied to the load. With the phase shifter, it is possible to achieve full control of power to the load device.

Only four components are used in the triac phase control circuit of Figure 8-21. Resistor R_1 and capacitor C_1 serve as a single-element phase shifter. The phase of the gate volage is shifted by R_1. The *diac* is a voltage-controlled trigger device. It will conduct in either direction when the applied voltage reaches a specific value. Diacs were purposely designed to trigger the gate of a triac. This particular diac will conduct when gate voltage reaches 28 V in either direction.

Operation of the single-element triac phase-shifter circuit is based on the time constant of R_1 and C_1. For the positive alternation, C_1 charges through R_1. When C_1 reaches the breakover voltage of the diac, it turns on C_1 then discharges through the diac and the gate of the triac. Gate current lowers the breakover voltage of the triac, and conduction occurs. Conduction continues for the remainder of the alternation.

During the negative alternation, C_1 charges in the reverse direction through R_1. When the reverse breakover voltage of the diac is reached, it goes into conduction. C_1 discharges through the diac and the gate of the triac, which triggers the triac into conduction. Conduction continues for the remainder of the negative alternation. The process then repeats itself for each succeeding alternation.

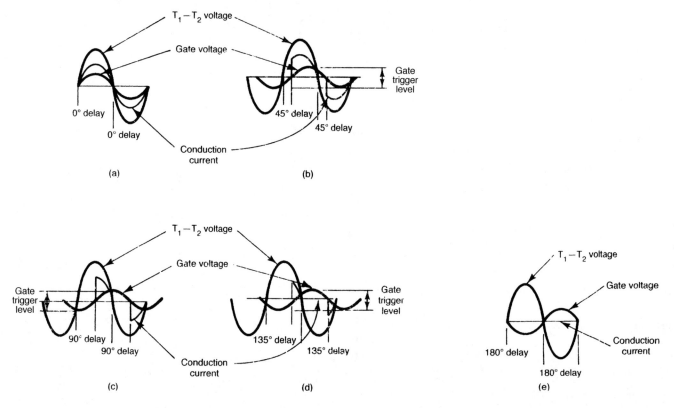

Figure 8-20. Operating states of a triac: (a) 100% power; (b) 75% power; (c) 50% power; (d) 25% power; (e) 0% power.

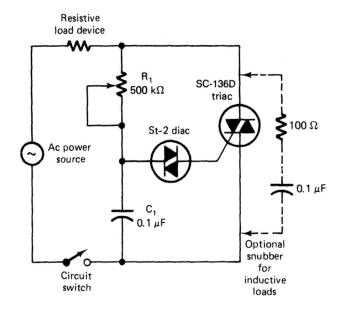

Figure 8-21. Basic triac phase shifter.

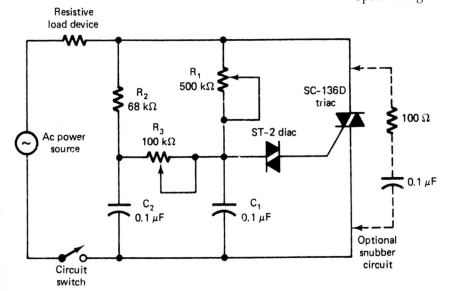

Figure 8-22. Two-stage phase shifter for a triac.

Delay of conduction is based on the time constant of R_1 and C_1. Control with a single-leg phase shifter is good only for the first 90° of an alternation. If conduction does not occur in the first 90°, it will remain off for the remainder of the alternation. Applications of this circuit are limited to lamp control, heater operation, and fan speed control.

An improved triac power control circuit is shown in Figure 8-22. This circuit employs a double-stage *RC* phase shifter. Control of this circuit is in the approximate range of 10° to 170°. Power control is from 16% to 95% of full power.

Operation of the double-stage *RC* phase shifter is based on alternate charge and discharge of two *RC* circuits. When power is applied, C_2 charges quickly through R_2. A short time later C_1 begins to charge from the developed voltage across C_2. When the charge on C_1 reaches the breakover voltage of the diac, it goes into conduction. C_1 discharges through the diac and the gate of the triac. This triggers the triac into conduction. Conduction then continues for the remainder of the alternation. The charge time of C_1 is based on adjustment of R_1 and R_3.

For the next alternation, C_2 recharges through R_2. C_1 begins to charge to the C_2 voltage a short time later through R_3. When the diac is triggered, C_1 discharges through the triac, increasing the gate current and causing the triac to be conductive. Conduction continues for the remainder of the alternation. The process then repeats itself for each succeeding alternation.

The double-stage *RC* phase shifter has an increased control range. When R_1 is large, C_1 charges to the developed voltage of C_2. R_3 should be adjusted so that the circuit drops out of conduction when R_1 is at its maximum value. R_3 is considered as a trimmer adjustment for the circuit.

DIAC POWER CONTROL

A diac is a special diode that can be triggered into conduction by voltage. This device is classified as a bidirectional trigger diode, meaning that it can be triggered into conduction in either direction. The word *diac* is derived from "*diode for ac.*" This device is used primarily to control the gate current of a triac. It will go into conduction during either the positive or the negative alternation. Conduction is achieved by simply exceeding the breakover voltage.

Figure 8-23 shows the crystal structure, schematic symbol, and *I-V* characteristics of a diac. Note that the crystal is similar to that of a transistor without a base. The N_1 and N_2 crystals are primarily the same in all respects. A diac will therefore go into conduction at precisely the same negative or positive voltage value. Conduction occurs only when input voltage exceeds the breakover voltage. A rather limited number of diacs are available today. The one shown here has a minimum V_{BO} of 28 V. This particular device is a standard trigger for triac control. Note that the voltage across the diac decreases in value after it has been triggered.

ELECTRONIC CONTROL CONSIDERATIONS

Electronic power control with an SCR or a triac is efficient when used properly. These devices are, however, attached directly to the ac power line. Severe damage to a load device and a potential electrical hazard may occur if this method of control is connected improperly. As a general rule, SCR and triac control should not be attempted for ac-only equipment such as fluorescent lamps, radios, TV receivers, induction motors, and other transformer-operated devices.

SCR and triac control can be used effectively to control resistive type loads such as incandescent lamps, soldering irons or pencils, heating pads, and electric blankets. It is also safe to use this type of control for universal motors. This type of motor is commonly used in portable power tools and some small appliances such as portable drills, saber saws, sanders, electric knives, and mixers. When in doubt about a particular device, check the manufacturer's instruction manual. It is also important that the wattage rating of the load not exceed the wattage of the control device. Wattage ratings are nearly always stamped on the device or indicated on a label. Check this rating before attempting to control the device with an SCR or a triac.

It is extremely important that all control circuits be placed in an insulated housing or container. The ac power line is connected directly to the control circuit. Touching any part of the control circuit may cause a severe electrical shock. A common practice is to place the control circuit in an insulated box that has no outside metal parts or knobs.

The procedure for servicing or analyzing the operation of an ac power control circuit is extremely important. Avoid touching all metal parts when in operation. Work with the circuit only on an insulated tabletop or bench. Do not attempt to connect electrically energized test instruments to the circuit. If at all possible, the control device should be energized by an isolation transformer to reduce the ground problems and some electrical hazards. Circuit testing should be done when the power control device is disconnected from the power line.

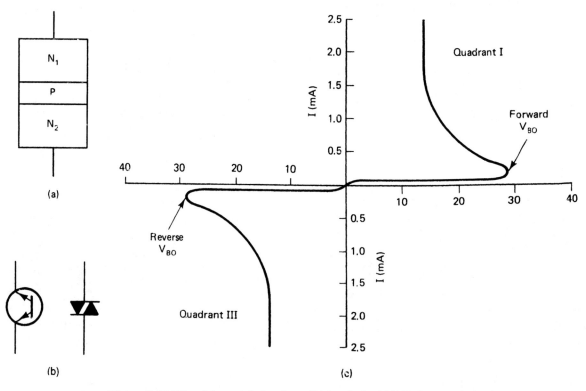

Figure 8-23. Diac: (a) crystal structure; (b) symbols; (c) I-V characteristics.

Appendices

Appendix A

Comprehensive Electronics Glossary

This glossary is an alphabetical list by topic of electronic terms. This glossary will provide you with brief discussions of electronics terms. The topics included are the following:

1. Basic electronics terms
2. Direct-current electronics terms
3. Magnetic and electrical power terms
4. Alternating-current electronics terms
5. Electronic measurement terms
6. Solid-state electronics terms
7. Electronic circuits terms
8. Communications electronics terms
9. Computer electronics terms
10. Industrial electronics terms

BASIC ELECTRONICS TERMS

Ac: abbreviation for alternating current.

Alternating current: current produced when electrons move first in one direction and then in the opposite direction.

Ampere: the electrical charge movement which is the basic unit of measurement for current flow in an electrical circuit.

Ampere-hour: a capacity rating of storage batteries which refers to the maximum current that can flow from the battery during a specific period of time.

Atom: the smallest particle that an element can be reduced to and still retain its characteristics.

Block diagram: a diagram used to show how the parts of a system fit together.

Cell: an electrical energy source that converts chemical energy into electrical energy.

Closed circuit: a circuit which forms a complete path so that electrical current can flow through it.

Component: An electrical device that is used on a circuit.

Compound: the combination of two or more elements to make some other material.

Conductor: a material that allows electrical current to flow through it easily.

Control: the part of an electrical system which affects what the system does; a switch to turn a light on and off is a type of control.

Conventional current flow: current flow that is assumed to be in a direction from high charge concentration (+) to low charge concentration (–).

Coulomb: a unit of electrical charge that represents a large number of electrons.

Current: the movement of electrical charge; the flow of electrons through an electrical circuit.

Dc: abbreviation for direct current.

Direct current: the flow of electrons in one direction from negative to positive.

Discharge: a condition reached by a battery when its chemical energy is released.

Electromotive force (EMF): the "pressure," or force, that causes electrical current to flow.

Electron: a small particle which is part of an atom that is said to have a negative (–) electrical charge; electrons cause the transfer of electrical energy from one place to another.

Electron current flow: current flow that is assumed to be in the direction of electron movement from a negative (–) potential to a positive (+) potential.

Electrostatic field: the area around a charge material in which the charge exists.

Element: the basic materials that make up all other materials; they exist by themselves (such as copper, hydrogen, carbon) or in combination with other elements (water is hydrogen and oxygen elements combined).

Energy: the capacity to do work.

Free electrons: electrons along the outer orbit which are removed easily to allow electrical current flow.

Fuse: an electrical overcurrent device that opens a circuit when it melts due to excess current flow through it.

Ground: of two types: *system grounds,* current-carrying conductors used for electrical power distribution; and *safety grounds,* not intended to carry electrical current but to protect individuals from electrical shock hazards.

Indicator: the part of an electrical system that shows if it is on or off or indicates some specific quantity.

Insulator: a material that offers a high resistance to electrical current flow.

Ion: an atom that has lost or gained electrons, making it a negative or positive charge.

Kinetic energy: energy that exists because of movement.

Load: the part of an electrical system that converts electrical energy into another form of energy, such as an electric motor which converts electrical energy into mechanical energy.

Matter: any material that makes up the world; anything that occupies space and has weight; can be either a solid, a liquid, or a gas.

Metallic bonding: the method in which loosely held atoms are bound together in metals.

Molecule: the smallest particle that a material can be reduced to before becoming its basic elements.

Neutron: a particle in the nucleus (center) of an atom that has no electrical charge or is neutral.

Nucleus: the center part of an atom, made of protons (positive charges) and neutrons (have no electrical charge).

Ohm: (symbol is Ω): the unit of measurement of electrical resistance.

Open circuit: a circuit that has a broken path so that no electrical current can flow through it.

Orbit: a circular path along which electrons travel around the nucleus of an atom. Path: the part of an electrical system through which the energy travels from a source to a load, such as the electrical wiring used in a building.

Piezoelectric effect: the property of certain crystal materials to produce a voltage when pressure is applied to them.

Potential energy: energy that exists due to position.

Potentiometer: (Rheostat) a variable-resistance component used as a control device in electrical circuits.

Power: the rate at which work is done.

Proton: a particle in the center of an atom which has a positive (+) electrical charge.

Resistance: the opposition to the flow of electrical current in a circuit; its unit of measurement is the ohm (Ω).

Scientific notation: a number written as a product of a number between 1 and 10 and a power of 10.

Semiconductor: a material that has electrical resistance somewhere between a conductor and an insulator.

Short circuit: a circuit that forms a direct path across a voltage source so that a very high and possibly unsafe electrical current flows.

Source: the part of an electrical system that supplies energy to other parts of the system, such as a battery that supplies energy for a flashlight.

Single phase: the type of electrical power that is supplied to homes and is in the form of a sine wave when it is produced by a generator.

Stable atom: an atom that will not release electrons under normal conditions.

Static charge: a charge on a material which is said to be either positive or negative.

Static electricity: electricity "at rest," which is due to positive negative electrical charges.

Switch: a control device used to turn a circuit on or off.

Symbol: used as a simple way to represent a component.

Three-phase ac: the type of electrical power that is generated at power plants and transmitted over long distances.

Valance electrons: electrons in the outer orbit of an atom.

Volt: the unit of measurement of electrical potential.

Voltage: electrical force, or "pressure," that causes current to flow in a circuit.

Voltage drop: the electrical potential (voltage) that exists across two points of an electrical circuit.

Watt: the unit of measurement of electrical power.

Work: the transforming or transferring of energy.

DIRECT-CURRENT ELECTRONICS TERMS

Battery: an electrical energy source that is two or more cells connected together.

Branch: a path of a parallel circuit.

Branch current: the current through a parallel branch of a circuit.

Branch resistance: the total resistance of a parallel branch of a circuit.

Branch voltage: the voltage across a parallel branch of a circuit.

Circuit: a path through which electrical current flows in a circuit.

Combination circuit: a circuit that has a portion connected in series with the voltage source and another part connected in parallel.

Complex circuit: another term for a combination circuit.

Current: the flow of electrons through a circuit path.

Difference in potential: the voltage across two points of a circuit.

Direct current (dc): the type of electrical power that is produced by batteries and power supplies; used mostly for portable and specialized power applications; the flow of electrons in one direction from negative (–)

to positive (+).

Directly proportional: when one quantity increases or decreases, causing another quantity to do the same.

Dry cell: a nonliquid cell that produces dc voltage by chemical action.

Electrode: the negative or positive element of a cell.

Electrolyte: the acid solution used in a cell that produces dc.

Equivalent resistance: a resistance value that would be the same value in a circuit as two or more parallel resistances of a circuit.

Hydrometer: an instrument used to measure the specific gravity, or "charge," of the electrolyte of a storage battery.

Inverse: the value of 1 divided by some quantity, such as 1IRT for finding parallel resistance.

Inversely proportional: when one quantity increases or decreases, causing another quantity to do the opposite.

Kirchhoff's current law: the sum of the currents flowing into any junction of conductors of a circuit is equal to the sum of the currents flowing away from that point.

Kirchhoff's voltage law: in any current loop of a circuit the sum of the voltage drops is equal to the voltage supplied to that loop.

Lead-acid cell: a secondary cell that has positive and negative plates made of lead peroxide and lead and has a liquid electrolyte of sulfuric acid mixed with water.

Nickel-cadmium cell: an alkaline cell that has a longer life than that of carbon-zinc dry cells.

Ohm's law: the law that explains the relationship of voltage, current, and resistance in electrical circuits.

Parallel circuit: a circuit that has two or more current paths.

Photovoltaic cell: a cell that produces dc voltage at the junction of two materials when light shines onto its surface.

Power (P): the rate of doing work in electrical circuits found by $P = I \times V$.

Primary cell: a cell that cannot be recharged.

Reciprocal: same as inverse.

Resistance: opposition to the flow of current in an electrical circuit.

Secondary cell: a cell that can be recharged by applying dc voltage from a battery charger.

Series circuit: a circuit that has one path for current flow.

Specific gravity: the weight of a liquid as compared to the weight of water, which has a value of 1.0.

Storage battery: another term used for lead-acid secondary cells.

Thermocouple: a device that has two pieces of metal joined together so that when its junction is heated, a voltage is produced.

Total current: the current that flows from the voltage source of a circuit.

Total resistance: the total opposition to current flow of a circuit, which may be found by removing the voltage source and connecting an ohmmeter across the points where the source was connected.

Total voltage: the voltage supplied by a source.

Voltage: electrical "pressure" that causes current to flow through a circuit.

Voltage drop: the voltage across two points of a circuit, found by $V = I \times R$.

Voltaic cell: a cell that produces voltage due to two metal elements which are suspended in an acid solution.

Watt: the unit of measurement of electrical power.

MAGNETIC AND ELECTRICAL POWER TERMS

Alnico: an alloy of aluminum, nickel, iron, and cobalt used to make permanent magnets.

Alternator: a rotating machine that generates ac voltage.

Ampere-turn: the unit of measurement of magnetic field strength; amperes of current times the number of turns of wire.

Armature: the movable part of a relay, rotating coils of a motor or the part of a generator into which current is induced.

Armature reaction: the effect in direct-current generators and motors in which current flow through the armature conductors reacts with the main magnetic field to cause distortion of the magnetic field and increase sparking between the brushes and commutator.

Back EMF: same as counter-electromotive force (CEMF).

Ballast: an inductive coil placed in a fluorescent light circuit to cause the light to operate.

Boiler: a chamber used in a steam power plant to heat water to convert it to steam.

Brush: a sliding contact made of carbon and graphite which touches the commutator of a generator or motor.

Capacitive heating: a method of heating nonconductive materials by placing them between two metal plates and applying a high-frequency ac voltage.

Coefficient of coupling: a decimal value that indicates the amount of magnetic coupling between coils.

Commutation: the process of applying direct current of the proper polarity at the proper time to the rotor windings of a dc motor through split rings and brushes; or the process of changing ac induced into the rotor of a dc generator into dc applied to the load circuit.

Commutator: an assembly of copper segments that provide a method of connecting rotating coils to the brushes of a dc generator.

Copper losses: heat losses in electrical machines due to the resistance of the copper wire used for windings; also called I_2R losses.

Core: laminated iron or steel that is used in the internal construction of the magnetic circuit of electrical machines.

Core saturation: when the atoms of a metal core material are aligned in the same pattern so that no more magnetic lines of force can be developed.

Counter electromotive force (CEMF): "generator action" that takes place in motors when voltage (EMF) is induced into the rotor conductors which opposes the source voltage (EMF).

Coupling: the amount of mutual inductance between coils.

Dc generator: a rotating machine that produces a form of direct-current (dc) voltage.

Delta connection: a method of connecting the stator windings of three-phase machines in which the beginning of one winding is connected to the end of the next winding; power input lines come into the junctions of the windings.

Domain theory: a theory of magnetism which assumes that groups of atoms produced by movement of electrons align themselves in groups called "domains" in magnetic materials.

Eddy currents: induced current in the metal parts of rotating machines which causes heat losses.

Efficiency: the ratio of output power and input power of a generator:

$$\text{Efficiency} = \frac{\text{power output (watts)}}{\text{power input (horsepower)}}$$

or the ratio of output power to input power; for a motor:

$$\text{Efficiency} = \frac{P_{\text{out}}}{P_{\text{in}}}$$
$$= \frac{\text{horsepower output} \times 746}{\text{watts input}}$$

Electromagnet: a coil of wire wound on an iron core so that as current flows through the coil it becomes magnetized.

Field coils: electromagnetic coils that develop the magnetic fields of electric machines.

Field poles: the laminated metal which serves as the core material for field coils.

Fluorescent lighting: a lighting method that uses lamps which are tubes filled with mercury vapor to produce a bright white light.

Flux (symbol is Φ): invisible lines of force that extend around a magnetic material.

Flux density: the number of lines of force per unit area of a magnetic material or circuit.

Fossil fuel system: a power system that produces electrical energy due to the conversion of heat from coal, oil, or natural gas.

Gauss: a unit of measurement of magnetic flux density.

Generator: a rotating electrical machine that converts mechanical energy into electrical energy.

Gilbert: a unit of measurement of magnetomotive force (MMF).

Ground fault: an accidental connection to a grounded conductor.

Ground-fault interrupter (GFI): a device used in electrical wiring for hazardous locations; it detects fault conditions and responds rapidly to open a circuit before shock occurs to an individual or equipment is damaged.

Horsepower: 1 horsepower (hp) = 33,000 ft-lb of work per minute or 550 ft-lb per second; it is the unit of measuring mechanical energy; 1 hp = 746 W.

Hot conductor: a wire that is electrically energized, or "live," and is not grounded.

Hydroelectric system: a power system that produces electrical energy due to the energy of flowing water.

Hysteresis: the property of a magnetic material that causes actual magnetizing action to lag behind the force that produces it.

Incandescent lighting: a method of lighting which uses bulbs that have tungsten filaments which when heated by electrical current, produce light.

Induced current: the current that flows through a conductor due to magnetic transfer of energy.

Induced voltage: the potential that causes induced current to flow through a conductor which passes through a magnetic field.

Induction motor: an ac motor that has a solid squirrel-cage rotor and operates on the induction (transformer action) principle.

Inductive heating: a method of heating conductors in which the material to be heated is placed inside a coil of wire and high-frequency ac voltage is applied.

Laminations: thin pieces of sheet metal used to construct the metal parts of motors and generators.

Laws of magnetism: (1) like magnetic poles repel; (2) unlike magnetic poles attract.

Left-hand rule: (1) to determine the *direction of the magnetic field* around a single conductor, point the thumb of the left hand in the direction of current flow (– to +) and the fingers will extend around the conductor in the direction of the magnetic field; (2) to determine the *polarity of an electromagnetic* coil, extend the fingers of the left hand around the coil in the direction of current and the thumb will point to the north polarity; (3) to determine the *direction of induced current flow* in a generator conductor, hold the thumb, forefinger, and middle finger of the left hand at right angles to one another, point the thumb in the direction of motion of the conductor, the forefinger in the direction of the magnetic field (N to S), and the middle finger will point in the direction of induced current.

Lenz's law: the induced voltage in any circuit is always in such a direction that it will oppose the force that produces it.

Lines of force: same as magnetic *flux*.

Lodestone: the name used for natural magnets used in early times.

Magnet: a metallic material, usually iron, nickel, or cobalt, which has magnetic properties.

Magnetic circuit: a complete path for magnetic lines of force from a north to a south polarity.

Magnetic field: magnetic lines of force that extend from a north polarity and enter a south polarity to form a closed loop around the outside of a magnetic material.

Magnetic flux: see Flux.

Magnetic materials: metallic materials such as iron, nickel, and cobalt that exhibit magnetic properties.

Magnetic poles: areas of concentrated lines of force on a magnet that produce north and south polarities.

Magnetic saturation: a condition that exists in a magnetic material when an increase in magnetizing force does not produce an increase in magnetic flux density around the material.

Magnetomotive force (MMF): a force that produces magnetic flux around a magnetic device.

Magnetostriction: the effect that produces a change in shape of certain materials when they are placed in a magnetic field.

MCM: a unit of measure for large circular conductors, equal to 1000 circular mils.

Motor: a rotating machine that converts electrical energy into mechanical energy.

Natural magnet: metallic materials that have magnetic properties in their natural state.

NEMA (National Electrical Manufacturers' Association): an organization that establishes standards for electrical equipment.

Neutral: a grounded conductor of an electrical circuit which carries current and has white insulation.

Neutral plane: the theoretical switching position of the commutator and brushes of a dc generator or motor which occurs when no current flows through the armature conductors and the main magnetic field has least distortion.

Nuclear fission system: a power system that produces electrical energy due to the heat developed by the splitting of atoms in a nuclear reactor.

Overcurrent device: a device such as a fuse or circuit breaker which is used to open a circuit when an excess current flows in the circuit.

Overload: a condition that results when more current flows in a circuit than it is designed to carry.

Permanent magnet: bars or other shapes of materials which retain their magnetic properties.

Permeability: the ability of a material to conduct magnetic lines of force as compared to air; the ability of a material to magnetize or demagnetize.

Polarities: same as *Magnetic poles*.

Power factor: the ratio of true power (watts) converted in a circuit or system to the apparent power (volt-amperes) delivered to the circuit or system.

Prime mover: a system that supplies the mechanical energy to rotate an electrical generator.

Pulsating dc: a dc voltage that is not in "straight line" or "pure" dc form, such as the dc voltage produced by a battery.

Reactive power (var): the quantity of "unused" power developed by reactive components (inductive or capacitive) in an ac circuit or system; this unused power is delivered back to the power source since it is not converted to another form of energy; the unit of measurement is the voltampere reactive (var).

Regulation: a measure of the amount of voltage change which occurs in the output of a generator due to changes in load.

Relay: an electromagnetic switch that uses a low current to control a higher current circuit.

Reluctance: the opposition of a material to the flow of magnetic flux.

Repulsion motor: a single-phase ac motor that has a wound rotor and brushes which are shorted together.

Residual magnetism: the magnetism that remains around a material after the magnetizing force has been removed.

Resistive heating: a method of heating which relies on the heat produced when electrical current moves through a conductor.

Retentivity: the ability of a material to retain magnetism after a magnetizing force has been removed.

Right-hand motor rule: the rule applied to find the direction of motion of the rotor conductors of a motor.

Rotating-armature method: the method used when a generator has dc voltage applied to produce a field to the stationary part (stator) of the machine and voltage is induced into the rotating part (rotor).

Rotating-field method: the method used when a generator has dc voltage applied to produce a field to the rotor of the machine and voltage is induced into the stator coils.

Rotor: the rotating part of an electrical generator or motor.

Running neutral plane: the actual switching position of the commutator and brushes of a dc generator or motor which shifts the theoretical neutral plane due to armature reaction.

Self-starting: the ability of a motor to begin rotation when electrical power is applied to it.

Shaded-pole motor: a single-phase ac motor that uses copper shading coils around its poles to produce rotation.

Short circuit: a fault condition that results when direct contact (zero resistance) is made between two conductors of an electrical system (usually by accident).

Single-phase ac generator: a generator that produces single-phase ac voltage in the form of a sine wave.

Single-phase motor: any motor that operates with single-phase ac voltage applied to it.

Slip rings: solid metal rings mounted on the end of a rotor shaft and connected to the brushes and the rotor windings; used with some types of ac motors.

Solenoid: an electromagnetic coil with a metal core which moves when current passes through the coil.

Speed regulation: the ability of a motor to maintain a steady speed with changes in load.

Split-phase motor: a single-phase ac motor which has start windings and run windings and operates on the induction principle.

Split rings: a group of copper bars mounted on the end of a rotor shaft of a dc motor and connected to the brushes and the rotor windings.

Squirrel-cage rotor: a solid rotor used in ac induction motors.

Starter: a resistive network used to limit starting current in motors.

Stator: the stationary part of a motor.

Steam turbine: a machine that uses the pressure of steam to cause rotation which is used to turn an electrical generator.

Synchronous motor: an ac motor that operates at a constant speed regardless of the load applied.

Three-phase ac generator: a generator that produces three ac voltages.

Torque: mechanical energy in the form of rotary motion.

True power (watts): the amount of power that is converted to another form of energy in a circuit or system; measured with a wattmeter.

Universal motor: a motor that operates with either ac or dc voltage applied.

Vapor lighting: a method of lighting that uses lamps which are filled with certain gases that produce light when electrical current is applied.

Voltage drop: the reduction in voltage caused by the resistance of conductors of an electrical distribution system; it causes a voltage less than the supply voltage at points near the end of an electrical circuit that is farthest from the source.

Watt: the basic unit of electrical power; the amount of power converted when 1 A of current flows under a pressure of 1 V.

Watt-hour meter: a meter that monitors electrical energy used over a period of time.

Wattmeter: a meter used to measure the actual electrical energy that is converted in a circuit or system.

Wye connection: a method of connecting the stator windings of three-phase machines in which the three *beginnings or ends of the windings are joined together to form a common (neutral) point and the other wires connect to the power lines.*

ALTERNATING-CURRENT ELECTRONICS TERMS

Ac: the abbreviation used for alternating current.

Admittance: the total ability of an ac circuit to conduct current, measured in siemens or mhos; the inverse of impedance $Y = 1/2$.

Air-core inductor: a coil wound on an insulated core or a coil of wire that does not have a metal core.

Amplitude: the vertical height of an ac waveform.

Angle of lead or lag: the angle between applied voltage and current flow in an ac circuit, in degrees; in an

inductive (L) circuit, voltage (E) leads current (I)—
ELI; in a capacitive (C) circuit, current (I) leads voltage (E)—ICE.

Apparent-power (volt-amperes): the total power *delivered* to an ac circuit; applied voltage times current.

Attenuation: a reduction in value.

Average voltage: the value of an ac sinewave voltage
which is found by this formula: $E_{avg} = E_{peak} \times$
0.636.

Band-pass filter: a frequency-sensitive ac circuit that allows incoming frequencies within a certain band to
pass through but attenuates frequencies that are below or above this band.

Bandwidth: the band (range) of frequencies that will pass
easily through a bandpass filter or resonant circuit.

Capacitance: the property of a device to oppose changes
in voltage due to energy stored in its electrostatic
field.

Capacitive reactance: the opposition to the flow of ac
current caused by a capacitive device (measured in
ohms).

Capacitor: a device that has capacitance and is usually
made of two metal plate materials separated by a
dielectric material (insulator).

Center-tap: a terminal connection made to the center of a
transformer winding.

Choke coil: an inductor coil used to block the flow of ac
current and pass dc current.

Condenser: another term used occasionally to mean capacitor.

Conductance: the ability of a resistance of a circuit to conduct current, measured in siemens or mhos; the inverse of resistance $G = 1/R$.

Cycle: a sequence of events that causes one complete pattern of alternating current from a zero reference, in a
positive direction, then back to zero, then in a negative direction and back to zero.

Decay: a term used for a gradual reduction in value of a
voltage or current.

Decay time: the time for a capacitor to discharge to a certain percentage of its original charge or the time required for current through an inductor to reduce to
a percentage of its maximum value.

Decibel: a unit used to express an increase or decrease in
power, voltage, or current in a circuit.

Delta connection: a method of connecting three-phase alternator stator windings in which the beginning of
one phase winding is connected to the end of the adjacent phase winding; power lines extend from the
junction of beginning-end connections. Dielectric:
an insulating material placed between the metal

plates of a capacitor.

Dielectric constant: a number that represents the ability of a dielectric to develop an electrostatic field, as
compared to air, which has a value of 1.0.

E: an abbreviation sometimes used for voltage; V is used
in this book.

Effective voltage: a value of an ac sinewave voltage that
has the same effect as an equal value of dc voltage:
$$V_{eff} = V_{peak} \times 0.707$$

Electrolytic capacitor: a capacitor that has a positive plate
made of aluminum and a dry paste or liquid used to
form the negative plate.

Electrostatic field: the field that is developed around a
material due to the energy of an electrical charge.

ELI: a term used to help remember that voltage (E) leads
current (I) in an inductive (L) circuit.

Farad: the unit of measurement of capacitance that is produced when a charge of 1 C causes a potential of 1
V to be developed.

Filter: a circuit used to pass certain frequencies and at-
tenuate all other frequencies.

Frequency: the number of ac cycles per second, measured
in hertz (Hz).

Frequency response: a circuit's ability to operate over a
range of frequencies.

Henry: the unit of measurement of inductance which is
produced when a voltage of 1 volt is induced when
the current through a coil is changing at a rate of 1
A per second.

Hertz: the international unit of measurement of frequency equal to 1 cycle per second.

ICE: a term used to help remember that current (I) leads
voltage (E) in a capacitive (C) circuit.

Impedance (Z): the total opposition to current flow in an
ac circuit which is a combination of resistance (R)
and reactance (X) in a circuit; measured in ohms:

$$Z = \sqrt{R^2 + X^2}$$

Inductance: the property of a circuit to oppose changes in
current due to energy stored in a magnetic field.

Inductive circuit: a circuit that has one or more inductors
or has the property of inductance, such as an electric
motor circuit.

Inductive reactance (X_L): the opposition to current flow
in an ac circuit caused by an inductance (L), measured in ohms: $X_L = 2\pi fL$.

Inductor: a coil of wire that has the property of inductance and is used in a circuit for that purpose.

In phase: two waveforms of the same frequency which
pass through their minimum and maximum values

at the same time and polarity.

Instantaneous voltage: a value of ac voltage at any instant (time) along a waveform.

Isolation transformer: a transformer with a 1:1 turns ratio used to isolate an ac power line from equipment with a chassis ground.

Lagging phase angle: the angle by which current *lags* voltage (or voltage *leads* current) in an inductive circuit.

Leading phase angle: the angle by which current *leads* voltage (or voltage *lags* current) in a capacitive circuit.

Maximum power transfer: a condition that exists when the resistance or impedance of a load (R_L) equals that of the source which supplies it (R_S).

Mho: ohm spelled backward; a unit of measurement sometimes used for conductance, susceptance, and admittance; being replaced by the *siemens*.

Mica capacitor: a capacitor made of metal foil plates separated by a mica dielectric.

Mutual inductance: when two coils are located close together so that the magnetic flux of the coil affect one another in terms of their inductance properties.

Parallel resonant circuit: a circuit that has an inductor and capacitor connected in parallel to cause response to frequencies applied to the circuit.

Peak-to-peak voltage: the value of ac sine-wave voltage from positive peak to negative peak.

Peak voltage (V_{peak}): the maximum positive or negative value of ac sine-wave voltage: $V_{peak} = V_{eff} \times 1.41$.

Period (time): the time required to complete one ac cycle: time 1/frequency.

Phase angle: the angular displacement between applied voltage and current flow in an ac circuit.

Power factor (pf): the ratio of power converted (true power) in an ac circuit and the power delivered (apparent power):

$$PF = \frac{\text{true power (watts)}}{\text{apparent power (volt-amperes)}}$$

Primary winding: the coil of a transformer to which ac source voltage is applied.

Quality factor: the "figure of merit" or ratio of inductive reactance and resistance in a frequency-sensitive circuit.

Radian: an angle that begins at the center of a circle and extends along the outside of the circle a distance equal to the radius of the circle: 1 radian = 57.3°; the circumference of a circle is equal to 2π (about 6.28 radians).

Reactance (X): the opposition to ac current flow due to inductance (X_L) or capacitance (X_C).

Reactive circuit: an ac circuit that has the property of inductance or capacitance.

Reactive power (VAR): the "unused" power of an ac circuit which has inductance or capacitance, which is absorbed by the magnetic or electrostatic field of a reactive circuit.

Resistance: the opposition to current flow due to a resistive device.

Resistive circuit: a circuit whose only opposition to current flow is resistance; a nonreactive circuit.

Resonant circuit: see *Parallel resonant circuit* and *Series resonant circuit*.

Resonant frequency: the frequency that passes most easily through a frequency-sensitive circuit when $X_L = X_C$ in the circuit:

$$f_r = \frac{1}{2\pi \times \sqrt{L \Sigma C}}$$

Root-mean-square (RMS) voltage: same as *Effective voltage*.

Saw tooth waveform: an ac waveform shaped like the teeth of a saw.

Secondary winding: the coil of a transformer into which voltage is induced; energy is delivered to the load circuit by the secondary winding.

Selectivity: the ability of a resonant circuit to select a specific frequency and reject all other frequencies.

Series resonant circuit: a circuit that has a resistor, inductor and capacitor connected in series to cause desired response to frequencies applied to the circuit.

Siemens: same as mho.

Signal: an electrical waveform of varying value which is applied to a circuit.

Sine wave: a waveform of one cycle of ac voltage.

Step-down transformer: a transformer that has a secondary voltage lower than its primary voltage.

Step-up transformer: a transformer that has a secondary voltage higher than its primary voltage.

Susceptance: the ability of an inductance (B_L) or a capacitance (B_C) to pass ac current; measured in siemens or mhos: $B_L = 1/X_L$ and $B_C = 1/X_C$.

Tank circuit: same as *parallel resonant circuit*.

Theta (Θ): a Greek letter used to represent the phase angle of an ac circuit.

Time constant (RC): the time required for the voltage across a capacitor in an RC circuit to increase to 63% *of its maximum value or decrease to 37% of its maximum value; time* = $R \times C$.

Time constant (RL): the time required for the current

through an inductor in an *RL* circuit to increase to 63% of its maximum value or decrease to 37% of its maximum value; time = I/R.

Transformer: an ac power control device that transfers energy from its primary winding to its secondary winding by mutual inductance and is ordinarily used to increase or decrease voltage.

True power (watts): the power actually *converted* to another form by an ac circuit; true power is measured with a wattmeter.

Turns ratio: the ratio of the number of turns of the primary winding (N_P) of a transformer to the number of turns of the secondary winding (N_S).

VAR (volt-amperes reactive): the unit of measurement of reactive power.

Vector: a straight line which indicates a quantity that has magnitude and direction.

Volt-ampere (VA): the unit of measurement of apparent power.

Waveform: the pattern of an ac frequency derived by looking at instantaneous voltage values that occur over a period of time; on a graph a waveform is plotted with instantaneous voltages on wave voltages that are separated in phase by 120 electrical degrees.

Wye connection: a method of connecting three-phase alternator stator windings in which the beginnings *or* ends of each phase winding are connected together to form a common or neutral point; power lines extend from the remaining beginnings or ends.

ELECTRONIC MEASUREMENT TERMS

Ammeter: a meter used to measure current (amperes).

Bridge circuit: a circuit with groups of components that are connected by a "bridge" in the center and used for precision measurements.

Cathode ray tube (CRT): a vacuum tube in which electrons are emitted from the cathode in the shape of a narrow beam and speeded up so that they will strike the screen and produce light.

Continuity check: a test to see if a circuit is an open or closed path.

D'Arsonval meter movement: a stationary magnet which has a moving electromagnetic coil inside that causes a meter needle to deflect.

Galvanometer: a meter that measures very small current values.

Kilowatt-hour (kWh): 1000 W per hour; a unit of measurement for electrical energy.

Loading: the effect caused when a meter is connected into a circuit that causes the meter to draw current from the circuit.

Megger: a meter used to measure very high resistances.

Moving coil: same as a d'Arsonval meter movement.

Multifunction meter: a meter that measures two or more electrical quantities, such as a VOM.

Multiplier: a resistance connected in series with a meter movement to extend the range of voltage it will measure.

Multirange meter: a meter that has two or more ranges to measure an electrical quantity.

Ohmmeter: a meter used to measure resistance (ohms).

Ohms per volt rating (WV): the rating that indicates the loading effect a meter has on a circuit when making a measurement.

Oscilloscope: an instrument that has a cathode ray tube to allow a visual display of voltages.

Polarity: the direction of an electrical potential (– or +) or a magnetic charge (north or south).

Precision resistor: a resistor used when much accuracy is needed.

Resistor: a component used to control the amount of current flow in a circuit.

Schematic diagram: a diagram used to show how the components of an electronic system fit together.

Sensitivity: same as the ohms per volt rating.

Shunt: a resistance connected in parallel with a meter movement to extend the range of current it will measure.

Voltmeter: a meter used to measure voltage.

Volt-ohm-milliammeter (VOM): a multifunction, multirange meter which is usually designed to measure voltage, current, and resistance.

Watt-hour: a unit of energy measurement equal to one watt per hour.

Watt-hour meter: a meter that measures the rate of power conversion.

Wattmeter: a meter used to measure power conversion (true power).

Wheatstone bridge: a bridge circuit that makes precision resistance measurements by comparing an "unknown" resistance with a known or standard resistance value. X axis: a horizontal line.

Y axis: a vertical line.

SOLID-STATE ELECTRONICS TERMS

Acceptor impurity: an element with less than four valence electronics that is used as a dopant for semiconductors. Acceptor atoms mixed with silicon makes an P material.

Active region: an area of transistor operation between cutoff and saturation.

Alternation: the positive or negative part of an ac sine wave.

Anode: the positive terminal or electrode of an electronic device such as a solid state diode.

Average value: an ac voltage or current value based on the average of all instantaneous values for one alternation $V_A = 0.637 \times V_{\text{mas}}$.

Barrier potential or voltage: the voltage that is developed across a P-N junction due to the diffusion of holes and electrons.

Base: a thin layer of semiconductor material between the emitter and collector of a bipolar transistor.

Beta: a designation of transistor current gain determined by $I_C I_B$

Bias (forward): voltage applied across a P-N function that causes the conducting of majority current carriers.

Bias (reverse): voltage applied across a P-N junction that causes little or no current flow.

Bipolar: a type of transistor with two P-N junctions. Current carriers have two polarities, including holes and electrons.

Cathode: the terminal of an electronic device that is responsible for electron emission.

Channel: the controlled conduction path of a field-effect transistor.

Circuit configuration: a method of connecting a circuit, such as common-emitter, common-source, and so on.

Collector: a section of a bipolar transistor that collects majority current carriers.

Compound: two or more elements combined together chemically.

Conduction band: the outermost energy level of an atom.

Covalent bonding: atoms joined together by an electron-sharing process that causes a molecule to be stable.

Current carriers: electrons or holes that support current flow in solid-state devices.

Cutoff: a condition or region of transistor operation where current carriers cease to flow through the device.

Depletion zone: an area near the P-N junction where it is void of current carriers.

Dielectric constant: a number that compares the insulating ability of a material to that of air. Air has a value of 1.

Donor impurity: an element with more than four valence electrons that is used as a dopant for semiconductors. Donor atoms mixed with silicon make an N

material.

Dopant: an impurity added to silicon or germanium.

Doping: the process of adding impurities to a pure semiconductor.

Drain: the output terminal of a field-effect transistor.

Dynamic characteristics of operation: electronic device operational characteristics that show how the device responds to ac or changing voltage values.

Electron: a negatively charged atomic particle.

Electron volt (eV): a measure of energy acquired by an electron passing through a potential of 1 V.

Electrostatic force: an interaction between negative and positive charged particles.

Emission: the release of electrons from the surface of a conductive material.

Emitter: a semiconductor section that is responsible for the release of majority current carriers.

Energy: something that is capable of producing work, such as heat, light, chemical, and mechanical action.

Enhancement: a conduction function where current carriers are pulled from the substrate into the channel of a MOSFET.

Extrinsic: a purified material that has a controlled amount of impurity added to it when manufactured.

Forbidden gap: an area between the valence and conduction band on an energy-level diagram.

Forward bias: a voltage connection method that produces current conduction in a P-N junction.

Gate: the control element of a field-effect transistor.

Hole: a charge area or void where an electron is missing.

Induced channel: the channel of an E -MOSFET that is developed by gate volt-age.

Intrinsic: a semiconductor crystal in a high level of purification.

Ion: an atom that has either lost an electron (positive ion) or gained an electron (negative ion).

Ionic bond: a binding force that joins one atom to another in a way that fills the valence band of an atom.

Junction: the point where P and N semiconductor materials are joined.

Kinetic energy: energy due to motion.

Light-dependent resistor (LDR): a photoconductive device that changes resistance with light energy.

Majority current carrier: electrons of an N material and holes of a P material.

Metallic bond: a force such as a floating cloud of ions holds atoms loosely together in conductor.

Minority current carrier: a conduction vehicle opposite to the majority current carrier, such as holes in an N material and electrons in the P material.

Negative resistance: a UJT characteristic where an increase in emitter current causes a decrease in emitter voltage.

Peak voltage point (P_V): a characteristic point of the unijunction transistor where the emitter voltage rises to a peak value.

Photon: a small packet of energy or quantum of energy associated with light.

Power dissipation: an electronic device characteristic that indicates the ability of the device to give off heat.

Q point: an operating point for an electronic device that indicates its dc or static operation with no ac signal applied.

Quanta: a discrete amount of energy required to move an electron into a higher energy level.

Quantum theory: a theory based on the absorption and emission of discrete amounts of energy.

Reverse bias: a voltage connection method producing very little or no current through a P-N junction.

Saturation: an active-device operational region where the output current levels off to a constant value.

Saturation region: a condition of operation where a device is conducting to its full capacity.

Secondary emission: the electrons released by bombardment of a conductive material with force.

Solid state: an area of electronics dealing with the conduction of current through semiconductors.

Source: the common of energy input lead of a field-effect transistor.

Stabilized atom: an atom that has a full complement of electrons in its valence band.

Static state: a dc operating condition of an electronic device with operating energy by no signal applied.

Substrate: a piece of N or P semiconductor material which serves as a foundation for others parts of a MOSFET.

Thermal stability: the condition of an electronic device that indicates its ability to remain at an operating point without variation due to temperature.

Thermionic emission: the release of electrons from a conductive material by the application of heat.

Unipolar: a semiconductor device that has only one P-N junction and one type of current carrier. FETs and UJTs are unipolar devices.

Valence band: the part of an energy-level diagram that deals with the outermost electrons of an atom.

Valence electrons: the outermost shell or layer of electrons in the structure of an atom.

Valley voltage: a characteristic voltage point of the UJT where the voltage value drops to its lowest level.

ELECTRONIC CIRCUITS TERMS

Amplification: the ability to control a large force by a smaller force. There is voltage, current, and power amplification. Astable multivibrator: a free-running generator that develops a continuous square-wave output.

Audio amplifiers: amplifiers that respond to the human range of hearing or frequencies from 15 to 15kHz.

Beta-dependent biasing: a method of biasing that responds in some way to the amplifying capabilities of the device.

Blocking oscillator: an oscillator circuit that drives an active device into cutoff or blocks its conduction for a certain period of time.

Bypass capacitor: a capacitor that provides an alternative path around a component or to ground.

Capacitance input filter: a filter circuit employing a capacitor as the first component of its input.

Capacitor divider: a network of series-connected capacitors, each having a voltage drop of ac.

Cascade: a method of amplifier connection in which the output of one stage is connected to the input of the next amplifier.

Center tap: an electrical connection point at the center of a wire coil or transformer. Comparator: an amplifier that has a reference signal input and a test signal input. When the test signal differs in value from the reference signal, a state change in the output occurs.

Complementary transistors: PNP and NPN transistors with identical characteristics.

Counter EMF: the voltage induced in a conductor through a magnetic field which opposes the original source voltage.

Coupling: a process of connecting active components.

Damped wave: a wave in which successive oscillators decrease in amplitude.

Darlington amplifier: two transistor amplifiers cascaded together.

Decibel: one tenth of a bel; used to express gain or loss.

Differentiator network: a resistor-capacitor combination where the output appears across the resistor. An applied square wave will produce a positive and negative spiked output.

Digital integrated circuit: an IC that responds to two-state (on-off) data.

Dual-in-line package (DIP): a packaging method for integrated circuits.

Efficiency: a ratio between output and input.

Electromagnetic field: the space around an inductor that changes due to current flow through a coil. The field expands and collapses with applied ac.

Electrostatic field: the space or area around a charged body where there is some reaction to the developed charge.

Feedback: transferring voltage from the output of a circuit back to its input.

Feedback network: a resistor combination that returns some of the output signal to the input of an IC.

Ferrite core: an inductor core made of molded iron particles.

Flip-flop: a multivibrator circuit having two stable states that can be switched back and forth.

Free running: an oscillator circuit that develops a continuous waveform that is not stabilized, such as an astable multivibrator.

Gain: a ratio of voltage or current output to voltage or current input.

Impedance: a form of opposition to alternating current measured in ohms. Electronic devices have input and output impedance values.

Impedance ratio: the ac resistance ratio between the input and output of a transformer.

Inductance: the property of an electric circuit that opposes a change in current flow.

Interbase resistance: the resistance between B_1 and B_2 of a unijunction transistor.

Linear amplifiers: amplifying circuits that increase the amplitude of a signal that is a duplicate of the input. These devices operate in the straight-line part of the characteristic curves.

Load line: a line drawn on a family of characteristic curves that shows how the device will respond in a circuit with a specific load resistor value.

Logarithm: a mathematical expression dealing with exponents showing the power to which a value called the base must be raised in order to produce a given number.

Monostable: a multivibrator with one stable state. It changes to the other state momentarily and then returns to its stable state.

Noninverting input: an input lead for an integrated circuit that does not shift the phase of a signal.

Nonsinusoidal: waveforms that are not sine waves, such as square, saw tooth, and pulsing waves.

Open loop: an operation amplifier connection method that causes full or maximum amplification.

Peak to peak (p-p): an ac value measured from the peak positive to the peak negative alternations.

Peak value: the maximum voltage or current value of an ac sine wave. $V_P = 1.414 \times$ RMS.

Piezoelectric effect: the property of a crystal material that produces voltage by changes in shape or pressure.

Pi filter: a filter with an input capacitor connected to an inductor-capacitor filter, forming the shape of the Greek capital letter pi.

Pulsating dc: a voltage or current value that rises and falls at a rate or frequency with current flow always in the same direction.

Push-pull: an amplifier circuit configuration using two active devices with the input signal being of equal amplitude and 1800 out of phase.

Read head: a signal pickup transducer that generates a voltage in a tape recorder.

Rectification: the process of changing ac into pulsating dc.

Regenerative feedback: feedback from the output to the input that is in phase so that it is additive.

Relaxation oscillator: a nonsinusoidal oscillator that has a resting or nonconductive period during its operation.

Resonant frequency: the frequency at which a tuned circuit oscillates.

Schematic: a diagram of an electronic circuit made with symbols.

Stability: the ability of an oscillator to stay at a given frequency or given condition of operation without variation.

Symmetrical: a condition of balance where parts, shapes, and sizes of two or more items are the same.

Synchronized: a condition of operation that occurs at the same time. The vertical and horizontal sweep of an oscilloscope are often synchronized or placed in sync.

Tank circuit: a parallel resonant *LC* circuit.

Threshold: the beginning or entering point of an operating condition. A terminal connection of the LM555 IC timer.

Time constant (*RC*): the amount of time required for the voltage of a capacitor in an RC circuit to increase 63.2% of the value or decrease to 36.7% of the value.

Toggle: a switching condition that changes back and forth.

Transducer: a device that changes energy from one form to another. A transducer can be on the input or output.

Triggered: a control technique that causes a device to change its operational state, such as a triggered monostable multivibrator.

Turns ratio: the ratio of the number of turns in the primary winding to the number of turns in the secondary winding.

Voltage follower: an amplifying condition where the input and output voltages are of the same value across different impedance.

COMMUNICATIONS ELECTRONICS TERMS

Amplitude modulation (AM): a method of transmitter modulation achieved by varying the strength of the RF carrier at an audio signal rate.

Audio frequency (AF): a range of frequency from 15 Hz to 15 kHz, the human range of hearing.

Beat frequency: a resulting frequency that develops when two frequencies are combined in a nonlinear device.

Beat frequency oscillator: an oscillator of a CW receiver. Its output beats with the incoming CW signal to produce an audio signal.

Bel (B): a measurement unit of gain that is equivalent to a 10:1 ratio of power levels.

Blanking pulse: a part of the TV signal where the electron beam is turned off during the retrace period. There is both vertical and horizontal blanking.

Buffer amplifier: an RF amplifier that follows the oscillator of a transmitter. It isolates the oscillator from the load.

Carrier wave: an RF wave to which modulation is applied.

Center frequency: the carrier wave of an FM system without modulation applied.

Chroma: short for "chrominance." Refers to color in general.

Compatible: a TV system characteristic in which broadcasts in color may be received in black and white on sets not adapted for color.

Continuous wave (CW): uninterrupted sine waves, usually of the radio-frequency type, that are radiated into space by a transmitter.

Crossover distortion: the distortion of an output signal at the point where one transistor stops conducting and another conducts, in class B amplifiers.

Crystal microphone: input transducer that changes sound into electrical energy by the piezoelectric effect.

Deflection: electron beam movement of a TV system that scans the camera tube or picture tube.

Deflection yoke: a coil fixture that moves an electron beam vertically and horizontally.

Diplexer: a special TV transmitter coupling device that isolates the audio carrier and the picture carrier signals from each other.

Dynamic microphone: an input transducer that changes sound into electrical energy by moving a coil through a magnetic field.

Field, even lined: the even-numbered scanning lines of one TV picture or frame.

Field, odd lined: the odd-numbered scanfling lines of one TV picture or frame.

Frame: a complete electronically produced TV picture of 525 horizontally scanned lines.

Ganged: two or more components connected together by a common shaft-a three-ganged variable capacitor.

Heterodyning: the process of combining signals of independent frequencies to obtain a different frequency.

High-level modulation: where the modulating component is added to an RF carrier in the final power output of the transmitter.

Hue: a color, such as red, green, or blue.

Interlace scanning: an electronic picture production process in which the odd lines are all scanned, then the even lines, to make a complete 525-line picture.

Intermediate frequency (IF): a single frequency that is developed by heterodyning two input signals together in a superheterodyne receiver.

I-signal (I): a color signal of a TV system that is in phase with the 3.58-MHz color subcarrier.

Line-of-sight transmission: an RF signal transmission that radiates out in straight lines because of its short wavelength.

Microphone: a device that changes sound energy into electrical energy.

Midrange: a speaker designed to respond to audio frequency in the middle part of the AF range.

Modulating component: a specific signal or energy used to change the characteristic of a RF carrier. Audio and picture modulation are very common.

Modulation: the process of changing some characteristic of an RF carrier so that intelligence can be transmitted.

Monochrome: black-and-white television.

Negative picture phase: a video signal characteristic where the darkest part of a picture causes the greatest change in signal amplitude.

Phasor: a line used to denote value by its length and phase by its position in a vector diagram.

Plumbicon: a television camera tube that operates on the photoconduction principle.

Q signal: a color signal of a TV system that is out of phase with the 3.58-MHz color subcarrier.

Radio frequency (RF): an alternating current (ac) frequency in excess of the human range of hearing which starts at 30,000 Hz.

Radio-frequency choke (RFC): an inductor or coil that offers high impedance to radio-frequency ac.

Retrace: the process of returning the scanning beam of a camera or picture tube to its starting position.

Saturation: the strength or intensity of a color used in a TV system.

Scanning: in a TV system, the process of moving an electron beam vertically and horizontally.

Selectivity: a receiver function of picking out a desired radio frequency signal. Tuning achieves selectivity.

Sidebands: the frequencies above and below the carrier frequency that are developed because of modulation.

Skip: an RF transmission signal pattern that is the result of signals being reflected from the ionosphere or the earth.

Sky-wave: an RF signal radiated from an antenna into the ionosphere.

Speaker: a transducer that converts electric current and/or voltage into sound waves. Also called a loudspeaker.

Stage of amplification: a transistor IC and all the components needed to achieve amplification.

Stereophonic (stereo): signals developed from two sources that give the listener the impression of sound coming into both ears from different points.

Stylus: a needle-like point that rides in a phonograph record groove.

Sync: an abbreviation for synchronization.

Synchronization: a control process that keeps electronic signals in step. The sweep of a TV receiver is synchronized by the transmitted picture signal.

Telegraphy: the process of conveying messages by coded telegraph signals.

Trace time: a period the scanning process where picture information is reproduced or developed.

Tweeter: a speaker designed to reproduce only high frequencies.

Vertical blocking oscillator: a TV circuit that generates the vertical sweep signal for deflection of the cathode-ray tube or picture tube.

Vestigial sideband: a transmission procedure where part of one sideband is removed to reduce bandwidth.

Videcon: a TV camera tube that operates on the photoconductive principle.

Voice coil: the moving coil of a speaker.

Voltage divider: a combination of resistors that produce different or multiple voltage values.

Wavelength: the distance between two corresponding points that represents one complete wave.

Woofer: a speaker designed to reproduce low audio frequencies with high quality.

Y signal: the brightness or luminance signal of a TV system.

Zero beating: the resulting difference frequency that occurs when two signals of the same frequency are heterodyned.

COMPUTER ELECTRONICS TERMS

Alphanumeric: a numbering system containing numbers and letters.

Analog: being continuous or having a continuous range of values.

Assembly language: a computer programming language that uses abbreviations in mnemonic form.

Base: the number of symbols in a number system. A decimal system has a base of 10.

Binary: a two-state or two-digit numbering system.

Binary-coded decimal: a code of 10 binary numbers for the decimal values from 0 to 9.

Central processing unit (CPU): a function part of a computer that achieves control, arithmetic logic, and signal manipulation.

Counter: a digital circuit capable of responding to state changes through a series of binary changes.

Decade counter: a counter that achieves 10 states or discrete values.

Decoding: a digital circuit that changes coded data from one form to a different form.

Digital: a value or quantity related to digits or discrete values.

Encoding: a converter that changes an input signal into binary data.

Fetch: a computer instruction that retrieves data from memory.

Firmware: permanently installed instructions in computer hardware that control operation.

Flip-flop: a circuit that changes back and forth between two states.

Gate: a circuit that performs special logic operations such as AND, OR, NOT, NAND, and NOR.

Hardware: electronic circuits, parts, and physical parts of a computer.

High-level language: a form of computer programming that uses words and statements.

Interface: a process or piece of equipment in a computer that brings two things together.

Keyboard: a device with keys or switches that are used in digital equipment.

Logic: a decision-making capability of computer circuitry.

Machine language: a form of computer programming that

involves direct coding operations and memory addresses.

Memory: the storage capability of a device or circuit.

Mnemonic: an abbreviation that resembles a word or reminds one of a word.

Negative logic: a system where low or "0" has voltage and high or "1" has little or no voltage.

Nonvolatile memory: a computer storage function that retains data without the use of electrical power. Disk and tape memory are nonvolatile.

Positive logic: a system where low, or 0, has no voltage and high, or 1, has voltage. Programmable read-only memory (PROM): a memory device that can be programmed by the user.

Radix: the base of a number system.

Reading memory: sensing binary information that has been stored on a magnetic tape, disk, or IC chip.

Read-only memory: a form of stored data that can be sensed or read. It is not altered by the sensing process.

Software: instructions, such as a program that controls the operation of a computer.

Toggle: a switching mode that changes between two states, such as on and off.

Truth table: a graph or table that displays the operation of a logic circuit with respect to its input and output data.

Volatile memory: stored binary data that are lost or destroyed when the electrical power is removed.

Writing memory: placing binary data into the storage cells of a memory device.

INDUSTRIAL ELECTRONICS TERMS

Alternation: half of an ac sine wave. There is a positive alternation and a negative alternation for each ac cycle.

Bidirectional triggering: the triggering or conduction that can be achieved in either direction of an applied ac wave. Diacs and triacs are of this classification.

Breakover voltage: the voltage across the anode-cathode of an SCR or terminal 1 and terminal 2 of a triac that causes it to turn on.

Circuit switching: an on-off control function. Conduction occurs during the on time with no conduction during the off time.

Commutation: the changing from on to off or from one state to the opposite state.

Conduction time: the time when a solid-state device is turned on or in its conductive state.

Contact bounce: when a mechanical switch is turned on, the contacts are forced together. This causes the contacts to open and close several times before making firm contact.

Delay time: the time that an alternation is held in an off state before it turns on.

Diac: a solid-state diode device that will trigger in either direction when its breakover voltage is exceeded. A bidirectional triggering device.

Forward breakover voltage: the voltage where an SCR or triac goes into conduction in quadrant I of its IV characteristics.

Gate: the control electrode of a triac, SCR, or FET.

Gate current (IG): the resulting current that flows in the gate when it is forward biased.

Holding current (IH): a current level that must be achieved when an SCR or triac latches or holds in the conductive state.

Internal resistance: the resistance between the anode-cathode of an SCR or T1-T2 of a triac. Also called the dynamic resistance.

Latch: a holding state where an SCR or triac turns on and stays in conduction when the gate current is removed.

Main 1 or 2: the connection points on a triac where the load current passes. Also called terminal 1 and terminal 2.

Off-state resistance: the resistance of an SCR or triac when it is not conducting.

Phase: a position or part of an ac wave that is compared with another wave. A wave can be in phase or out of phase with respect to another.

Phase shifter: a circuit that changes the phase of an ac waveform.

Quadrant: one fourth, or a quarter part, of a circle or a graph.

Reverse breakdown: the conduction that occurs in quadrant III of the IV characteristic of a triac.

Rheostat: the two-terminal variable resistor.

Snubber: a resistor-capacitor network placed across a triac to reduce the effect of an inductive voltage.

Switching device: an electronic device that can be turned on and off very rapidly. This type of device is capable of achieving control by switching.

Terminal (T_1 or T_2): the connecting points where load current pass through a triac. These two connection points are also called main 1 or 2.

Triggering: the process of causing an NPNP device to switch states from off to on.

Turn-on time: the time of an ac waveform when an alternation occurs.

Appendix B

Electronic Symbols

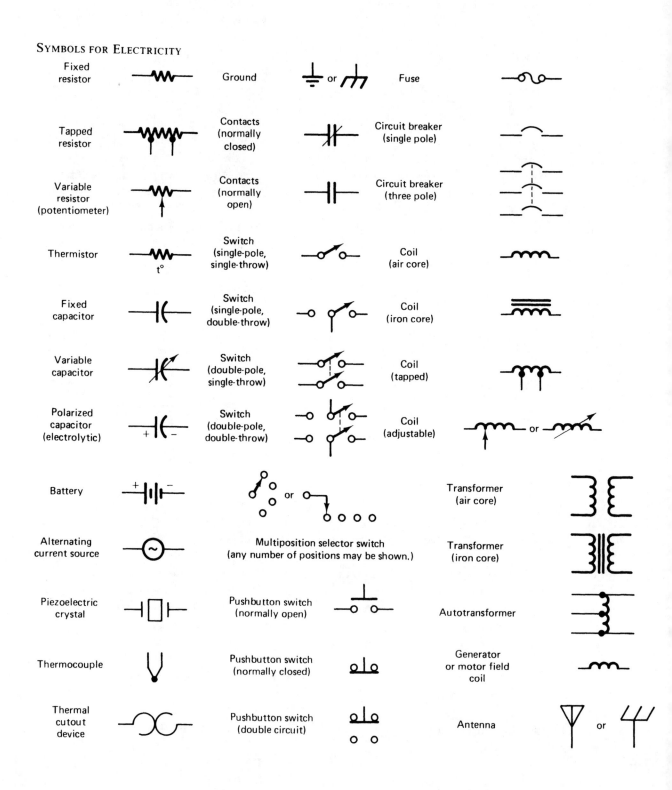

Symbols for Electricity

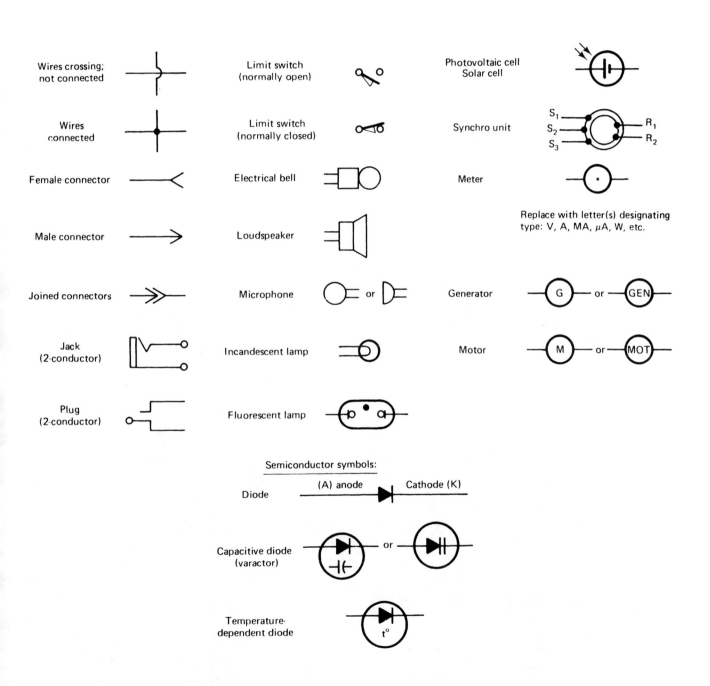

Wires crossing; not connected

Wires connected

Female connector

Male connector

Joined connectors

Jack (2-conductor)

Plug (2-conductor)

Limit switch (normally open)

Limit switch (normally closed)

Electrical bell

Loudspeaker

Microphone — or —

Incandescent lamp

Fluorescent lamp

Photovoltaic cell Solar cell

Synchro unit

S_1 S_2 S_3 R_1 R_2

Meter

Replace with letter(s) designating type: V, A, MA, μA, W, etc.

Generator — G — or — GEN —

Motor — M — or — MOT —

Semiconductor symbols:

Diode — (A) anode — Cathode (K)

Capacitive diode (varactor) — or —

Temperature-dependent diode — $t°$ —

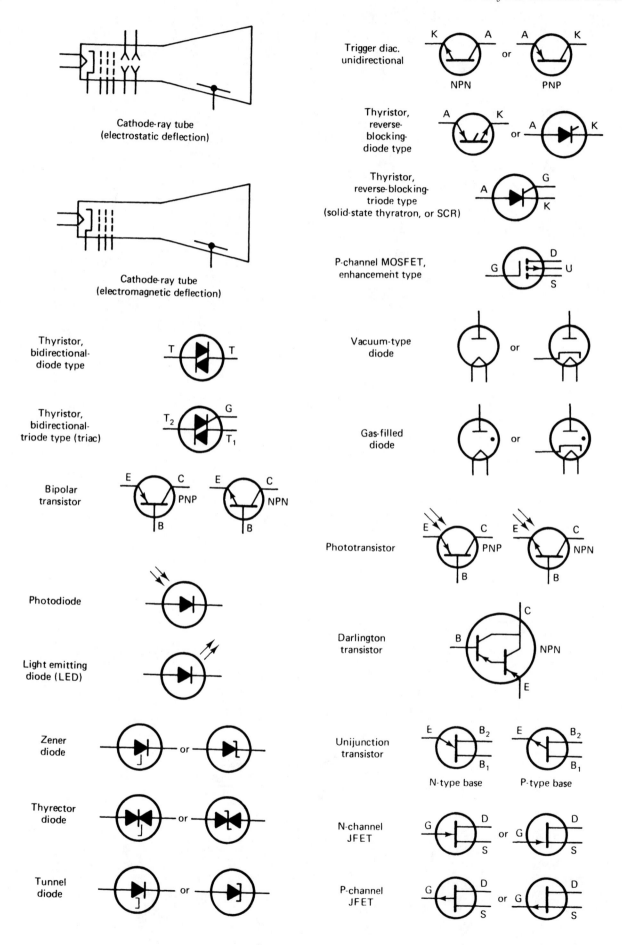

Cathode-ray tube
(electrostatic deflection)

Cathode-ray tube
(electromagnetic deflection)

Thyristor,
bidirectional-
diode type

Thyristor,
bidirectional-
triode type (triac)

Bipolar
transistor

Photodiode

Light emitting
diode (LED)

Zener
diode

Thyrector
diode

Tunnel
diode

Trigger diac,
unidirectional

Thyristor,
reverse-
blocking-
diode type

Thyristor,
reverse-blocking-
triode type
(solid-state thyratron, or SCR)

P-channel MOSFET,
enhancement type

Vacuum-type
diode

Gas-filled
diode

Phototransistor

Darlington
transistor

Unijunction
transistor

N-channel
JFET

P-channel
JFET

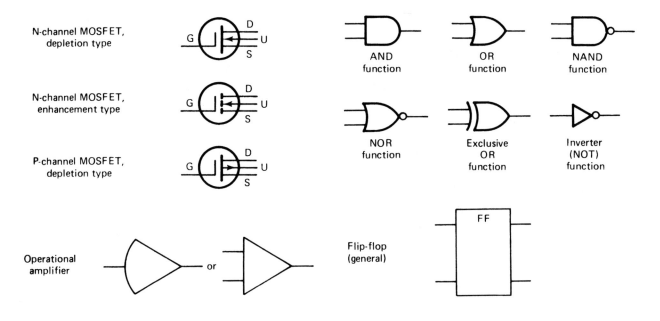

Appendix C

Electrical Safety

Electrical safety is very important. Many dangers are not easy to see. For this reason, safety should be based on understanding basic electrical principles. Common sense is also important. The physical arrangement of equipment in the electrical lab or work area should be done in a safe manner. Well designed electrical equipment should always be used. For economic reasons, electrical equipment is often improvised. It is important that all equipment be made as safe as possible. This is especially true for equipment and circuits that are designed and built in the lab or shop.

Work surfaces in the shop or lab should be covered with a material that is nonconducting and the floor of the lab or shop should also be nonconducting. Concrete floors should be covered with rubber tile or linoleum. A fire extinguisher that has a nonconducting agent should be placed in a convenient location. Extinguishers should be used with caution. Their use should be explained by the teacher.

Electrical circuits and equipment in the lab or shop should be plainly marked. Voltages at outlets require special plugs for each voltage. Several voltage values are ordinarily used with electrical lab work. Storage facilities for electrical supplies and equipment should be neatly kept. Neatness encourages safety and helps keep equipment in good condition. Tools and small equipment should be maintained in good condition and stored in a tool panel or marked storage area. Tools that have insulated handles should be used. Tools and equipment plugged into convenience outlets should be wired with three-wire cords and plugs. The purpose of the third wire is to prevent electrical shocks by grounding all metal parts connected to the outlet.

Soldering irons are often used in the electrical shop or lab. They can be a fire hazard. They should have a metal storage rack. Irons should be unplugged while not in use. Soldering irons can also cause burns if not used properly. Rosin-core solder should always be used in the electrical lab or shop.

Adequate laboratory space is needed to reduce the possibility of accidents. Proper ventilation, heat, and light also provide a safe working environment. Wiring in the electrical lab or shop should conform to specifications of the National Electrical Code (NEC). The NEC governs all electrical wiring in buildings.

LAB OR SHOP PRACTICES

All activities should be done with low voltages whenever possible. Instructions should be written, with clear directions, for performing lab activities. All lab or shop work should emphasize safety. Experimental circuits should always be checked before they are plugged into a power source. Electrical lab projects should be constructed to provide maximum safety when used.

Electrical equipment should be disconnected from the source of power before working on it. When testing electronic equipment, such as TV sets or other 120-V devices, an isolation transformer should be used. This isolates the chassis ground from the ground of the equipment and eliminates the shock hazard when working with 120-V equipment.

ELECTRICAL HAZARDS

A good first-aid kit should be in every electrical shop or lab. The phone number of an ambulance service or other medical services available should be in the lab or work area in case of emergency. Any accident should be reported immediately to the proper school officials. Teachers should be proficient in the treatment of minor cuts and bruises. They should also be able to apply artificial respiration. In case of electrical shock, when breathing stops, artificial respiration must be immediately started. Extreme care should be used in moving a shock victim from the circuit that caused the shock. An insulated material should be used so that someone else does not come in contact with the same voltage. It is not likely that a high-voltage shock will occur. However, students should know what to do in case of emergency.

Normally, the human body is not a good conductor of electricity. When wet skin comes in contact with an electrical conductor, the body is a better conductor. A slight shock from an electrical circuit should be a warning that something is wrong. Equipment that causes a shock should be immediately checked and repaired or replaced. Proper grounding is important in preventing shock.

Safety devices called ground-fault circuit interrupters (GFIs) are now used for bathroom and outdoor power receptacles. They have the potential of saving many lives by preventing shock. GFIs immediately cut off power if a shock occurs. The National Electric Code specifies where GFIs should be used.

Electricity causes many fires each year. Electrical wiring with too many appliances connected to a circuit overheats wires. Overheating may set fire to nearby combustible materials. Defective and worn equipment can allow electrical conductors to touch one another to cause a short circuit, which causes a blown fuse. It could also cause a spark or arc which might ignite insulation or other combustible materials or burn electrical wires.

FUSES AND CIRCUIT BREAKERS

Fuses and circuit breakers are important safety devices. When a fuse "blows," it means that something is wrong in the circuit. Causes of blown fuses could be:
1.   A short circuit caused by two wires touching
2.   Too much equipment on the same circuit
3.   Worn insulation allowing bare wires to touch grounded metal objects such as heat radiators or water pipes

After correcting the problem a new fuse of proper size should be installed. Power should be turned off to replace a fuse. Never use a makeshift device in place of a new fuse of the correct size. This destroys the purpose of the fuse. Fuses are used to cut off the power and prevent overheated wires.

Circuit breakers are now very common. Circuit breakers operate on spring tension. They can be turned on or off like wall switches. If a circuit breaker opens, something is wrong in the circuit. Locate and correct the cause and then reset the breaker.

Always remember to use common sense whenever working with electrical equipment or circuits. Safe practices should be followed in the electrical lab or shop as well as in the home. Detailed safety information is available from the National Safety Council and other organizations. It is always wise to be safe.

Appendix D
Soldering Techniques

Soldering is an important skill for electrical technicians. Good soldering is important for the proper operation of equipment.
Solder is an alloy of tin and lead. The solder that is most used is 60/40 solder.   This means that it is made from 60% tin and 40% lead. Solder melts at a temperature of about 400°F.

For solder to adhere to a joint, the parts must be hot enough to melt the solder. The parts must be kept clean to allow the solder to flow evenly. Rosin flux is contained inside the solder. It is called rosin-core solder.

A good mechanical joint must be made when soldering. Heat is then applied until the materials are hot. When they are hot, solder is applied to the joint. The heat of the metal parts (not the soldering tool) is used to melt the solder. Only a small amount of heat should be used. Solder should be used sparingly. The joint should appear smooth and shiny. If it does not, it could be a "cold" solder joint. Be careful not to move the parts when the joint is cooling. This could cause a "cold" joint.

When parts that could be damaged by heat are soldered, be very careful not to overheat them. Semiconductor components, such as diodes and transistors, are very heat sensitive. One way to prevent heat damage is to use a *heat sink,* such as a pair of pliers. A heat sink is clamped to a wire between the joint and the device being soldered. A heat sink absorbs heat and protects delicate devices. Printed circuit boards are also very sensitive to heat. Care should be taken not to damage PC boards when soldering parts onto them.

Below are some rules for good soldering:
1. Be sure that the tip of the soldering iron is clean and tinned.
2. Be sure that all the parts to be soldered are heated. Place the tip of the soldering iron so that the wires and the soldering terminal are heated evenly.
3. Be sure not to overheat the parts.
4. Do not melt the solder onto the joint. Let the solder flow onto the joint.

5. Use the right kind and size of solder and soldering tools.
6. Use the right amount of solder to do the job but not enough to leave a "blob."
7. Be sure not to allow parts to move before the solder joint cools.

Appendix E
Power Supply Construction Project

The following is a step-by-step procedure for constructing a 9-V power supply using printed-circuit (PC) board. If PC board is not available, the parts may be soldered together using any type of circuit board. Materials for circuit board construction are available through many electronic parts dealers.

Power-Supply Project
1. To construct a regulated power supply, using the circuit of Figure E-1.
2. A representative PC-board layout is shown in Figure E-2. The circuit board may require some modification according to your part selection.

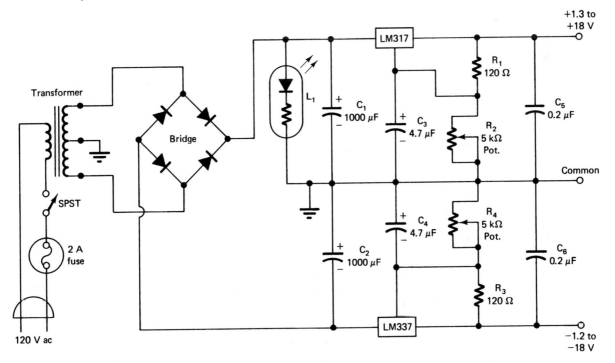

Figure E-1

Parts

C_1 and C_2	1000 µF, 35 V
C_3 and C_4	4.7 µF, 35 V
C_5 and C_6	0.2 µF, 50 V
R_1 and R_3	120 Ω, 1/4 W
R_2 and R_4	5 kΩ Pot.
L_1	Indicator lamp LED and resistor (12 V)

1 LM317 (+ Reg)

1 LM337 (− Reg)

1 Bridge rectifier

1 Transformer 25 V, 2 A, CT secondary

1 Fuse/holder (2 A)

1 SPST switch

1 Line cord/plug

3 Binding posts

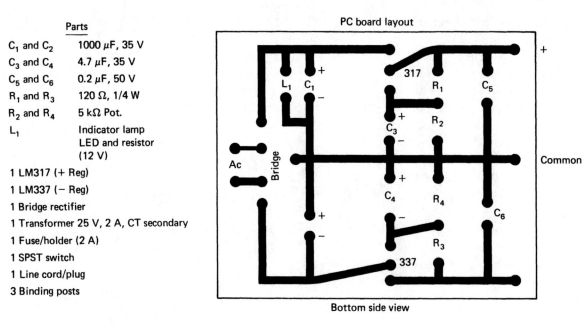

Figure E-2

Make a prototype layout of the PC board on a piece of card stock. Use the actual dimensions of the parts being used.

3. Lay out the PC board using tape and PC resist dots.

4. Etch the board.

5. Drill holes at the designated locations.

6. Solder components in place.

7. Wire the external circuitry.

8. Assemble the external circuitry and PC board.

9. Test the operation of the circuit with a voltmeter and/or oscilloscope. The output should be 18 V DC.

10. Assemble the entire unit in a cabinet and test it again for operation.

Index